Textiles for
Industrial Applications

Textiles *for* Industrial Applications

R. Senthil Kumar

Department of Textile Technology
Kumaraguru College Of Technology
Coimbatore, Tamilnadu, India

CRC Press
Taylor & Francis Group
Boca Raton London New York

CRC Press is an imprint of the
Taylor & Francis Group, an **informa** business

A CHAPMAN & HALL BOOK

CRC Press
Taylor & Francis Group
6000 Broken Sound Parkway NW, Suite 300
Boca Raton, FL 33487-2742

First issued in paperback 2018

ISBN-13: 978-1-4665-6649-1 (hbk)
ISBN-13: 978-1-138-37476-8 (pbk)

Visit the Taylor & Francis Web site at
http://www.taylorandfrancis.com

and the CRC Press Web site at
http://www.crcpress.com

Contents

Preface

Industrial textiles are the fastest growing and promising segments of the textile industry. According to the end use of the product, the textile industry can be simply classified into three categories—clothing, home textile, and industrial textile—and the proportion of these categories may reflect the level of textile technology development. Industrial textiles, also known as technical textiles, are textile products wherein functionality is the primary concern. The industrial textiles industry has immense potential in the developing countries. Depending on the characteristics of the products, functional requirements, and end use applications, the highly diversified range of industrial textile products has been grouped into several categories.

The industrial textile industry has the following characteristics and advantages: intensive capital, high-end technology, low labor requirements qualified workforce, and a huge domestic market. These are totally different from the traditional textile industry, which has intensive labor and poor technology. In the last decade, industrial textile has experienced rapid growth. Compared with the other products of the textile industry, the index of competitive ability for industrial textile is still relatively low. However, the tendency for growth has been more prominent than others. The industrial textile industry has enormous potential for growth if it optimizes and upgrades the industrial structure.

The design and development of the industrial textile product need basic understanding and application of textile science and technology. Technology advances in the industry are driven by forces outside the pure textile sector, that is, polymer and fiber producers and, in some cases, the machinery producers of fabric manufacturing techniques. There is a growing need for nontextile application know-how in many segments of the industrial textiles market. Textile technologists are needed who understand the various engineering aspects of potential industrial applications so that suitable textile structures can be produced. The producer of the automotive textiles should have intricate knowledge of the various aspects of automobiles.

This book, *Textiles for Industrial Applications*, consists of 13 chapters covering a wide range of sectors such as filtration; belting; ropes; medical, automotive, and civil composites; and miscellaneous products. The chapters are specially crafted to cover the various aspects of product development and current developments in the industrial textiles sector. Chapter 12, "Testing of Industrial Textiles," provides the readers with a broad outlook on testing methods.

I extend my gratitude to the management and colleagues of Kumaraguru College of Technology, Coimbatore, for their continuous support throughout this work. I am grateful to Mr. C. Vigneswaran, who coauthored the book by contributing Chapters 1 through 3, 11, and 12. I also thank my family members for their patience and support during the preparation of this book.

Mr. R. Senthil Kumar
Assistant Professor (Senior Grade)
Department of Textile Technology
Kumaraguru College of Technology
Coimbatore, India

Author

R. Senthil Kumar completed his postgraduation in textile engineering from the Indian Institute of Technology, Delhi, India. He contributed the chapter "Mechanical Finishing Techniques for Technical Textiles" to the book *Advances in Dyes, Chemicals and Finishes for Technical Textiles* (Woodhead Publication, 2013). He has published 30 technical and management-related papers in various international and national journals; has prepared manuals on nonwoven technologies for international and local textile industries; has rich experience in the yarn manufacturing industry with over six years in production and R&D; is actively engaged in consultancy to various textile industries; and has prepared and published nearly 200 technical and management-related textile study materials in an online portal (www.scribd.com/sen29iit), with more than 270,226 users accessing the website at the time of writing. His areas of specialization are fiber science, textile composites, technical textiles, nonwoven technologies, and textile testing. He has been working as an assistant professor (senior grade) in the Department of Textile Technology, Kumaraguru College of Technology, Coimbatore, India, since 2009.

1

Introduction

1.1 Industrial Textiles: An Overview

The word "textile" used in the literal meaning/use of it a person thinks of is the apparel fabric, but basically "textile" is beyond that. The need of mankind has increased day by day; also due to growth and development of industries, the demand of functional products has increased. It was beyond the reach of traditional products to serve these demands. Technical textiles are different from the conventional textiles. Technical textiles are semifinished or finished textiles and textile products manufactured for performance characteristics. Unlike conventional textiles used traditionally for clothing or furnishing, technical textiles are used basically on account of their specific physical and functional properties and mostly by other user industries. Technical textile product can exist and be used in various forms of fibrous structures from simple filament to a complex end product. The most common textile products in this category include high-performance fibers, ropes, webbings, tapes, filter media, paper-making felts and fabrics, heat and sound insulators, coated fabrics, protective clothing, composites, agro textiles, geotextiles, and medical and hygienic products.

The technical textiles industry has immense potential in the developing countries. Asia is now emerging as a powerhouse of both production and end-use consumption of technical textiles. Depending on the product characteristics, functional requirements, and end-use applications, the highly diversified range of technical textile products has been grouped into several categories. Among these, industrial textile is the promising and emerging sector that fills the areas of applications in heavy-duty and functional textiles in engineering field. Global market volume of industrial textiles varies depending on the type of end-use applications. Higher-value products exist at the upper end of price level at lower volumes, and these are used in much specialized products where the performance and not the price is the determining factor. The industrial textile sector is concentrating on products such as filter fabrics, hoses, ropes and cordages, belts, sound absorption materials, printed circuit boards (PCBs), composite materials, etc.

Technical textile products are classified according to the application in 12 main fields: *agrotech* (agriculture, gardening, and forestry); *buildtech* (construction); *clothtech* (technical components for clothing and shoes); *geotech* (geotextile and road construction); *hometech* (furniture components, upholstery, and floor coverings); *indutech* (filtration, cleaning, and other industrial applications); *meditech* (medical and hygienic textile); *mobiltech* (automotive industry, marine construction, railroad, and aviation); *oekotech* (products for environment protection); *packtech* (packing); *protech* (protection of people and property); and *sporttech* (sport and leisure).

Indutech: These are the industrial textiles, also known as indutex, used in different ways by many industries for activities such as separating and purifying industrial products, cleaning gases and effluents, transporting materials between processes, and acting as substrates for abrasive sheets and other coated products. They range from lightweight nonwoven filters, knitted nets and brushes, to heavyweight coated conveyor belts.

1.2 Industrial Textiles: Definition and Scope

1.2.1 Definition

The word "industrial textiles" are defined as specially designed and engineered structures that are used in products, processes, or services of mostly nontextile industries. The term "industrial textiles" is the most widely used term for nontraditional textiles.

1.2.2 Scope

According to the end use of the product, textile industry can be simply classified into three main categories of clothing, home textile, and industrial textile, and the proportion of these categories may reflect the level of textile technology development. The field of industrial textiles comprises such diverse products as filter fabrics and battery separators, sound absorption materials and PCBs, composites and cigarette filters, or more commodity products like safety belts, conveyor/transmission belts, hoses, and ropes. The development of industrial textile with more requirements of high technology implies the reasonable structure of textile industry. The increasing percentage shows the optimized industrial structure of textile; however, big differences still exist as compared with developed countries.

In the last decade, industrial textile has already experienced rapid increase. Compared with the other products of textile industry, the index of competitive ability for industrial textile is still relatively lower. However, the tendency of increase is more obvious than others. It is worth to mention that half amount of the export of textile products belongs to processing and assembling trade, which means lower profits along with greater export amount, larger market share, and higher competitive index. Hence, accurate assessment of competitive level of textile industry requires the comprehensive consideration of trading amount, price, and added value to explore the actual competitive ability of industry.

New technologies for producing microfibers have also contributed toward the production of high-tech articles. By designing new processes for fabric preparation and finishing, and due to advances in technologies for production and application of suitable polymeric membranes and surface finishes, it is now possible to successfully combine the consumer requirements of aesthetics, design, and function in protective clothing for different end-use applications.

Industrial textiles find many areas of industrial needs and fulfilling the requirements such as filtration, hose, ropes, belts, composites, needled fabrics, air bags, and other textile-based industrial products.

1.2.3 Filtration Textiles

Filtration plays a critical role in our day-to-day life by providing healthier and cleaner products and environment. Textile materials are used in the filtration of air and liquids, in food particles, and in industrial production. Filtration fabrics are used widely in vacuum cleaners, power stations, petrochemical plants, sewage disposal, etc. Textile materials, particularly woven and nonwoven, are suitable for filtration because of their complicated structure and thickness. Dust particles have to follow a tortuous path around textile fibers. Thus, due to their structure, they have high filtration efficiencies. The usage of the filter fabrics varies according to their end use. This depends on the properties, the filters, which ultimately depend on the characteristics of the raw material used for the manufacturing of the filter fabric.

1.2.4 Hose, Ropes, and Belts

Hose is a hollow tube designed to carry fluids from one location to another. Hoses cover a wide range of sectors from domestic to industrial applications. Ropes are the twisted components that find application in fields as diverse as construction, seafaring, exploration, sports, and communications and have been since prehistoric times. Belts can be classified as transmission belts and conveyor belts, both of play a predominant role in today's manufacturing sector. Textile structures are reinforced in these components, which act as tension member contributing to mechanical property enhancement. These three components are prevalent in many machines of various manufacturing processes.

1.2.5 Composites

In the last two decades, the uses of textile structures made from high-performance fibers are finding increasing applications in composites.

High-performance textile structures may be defined as materials that are highly engineered fibrous structures having high specific strength, high specific modules, and designed to perform at high temperature and high pressure (loads) under corrosive and extreme environmental conditions. Significant developments have taken place in fibers, matrix polymers, and composite-manufacturing techniques. Composites that are a part of industrial textiles have a significant role in many applications especially in automobiles and aerospace applications.

1.2.6 Technical Needled Fabrics

The range of specialty products and markets for technical needled fabric is extensive. Some major technical areas for needled felts and fabrics are thus: automotive, furniture and bedding, filtration, geotextiles, and roofing. Needled fabrics in automotive application are used not only as interior coverings with aesthetic values but also as fuel and air filters, packings, dampers, etc., with performance value. The needled structure is an ideal media for air filtration and particulate emission control. Needled geotextiles are the key geotextile products because of some ideal functional properties such as bulk, toughness, and permeability.

1.2.7 Air Bags

The opportunities and challenges for the textile and making-up industries are great in the area of air bag production. This is so because of the great demand especially in view of the legalization that is already enforced in many countries, and other countries are also going to follow this action sooner or later. Approximately 1.42 m^2 of fabric is required to make a driver-side air bag, and $2.5–4.18 \text{ m}^2$ fabric is required to make passenger-side air bags or driver-side bags on light trucks. These figures point out that the air bag market is of great importance for the use of technical textiles. Air bags are usually made of coated or uncoated fabrics made of PA6.6 yarns, with minimum air permeability. The trend toward uncoated fabrics is expected to continue and so is the increased trend toward more air bags per car and full-size bags. There is also technical challenge of manufacturing the bag using more rational techniques and according to the tough specifications formulated by the automotive industry.

1.2.8 Other Textile-Based Industrial Products

Industrial textiles cover wide range of products other than mentioned earlier. Some of the products based on textile structures are acoustic materials, PCBs, decatizing cloth, bolting cloth, coated abrasives, battery separators, etc. The manufacturing technologies adopted to produce the various textile structures used in industrial textiles are weaving, knitting, nonwovens, braiding, and felting. Sometimes the textile material may be used in fiber or yarn form in industrial textiles. The industry of industrial textile has characteristics and advantages of intensive capital, high technology, low labor force, qualified worker, and huge domestic market, which are totally different from traditional textile industry with intensive labor and poor technology. At present, the exported amount of industrial textile is relatively low, and most of the products are still sold in domestic market, which leads to the low competitive index of international trade. In the long run, the industry of industrial textile with high added value and advanced technology will have huge potential market and great competitive ability. Accelerating the optimization and upgrading of the industrial structure, the industrial textile still has enormous potential room for growth under stable economic development, especially for filtration materials, composites, and coated textiles. The filtration materials are most likely to fluctuate with the national environment policy. In recent years, the higher environmentally friendly requirement for steel, metallurgy, chemical engineering, cement, and coal industry and definite requirement and norm for filter bag and dust settling pocket make the application of filter material more standardized and offer more potential market for industrial textile. With the pace of industrial upgrading, some traditional large textile manufacturers have become involved in industrial textiles step-by-step, encouraging new vitality for industrial textiles on the basis of their advantages of human resources, finance technology, and management.

1.3 Industrial Textiles: Market Scenario

Technical textiles are one of the faster-growing sectors of the global textile industry. The world textile industry is moving rapidly toward the manufacture of high-added-value textile structures and products such as medical textiles, protective textiles, and smart textiles. Market size of the global consumption of technical textiles in 2000 was consumed nearly 16.7 million tons of fibers and polymers; finished product value $92.9 billion including nonwovens. According to "Technical Textiles World Market Forecast" (David Rigby Associates, 2012), the global market size for technical textiles is expected to reach $127 billion by 2013 at an annual growth rate of 3.5%. India contributes 8% to the global market size of technical textiles. Asia leads the world in terms of textile consumption with 8.5 million tonnes followed by the United States and Europe with 5.8 and 4.8 million tonnes, respectively. According to statistics, the global textile market is worth of more than $400 billions at present. In a more liberalized environment, the industry is facing competition as well as opportunities. It is predicted that global textile production will grow up to 50% by 2014. The world textile and apparel industry has gone into a phase of transformation since the elimination of quota in 2005. It is expected that China will represent around 40% of global trade by 2014. In spite of its significant growth trend, China's rising costs and perceived risks are creating more opportunities for other low-cost countries. It is also expected that India will represent around 18% of global trade by 2014. India is rapidly expanding its role with new capacity buildup in the management control of textile trades through vertical integration. The advantage of vertical integration is that it avoids the holdup problem. Pakistan, Vietnam, Cambodia, and Bangladesh are relying on their low manufacturing costs. Thus, they are building up more capacity in textile manufacturing (Figure 1.1).

1.4 Future Trends of Industrial Textiles

The market for industrial textiles continues to grow, and the future of this sector embraces a wider economic sphere of activity than just manufacturing of textiles. Companies that recognize strategic and tactical implications of this and take action appropriate to their situation have good chances of continuing to be successful. As per some of the published statistics, the consumption by volume of industrial textiles in the world market appears to have continued to grow over recent years. Until recent times, the major driver of innovation in the growth of industrial textile consumption was "technology push," which has been now replaced in many market sectors by "market pull." In "technology push," new materials and textile processes lead to new or extended performance capabilities and product possibilities. These in turn lead to new and extended end-use ideas that can be compared with the actual needs of customers, leading to specific products to meet them as nearly as possible. The main spur for product innovation becomes the customer's problem rather than the available technologies.

Requirements to be met by technical yarns and fibers that include high tenacity, low elongation at break, high modulus, low thermal shrinkage, high thermal stability, high resistance to corrosive chemicals, etc., have placed great challenges to the R&D people at the major fiber producers. In many countries, the debate regarding environmental loading of oil-based polymers has also influenced the development of materials and products. The driving technological force in industrial textiles has thus far been materials development spearheaded by advances in fibers, polymers, and chemical technology. The mechanical processing has not played an important role so

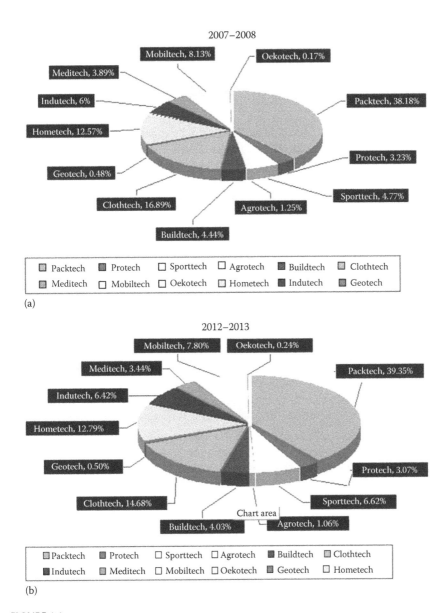

FIGURE 1.1
World consumption of technical textiles: (a) 2007–2008 and (b) 2012–2013 (projected).

far in the sense that very few nonconventional and technical textile-specific machineries have been put in the market. The conventional spinning, weaving, knitting, and nonwoven techniques have been used for producing the majority of items. The coating technology used is also that applicable to apparel and household textiles. This means that in terms of technology, the industry is very flexible in its ability to switch from conventional textiles to industrial textiles.

Industrial textile products and innovations: Innovations enable a company to differentiate its products and take competitive advantage in the market. Many innovations have been made in the field of textile manufacturing:

- *Smart or intelligent textiles* can think for themselves. They can sense and react to external conditions and also retain the aesthetic and technical properties of textile material. They are used to measure strain, temperature, pressure, electric currents, magnetic fields, etc., in defense, aerospace, science and research, nuclear plants, etc.

- *Ultrafine textiles* use tightly woven fabrics that have high resistance to dust, water, and wind. They are also used in the medical field for making bandages, sheets, patients' gowns, curtains, and bedsheets.

- *Electronic textiles or e-textiles* are used for manufacturing fabrics that have electronic interconnections within them. These clothings measure blood pressure, heart rate, body temperature, etc., and relay the data to a computer, cell phone, or other device that could signal for help if the wearer experiences a health problem. Another variation of it may have an MP3 player fitted in it.

- *Antiallergic and antibacterial textiles* can reduce all types of bacterial and fungal allergies like colds and flues; improve sleep, meditation, and relaxation; increase lung capacity, absorption of vitamins B and C, and relief from migraine, respiratory tracks, nose disorders, stress, etc.

- *Antimagnetic textiles and antiradiation fabrics* offer protection against magnetic pull in areas with active magnetic field and from ultraviolet radiations. These textiles are used in industries such as aerospace, aviation, petrochemical, textile, electronics, machinery, and environmental protection.

- *Soluble textiles* can dissolve in water at a temperature ranging from 37°C to 40°C depending on their composition. Sterile hygienic materials can be used to protect patients and medical staff from infections, surgical garments and drapes, face masks, and shoe covers; are used in food industry including food science, agriculture, and ceramics; paper and ink technology; and explosives.

1.5 Technology and Material Trends

The driving technological force in industrial textiles has thus far been materials development spearheaded by advances in fibers, polymers, and chemical technology. Mechanical processing has also played an important role so far in the sense that very few nonconventional and technical textile-specific machineries have been put in the market. Conventional spinning, weaving, knitting, and nonwoven techniques have been used for producing the majority of items. The coating technology used is also that applicable to apparel and household textiles. This means that in terms of technology, the industry is very flexible in its ability to switch from conventional textiles to industrial textiles.

Technology advances in the industry are driven by forces outside the pure textile sector, that is, polymer and fiber producers and in some cases the machinery producers of fabric fabrication techniques. There is a growing need for nontextile application know-how in many segments of the technical textile market. Textile technologists are needed who understand the civil engineering aspects of potential geotextile applications so that suitable textile structures can be produced. Technologists have to understand the mechanical and production engineering aspects of fiber composites in automotive and aeronautical applications to be able to design a suitable textile or fiber-reinforced composite components. Textile engineers have also started using CAD, CAM, and CAE tools not only for designing suitable textile reinforcements but also to have a common language necessary for fruitful cooperation with the design engineers working at car companies. Textile technologists do not always understand the functional requirements of particular application, and often for textile industry, the newer customers do not recognize the particular requirements of the textile company with regard to specifications, tolerances, etc.

Bibliography

David Rigby Associates, Technical textiles and nonwovens: World market forecasts to 2010, Manchester, U.K.

Mauretti, G. J., The outlook for technical textiles worldwide presented to Korean Federation of Textile Industries (KOFOTI) July 13, 2006, Korea.

Medical plastics and pharmaceutical industry, Gujarat, www.medicalplasticsindia.com/mpds/2008/may/coverstory1.htm (accessed May 25, 2013).

Ministry of Industry, Commerce, and Consumer Protection (Mauritius), www.industryobservatory.org/technical.php (accessed May 25, 2013).

Mishra, N. D. and J. N. Sheik, The scope of technical textile in India, *Bombay Technologist*, 55, 66–69 (2005).

Technical Textiles, The future is being woven (or nonwoven) in India, *Textile Review*, (2006) 13–18.

Texline.com, www.teonline.com/industry-overview.html (accessed May 25, 2013).

Texline.com, www.teonline.com/knowledge-centre/study-technical-textiles.html (accessed May 25, 2013).

2

Fiber, Yarn, and Fabric Structures Used in Industrial Textiles

2.1 Fibers in Used Industrial Textiles

During the last decade, there was a remarkable growth of the production of industrial textiles, which definitively promoted this branch as one of the most potential and dynamical fields for the development of the textile industry. The ranges of industrial textile products are very broad, and it comprises various products such as industrial filters, hose and belts, ropes, acoustic materials, battery separators, and composites. Today, we have access to a wide variety of fibers and yarns showing the appropriate characteristics required for producing high-tech textiles. The effects of different engineering and technological parameters on mechanical properties of high-modulus and high-strength polymer fibers and yarns are very important when designing industrial textiles. The fact is that only a small quantity of products (about 2%–3% by volume) utilize high-performance fibers, where as the bulk of the products use conventional textile materials such as polyethylene (PE), polypropylene (PP), polyamide (PA), viscose, cotton, jute, and glass. The technology of industrial textile is not always confined to the products themselves or to the production technology, but it incorporates the "know-how" for the application of these products for a broad range of end users. The development of industrial textile is closely associated to advancements in fiber production.

2.1.1 Natural Fibers

Natural fibers are obtained from plants, animals, and geological processes. Plant fibers include cotton, flax, jute, bamboo, ramie, kapok, hemp, and sisal among others. Flax is the oldest known fiber crop and is used for linen production. Jute fiber is the cheapest and strongest of all natural fiber and ranks second in production after cotton. It is used in traditional packaging fabrics, carpet backing, mats, bags, tarpaulins, ropes, twines, etc. Manila is a bast fiber, which finds application in ropes. Jute grids are used in drainage applications to separate contaminants from water. Cotton was used conventionally in hose manufacturing and low-concentration filtration process. Decatizing cloth made up of cotton-blended fabric is used in decatizing machine, which is a part of the mechanical finishing process of textiles. Bast fibers such as jute, linen, and hemp are used in composites for short-term applications owing to its biodegradability. It is typical that the natural fibers are heavyweight products with limited resistance to moisture, microbes, fungus, and low flame resistance. These limitations reduce the possibility of use of natural fibers in industrial textiles.

2.1.2 Synthetic Fibers

2.1.2.1 Viscose Rayon

Viscose rayon is the first commercial man-made fiber used as reinforcing material for tires and other rubber products like safety belts, conveyor belts, and hoses. The fiber has relatively high uniformity, tenacity (16–30 cN/tex), and modulus, especially if impregnated with rubber. The viscose fiber obtained by a special process of spinning has tenacity up to 40 cN/tex and elongation of 11%–17%. The tires designed for high-quality roads still employ viscose fiber due to better thermal resistance. Majority of all rayon for polymer reinforcement is used as continuous filament, but there is still some use of spun staple rayon, where the main property required is bulk rather than strength. Viscose rayon is used as backing cloth in the coated abrasives owing to tensile strength and flexibility.

2.1.2.2 Polyamide

PA fiber is characterized by high tenacity, elasticity, resistance to abrasion, and moisture. Capability of energy resilience is a condition for an application in manufacturing climbing ropes and linen for parachutes and sail fabrics. The typical application of PA is for reinforcing tires for use at low-quality roads and of road vehicles. Nylon is used as facing fabric for the transmission belts due to its better abrasion property. The usage of PA fibers, due to its poor acid resistance, is limited in industrial filtration. PAs are used as carcass material in conveyor as well as transmission belts due to its better adhesion property. The better abrasion resistance of PAs finds its application in the manufacturing of industrial brushes. Bolting fabrics made up of nylon is used as screen in screen printing operation. The nylon 6 yarn is woven into a fabric, which is then cut into the required size for making computer printer ribbons. The reasons attributed for the selection of nylon 6 are tensile strength, capillary action, scratch resistance, and heat resistance.

2.1.2.3 Polyester

Production of polyester made it possible to get fibers for technical applications at a lower price compared to PA and viscose fibers. Polyester is used as reinforcement material in tire cord, hose, and conveyor belts because of its superior mechanical properties. Polyester has good resistance toward acid and moisture, which finds its application in liquid filtration process of highly acidic in nature. Due to its excellent moisture resistance, it is used as the reinforcement fabric in water hoses. Polyester cords are widely used as the carcass material in the transmission belts. The lower modulus, elasticity, and better shock absorption property of polyester enable the belt to rotate smoothly over small-diameter pulleys. Polyester is used as the backing cloth in coated abrasives. Polyester fabrics are used as paper-making fabrics due to its good drainability, abrasion resistance, and moisture resistance.

2.1.2.4 Polyolefin Fibers

PE and PP are the polyolefin fibers that contribute significantly to the industrial textiles. Advantages of polyolefin fibers are low price, low specific gravity, good abrasion resistance, and low moisture content. These properties have determined their usage in a range of technical applications such as ropes, filter fabrics, nets, etc. PE is used in battery separators due to its excellent chemical resistance.

2.1.2.5 High-Performance Fibers

The development of carbon fibers and aramid fibers in the 1960s triggered many developments of high-performance fibers and yarns. Today access to a wide variety of fibers and yarns showing the appropriate characteristics is required for producing high-tech textiles. These include high modulus/ high tenacity, heat resistance, and stability to chemicals even at elevated temperatures. Aramid is used as reinforcement material in the specialty conveyor belt where high strengths and modulus are required. The typical properties of aramid fiber are low density, high strength, good impact resistance, good abrasion resistance, good chemical resistance, good resistance to thermal degradation, and compressive strength similar to E-glass fibers. The characteristic property of aramid fiber is a high melting temperature of 370°C (compared to 248°C for conventional PA). Due to such a property, the use of aramid fiber is extended to high-temperature applications. PE processed by extended highly oriented chain structure has got much higher strength. The extension of polymer chains and high longitudinal orientation is a precondition for accomplishing high mechanical properties. The result of this treatment is the production of high-performance

PE fiber (HPPE), of so far the highest strength of 400 cN/tex, that is, two times higher than aramid fiber.

The fiber is later commercialized under several versions of which the most known are Dyneema® and Spectra. The advantages of HPPE are low specific gravity (0.396 g/cm³), almost about half less than a high-modulus carbon fiber, and about one-third less than the aramid fiber. The fiber has a low melting temperature (~150°C), which restricts the possibility for the high-temperature application. Carbon fiber can be manufactured from several precursors, of which rayon and acrylic are the most usually employed. Carbon-reinforced composites find application in civil aviation, special sport, and industrial goods, such as turbine parts for generators and reinforced fuel tanks. Glass fiber is at great extent accepted in the production of high-performance composite materials, including protective materials, various filters, protective clothing, and packing. Nonwoven glass mats are used as battery separators due to its superior resistance to chemical, oxidation, and contaminants. Dimensional stability and high strength-to-weight ratio of glass fiber are also a reason for its use in battery separators and many composite applications. Glass fiber prepregs are used in printed circuit board owing to its uniform dielectric constant, lower dissipation factor, and dimensional stability.

The development of technical textile is closely associated to advancements in fiber production. The creation of PA fiber (Carothers, 1930) gave a direction for the development of polymer technology, followed by the invention of polyester, PE, PP, and carbon fibers. In recent times, high-performance fibers such as aramid, UHMW-PA, and HP-PE that had an extraordinary significant influence for the development of industrial textile were obtained. It is to expect that the knowledge gained so far from manufacturing high-performance fiber would be of benefit in realizing the predetermined goal of processing technical fiber of fantastic tenacity of 900 cN/tex (100 g/denier). This would mean that a great number of products, the metals and other traditional constructive materials, would be replaced. The typical properties of some high-performance fibers are depicted in Table 2.1.

Numerous multifunctional fibers are nowadays available on the market, offering a diversity of improved functional properties. In addition to thermally adaptable fibers (e.g., hollow high-loft PET fibers, PP/PET blend fibers, and hollow fibers containing water-soluble phase change materials), a new generation of fibers based on a multiproperty holistic concept are developed for the use in automotive interiors, battery warmer, outdoor architectural structures, protective clothing for bullet-proof vests, geotextile, agriculture, etc.

2.1.2.6 Specialty Fibers

Bicomponent technologies have made significant advancements since their introduction in the mid-twentieth century. Fibers with different cross sections

TABLE 2.1

Mechanical Characteristics of High-Performance Fibers

Fiber Type	Tensile Strength (MPa)	Tensile Modulus (GPa)	Density (g/cm³)	Failure Strain (%)
Aramid	3450–3620	112–179	1.44–1.47	1.9–6.2
E glass	3448	72.4	2.54	3.5
Vectran HS	3210	135	1.41	2.3
Carbon	2900–4800	230–390	1.74–1.81	0.7–1.8
HPPE	3090	172	0.97	1.8

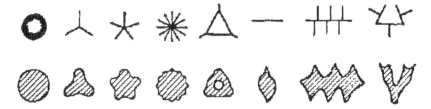

FIGURE 2.1
Polyamide fibers of various cross sections.

to enlarge the fiber surface enhancing the performance are increasingly being used that are shown in Figure 2.1. Fibers that show a profiled cross section possess a wider specific surface that makes the separation of particle smaller than 5 μ more effective. Factors influencing these characteristics are the cross section, shape, and microfibrillation of such materials. Both finest fibers and microfibers are, because of their filter surface, preferably arranged and used to enlarge the effective filter surface.

2.2 Yarn Formation

Yarn is a continuous strand that is made up of filaments or fibers. It is used to make textiles of different kinds. Yarn formation methods were originally developed for the spinning of natural fibers including cotton, linen, wool, and silk. Since the overall physical characteristics of the fibers and processing factors needed differed from fiber to fiber, separate processing systems were developed. As synthetic fibers were introduced, synthetic spinning systems for texturized and untexturized cut staple were developed as modifications of existing staple systems, whereas spinning systems for texturized and untexturized filament were developed separately. Yarn has different forms such as staple fibers, monofilament, multifilament, tow, and textured yarn.

2.2.1 Staple Fiber Yarns

Staple fiber yarns can be manufactured in short-staple and long-staple spinning systems. In natural fiber spinning system, fiber bale is opened, cleaned, and converted to sliver form. The sliver is then drafted and twisted into roving and subsequently into yarn form. The object of spinning and of the processes that precede it is to transform the single fibers into a cohesive and workable continuous-length yarn. Yarn making from staple fibers involves picking (opening, sorting, cleaning, blending), carding and combing (separating and aligning), drawing (reblending), drafting (drawing into a long strand), and spinning (further drawing and twisting), Staple yarns, made from shorter fibers, require more twist to provide a sufficiently strong yarn; filaments have less need to be tightly twisted. For any fiber, yarns with a smaller amount of twist produce fabrics with a softer surface; yarns with considerable twist, hard-twisted yarns, provide a fabric with a more wear-resistant surface and better resistance to wrinkles and dirt, but with a greater tendency to shrinkage. Hosiery and crepe fabrics are made from hard-twisted yarns. Short-staple spinning is the logical development of the cotton spinning history, but the range of fibers has increased dramatically in this century. Short-staple spun yarns can be manufactured by several spinning techniques such as ring, rotor, air jet, friction, and twistless spinning. The main advantage of staple fiber spinning system is the possibility of altering different fiber blend ratio to achieve the desirable properties in the yarn.

Staple fiber yarn structure is predominantly used in filtration process that needs high dust retention capacity. Manufactured fibers used in textile manufacture come from both natural and man-made sources. Natural sources are either organic or inorganic. Organic materials include those from plant cellulose or rubber and those from manufactured polymers. Those from polymers, derived primarily from petroleum, coal, and natural gas, include polyesters, acrylics, nylon, PE, PP, polyvinylchloride, polyurethane, and synthetic rubbers. Synthetic fibers made from cellulose include rayon, acetate, and triacetate. Inorganic fiber materials include metal and glass. Man-made staple fibers are made from tow, which is extruded in the same basic way as with filament yarns; however, the number of filaments involved is vastly larger. Filament tow is cut into staple fibers according to the end-use requirement. Synthetic staple fibers usually arrive at the mill in compacted bales containing about 500 lb of fiber. Synthetic staple fibers are blended with other fibers in blow room or draw frame and then subsequently processed in speed frame and ring frame machines. Spinning process involves drafting, twisting, and winding of the yarn into the cop.

2.2.2 Filament Production

The term "spinning" is also used to refer to the extrusion process of making synthetic fibers by forcing a liquid or semiliquid polymer through small holes in an extrusion die, called a spinneret, and then cooling, drying, or coagulating the resulting filaments. The fibers are then drawn to a greater length to align the molecules to increase their strength. The monofilament fibers may be used directly as-is, or may be cut into shorter lengths, crimped into irregular shapes, and spun with methods similar to those used with natural fibers. Extruded filament yarn manufacture is a short mechanical process involving only one or two steps. The term "spinning" here defines the extrusion process through spinnerets of fluid polymer masses that are able to solidify in a continuous flow. The polymer processing from the solid to the fluid state can take place by two methods:

1. *Melting:* this method can be applied on thermoplastic polymers that show stable performances at the processing temperatures.
2. *Solution:* the polymer is dissolved in variable concentrations of solvent according to the kind of polymer. Solvent evaporation is carried out by either coagulation or evaporation.

The three main stages of melt spinning polymers are preparation of the melt, extrusion, and winding. A molten polymer is forced through a spinneret orifice at a given temperature, pressure, and rate. The flow is collected at a different velocity at the site of "take-up." The distance between the spinneret and take-up is variable. Once the polymer reaches the take-up area, the process of initial fiber formation through solidification and cooling is finished. Solution spinning is classified into wet spinning and dry spinning. In the wet-spinning process, dissolved polymer solution is pumped through a spinneret that is submerged in a coagulating bath. The bath contains the solvent and water. As the polymer solution passes through the coagulant or nonsolvent, a phase change occurs whereby the solvent diffuses out and the nonsolvent diffuses in. The newly formed fibers emerge in a gel form from the bath where they are later subjected to a series of after-treatments. In contrast to wet spinning where nonsolvents diffuse into and solvents diffuse out of the dope, formation of fibers by dry spinning occurs from solvent evaporation. Evaporation is encouraged through a hot inert gas as the gel passes out of the spinnerets.

2.2.3 Textured Yarn Production

Texturization is the process of imparting crimp to the man-made fiber. In this way, the yarn contains the many air pockets needed to produce insulation properties, permeability, and softness. Furthermore, the yarn now occupies a greater volume; the greater the bulk, the better the cover. Also the yarn becomes more extensible, and this, too, is an added attraction. It is possible to get various combinations of stretch and bulk. Texturization process is done using several techniques such as false twist, air jet, knit-de-knit, gear, stuffer box, and draw texturization.

2.2.4 Doubled Yarn Production

For sewing threads, as well as certain specialty and industrial yarns, it is necessary to ply (i.e., to double or fold) the yarns to give them a smoother and less hairy character. Doubling improves the evenness; plying balances torque if carried out correctly and binds some of the hairs on the component yarns. The traditional methods include assembly winding to place the single yarns parallel to each other as a closely spaced pair (or group) of yarns on an intermediate package. The new package is then used as a feed for a twisting machine, and the output is a plied yarn. However, the cost of assembly winding approaches 25% of the total winding costs, and the system is prone to problems.

2.3 Fabric Structures

2.3.1 Fabric Formation

Fabric formation technology can be broadly classified into weaving, knitting, and nonwoven.

2.3.1.1 Weaving

The conversion of yarn into woven fabric is accomplished by interlacing warp and the weft on a weaving machine or loom. In a weaving machine, the warp yarns are passed from a warp beam to the fabric beam. During this process, each warp yarn is led through an eye on a heald attached to a harness. The harness lifts some of the warp yarns and depresses the remainder to form a gap between them known as a "shed" through which the weft is inserted. This operation is known as shedding, and the insertion of weft is called "picking." The sley beats up the weft to the edge of the woven fabric.

Weaving machines are classified according to the type of weft insertion system, which at present includes shuttles, rapiers, projectiles, air jets, and water jets. Woven fabrics find application in high-velocity filtration where mechanical properties of filter fabric are crucial in deciding life of filter fabric. Woven fabrics are used as reinforcements in hoses, conveyor and transmission belts, and composite materials owing to its superior mechanical properties.

2.3.1.2 Woven Structures

Woven structures have the greatest history of application in textile manufacturing. Woven fabrics are made on looms in a variety of weights, weaves, and widths. Woven structures are bidirectional, providing good strength in the direction of warp yarn. The threads that run along the length of the fabric are called warp or ends, while the threads that run along the width of the fabric from selvedge to selvedge are referred as weft or picks. Woven fabrics are produced by the interlacing of warp (0°) fibers and weft (90°) fibers in a regular pattern or weave style. The fabric's integrity is maintained by the mechanical interlocking of the fibers. Mechanical properties of woven fabrics, which are especially important for industrial textile, depend on the type of raw materials, type and count of warp and weft yarns, yarn density, and the type of weave structure. Drape (the ability of a fabric to conform to a complex surface), surface smoothness, and stability of a fabric are controlled primarily by the weave style. However, the tensile strength of woven fabrics is compromised to some degree because fibers are crimped as they pass over and under one another during the weaving process. The crimp influences the fiber volume fraction, fabric thickness, and fabric mechanical properties. Fiber volume fraction and fabric thickness are the important process parameters influencing the properties of textile composites. The cloths are lighter in weight, typically from 6 to 10 oz/square yard, and require about 40–50 plies to achieve a 1 in. thickness. Impact resistance is enhanced because the fibers are continuously woven. Fabric area density and cover factor influence strength, thickness, stiffness, stability, porosity, filtering quality, and abrasion resistance of fabrics. The woven structures can be broadly classified as given in the following sections (Figures 2.2 through 2.4).

2.3.1.2.1 Plain Weave

The structure where warp yarns alternatively lift and go over across one weft yarn and vice versa is the simplest woven structure called plain weave. However, it is the most difficult of the weaves to drape, and the high level of fiber crimp imparts relatively low mechanical properties compared with the other weave styles. With large fibers (high tex), this weave style gives excessive crimp, and therefore it tends not to be used for very heavy fabrics.

2.3.1.2.2 Twill Weave

Twill is a weave that produces diagonal lines on the face of a fabric. One or more warp yarns alternately weave over and under two or more weft yarns in a regular repeated manner. The direction of the diagonal lines viewed along the warp direction can be from upward to the right or to the left making Z or S twill. This produces the visual effect of a straight or broken diagonal "rib" to the fabric. Compared to plain weave of the same cloth parameters, twills have longer floats, fewer intersections, and a more open construction. Superior wet-out and drape are better in the twill weave over the plain weave.

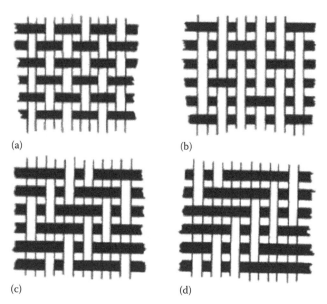

FIGURE 2.2
Typical traditional woven fabric constructions: (a) plain weave, (b) satin weave, (c) 2/2 twill weave, and (d) 3/3 twill weave.

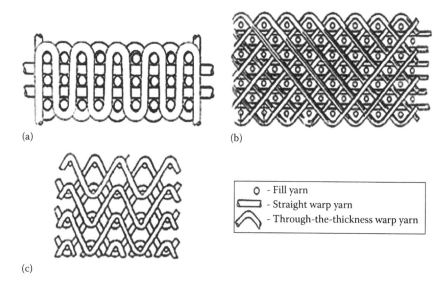

O - Fill yarn
⬭ - Straight warp yarn
⋀ - Through-the-thickness warp yarn

FIGURE 2.3
Constructions of multidirectional woven fabrics: (a) 3-D orthogonal, (b) full-depth warp interlock, and (c) angle interlock.

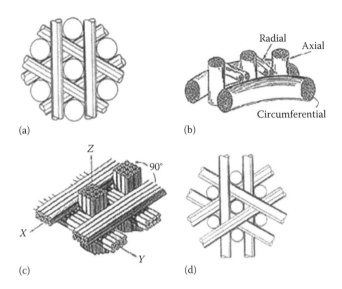

FIGURE 2.4
Advanced woven structures for industrial textiles: (a) 4-D in plane, (b) 3-D cylindrical, (c) 3-D orthogonal, and (d) 4-D pyramidal.

2.3.1.2.3 Satin Weave

A weave where binding places arranged to produce a smooth fabric surface free from twill lines is called satin. The "harness" number used in the designation (typically 4, 5, and 8) is the total number of yarns crossed and passed under, before the yarn repeats the pattern. The five-end satin is most frequently used for technical applications for providing firm fabric although having moderate cover factor. Satin weaves are very flat, have good wet-out, and a high degree of drape. The low crimp gives good mechanical properties. Satin weaves allow fibers to be woven in the closest proximity and can produce fabrics with a close "tight" weave. However, the style's low stability and asymmetry need to be considered.

2.3.1.2.4 Leno Weave

A form of plain weave in which adjacent warp yarns are twisted around consecutive weft yarns to form a spiral pair, effectively "locking" each weft in place, thus securing a firm hold on the filling yarn and preventing them from slipping out of position. It is also called the gauze weave. Leno weave improves the stability in "open" fabrics that have a low fiber count (Figure 2.5).

2.3.2 Noncrimp Fabrics

In noncrimp fabrics, yarns are placed parallel to each other and then stitched together using polyester thread. Warp unidirectional fabric is used when

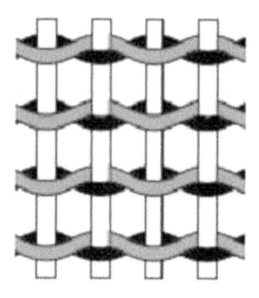

FIGURE 2.5
Leno weave.

fibers are needed in one direction only, for example, in stiffness-critical applications such as water ski applications where the fabric is laid along the length of the ski to improve resistance to bending (Figure 2.6). Noncrimp fabrics offer greater flexibility compared to woven fabrics. Noncrimp fabrics offer greater strength because fibers remain straight, whereas in woven fabrics, fibers bend over each other. Noncrimp fabrics are available in a thick layer and thus an entire laminate could be achieved in a single-layer fabric.

FIGURE 2.6
Noncrimp fabric.

2.3.3 Knitted Fabrics

Knitting is the second most frequently used method, after weaving, that turns yarns or threads into fabrics. It is a versatile technique that can make fabrics having various properties such as wrinkle resistance, stretchability, better fit, particularly demanded due to the rising popularity of sportswear, and casual wears. As of present day, knitted fabrics are used widely for making hosiery, underwear, sweaters, slacks, suits, and coats apart from rugs and other homefurnishings. A knitted fabric may be made with a single yarn, which is formed into interlocking loops with the help of hooked needles. According to the purpose of the fabric, the loops may be loosely or closely constructed. Knitted fabrics are textile structures assembled from basic construction units called loops. There exist two basic technologies for manufacturing knitted structures: weft- and warp-knitted technology. Knitted fabrics are easy to handle and can be cut without falling apart. A knitted reinforcement is constructed using a combination of unidirectional reinforcements that are stitched together with a nonstructural synthetic such as polyester. A layer of mat may also be incorporated into the construction. The process provides the advantage of having the reinforcing fiber lying flat versus the crimped orientation of woven structure. Knitted fibers are most commonly used to reinforce flat sections or sheets of composites, but complex three-dimensional (3-D) performs have been created by using prepreg yarn.

2.3.3.1 Weft-Knitted Fabrics

The feature of the weft-knitted fabric is that the loops of one row of fabric are formed from the same yarn. The horizontal row of loops in a knitted fabric is called a course and the vertical row of loops is called a wale. The stitch density is the number of stitches per unit area in the knitted fabric. The stitch length is the length of a yarn in a knitted loop and is an important factor that determines the properties of the weft-knitted fabric. The cover factor is a number that indicates the extent to which the area of a knitted fabric is covered by the yarn. The higher cover factor indicates a more tight structure and vice versa. The fabric area density is a measure of the mass per unit area of the fabric. In weft-knitted fabrics, the loops are formed successfully along the fabric width. The yarn is introduced more or less under the right angle regarding the direction of the fabric formation. The feature of the weft-knitted fabric is that the neighboring loops of one course are created of the same yarn (Figure 2.7). The simplest weft-knit structure produced by the needles of one needle-bed machine is called plain knit or jersey knit. The plain knit has a different appearance on both sides of the fabric. A structure produced by the needles of both needle beds is called rib structure or double jersey having the same appearance on both sides of the fabric. Weft-knitted fabric can be produced on a number of

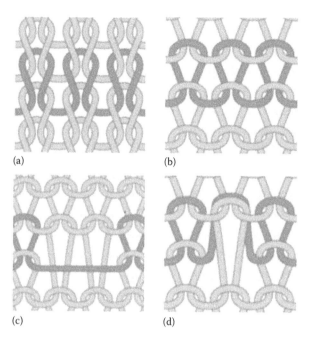

FIGURE 2.7
Weft-knitted fabric structures: (a) knit stitch, (b) purl stitch, (c) missed stitch, and (d) tuck stitch.

different types of knitting machines. Circular or flat bar machines using a latch needle can produce both fabrics and knitting garments. Straight bar or circular machines using a bearded needle can produce shaped knitwear. Many machines can produce a double fabric structure with differing knitted structure on each fabric face. Spacer yarns can be inserted between the front and back fabrics, thus creating a complex three-layer structure. The properties of each layer are determined by the fiber, yarn properties, and structure of that layer. These structures can be tailored for specific applications and are useful in the protective textile field (Figure 2.8).

In the more common weft knitting, the wales are perpendicular to the course of the yarn. In warp knitting, the wales and courses run roughly parallel. In weft knitting, the entire fabric may be produced from a single yarn, by adding stitches to each wale in turn, moving across the fabric as in a rasterscan. By contrast, in warp knitting, one yarn is required for every wale. Warp-knitted fabrics such as tricot and milanese are resistant to runs and are commonly used in lingerie.

2.3.3.2 Warp-Knitted Fabrics

In warp-knitted technology, every loop in the fabric structure is formed from a separate yarn called warp, which is mainly introduced in the longitudinal

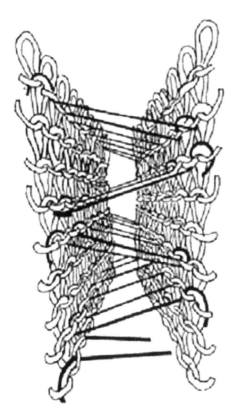

FIGURE 2.8
Double-layered weft-knitted fabric construction.

fabric direction. The most characteristic feature of the warp-knitted fabric is that neighboring loops of one course are not created from the same yarn. To accomplish the warp-knitted structure, every needle along the width of the fabric must receive yarn from the individual guide. The function of the guide is to lead and wrap the warp yarn around the knitting needle during the knitting process. The loop structure in the warp-knitted and weft-knitted structure is similar in appearance (Figure 2.9). The warp-knitted structure is very flexible, and regarding construction, it can be elastic or inelastic. The mechanical properties are in many cases similar to those of woven structures. The best description of warp-knitted fabrics is that they combine the technological, production, and commercial advantages of woven and weft-knitted fabrics. Warp-knitted fabrics can be produced on a number of different types of knitting machine. Raschel machines using a latch or compound needle produce high pile upholstery fabrics, industrial furnishing fabrics, and bags for vegetables. Tricot machines using bearded or compound needle produce lace, nets, and outerwear fabrics. Weft insertion with, for example, elastic yarns or fleeces can produce directionally orientated fabrics. Warp-knitted fabrics

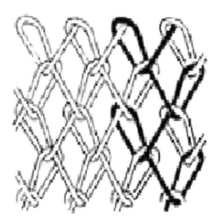

FIGURE 2.9
Warp-knitted fabric construction.

are commonly used in linings for protective clothing and laminated with polyurethane foams to provide a strong flexible base for the foam.

2.3.3.3 Warp Knitting versus Weft Knitting

Warp knits, which generally have a flat, smooth surface (though they can also be made with a pile), have little or no vertical stretch and varying degrees of crosswise stretch. Produced in a large variety of weights in a wide range of fiber types, warp knits are run-resistant and do not ravel. With a few exceptions, weft knits have moderate to great amounts of crosswise stretch and some lengthwise stretch (some jerseys, however, have little or no crosswise or lengthwise stretch). On many weft knits, the edges may curl. As with warp knits, weft knits are made from many different fibers and come in many weights. If a stitch in a weft knit is broken, the fabric will tend to run, but a weft knit ravels only from the yarn end knitted last.

2.3.4 Multiaxial Fabrics

In recent years, multiaxial fabrics have begun to find favor in the construction of composite components. The main fibers can be any of the structural fibers available in any combination (Figure 2.10). Multiaxials are nonwoven fabrics made with unidirectional fiber layers stacked in different orientations and held together by through-the-thickness stitching, knitting, or a chemical binder.

The proportion of yarn in any direction can be selected at will. In multiaxial fabrics, the fiber crimp associated with woven fabrics is avoided because the fibers lie on top of each other, rather than crossing over and under. This makes better use of the fibers' inherent strength and creates a fabric that

FIGURE 2.10
Multiaxial fabric.

is more pliable than a woven fabric of similar weight. Super-heavyweight nonwovens are available (up to 200 oz/yd^2) and can significantly reduce the number of plies required for a layup, making fabrication more cost-effective, especially for large industrial structures. High interest in noncrimp multiaxials has spurred considerable growth in this reinforcement category.

2.3.5 Braided Fabrics

Braid is a rope-like thing, which is made by interweaving three or more strands, strips, or lengths, in a diagonally overlapping pattern. Braiding is a predominant manufacturing technique for developing textile-reinforced hoses and ropes. Braiding is one of the major fabrication methods for composite reinforcement structures. Braiding is probably the simplest way of fabric formation. Diagonal interlacing of yarns forms a braided fabric. Each set of yarns moves in an opposite direction. Braided fabrics are continuously woven on the bias and have at least one axial yarn that is not crimped in the weaving process. The braid's strength comes from intertwining three or more yarns without twisting any two yarns around each other. This unique architecture offers, typically, greater strength-to-weight ratio than woven fabrics. It also has natural conformability, which makes braid especially suited for production of sleeves and preforms because it readily accepts the shape of the part that it is reinforcing, thereby obviating the need for cutting, stitching, or manipulation of fiber placement. Braids also are available in flat fabric form. These can be produced with a triaxial architecture, with fibers oriented at 0°, +60°, and −60° within one layer. This quasi-isotropic architecture within a single layer of braided fabric can eliminate problems associated with the layering of multiple 0°, +45°, −45°, and 90° fabrics.

Furthermore, the propensity for delamination (layers of fiber separating) is reduced dramatically with quasi-isotropic braided fabric. Its 0°, +60°, and −60° architecture gives the fabric the same mechanical properties in every direction, so the possibility for a mismatch in stiffness between layers is eliminated. In both sleeve and flat fabric forms, the fibers are continuous and mechanically interlocked. Because all the fibers in the structure are involved in a loading event, the load is evenly distributed throughout the structure. Therefore, braid can absorb a great deal of energy as it fails. Braiding is the most widely used manufacturing technology in the development of rope and hose. Braid's impact resistance, damage tolerance, and fatigue performance have attracted composite manufacturers in a variety of applications, ranging from hockey sticks to jet engine fan cases. Three-dimensional braiding is a relatively new topic and mainly developed for industrial composite materials.

Braiding is done by intertwining of yarns in whatever direction suited to the manufacturer's purpose. Braiding can be classified as 2-D and 3-D braiding. Two-dimensional braid structure can be circular or flat braid. They are formed by crossing a number of yarns diagonally so that each yarn passes alternately over and under one or more of the others. Two-dimensional braids are produced through circular braiding machine and rotary machine (Figure 2.11).

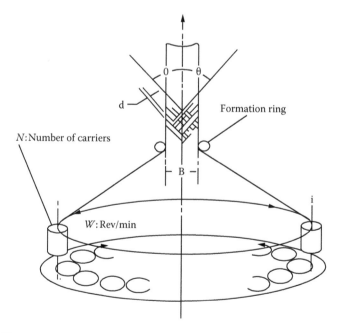

FIGURE 2.11
Braided structure formation in circular braiding machine.

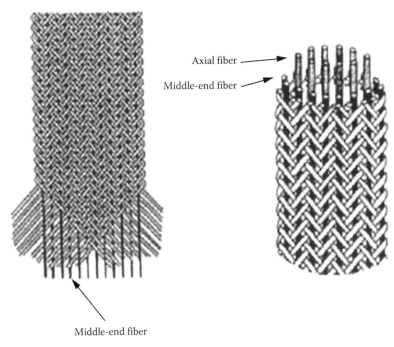

Axial fiber

Middle-end fiber

Middle-end fiber

FIGURE 2.12
Braided structure.

Three-dimensional braiding is relatively new and was developed mainly for composite structures. In it, a 2-D array of interconnected 2-D circular braids is created on two basic types of machines: the horn gear and cartesian machines. Braided fiber architecture resembles a hybrid of filament winding and woven material. Like filament winding, tubular braid features seamless fiber continuity from end to end of a part. Like woven material, braided fibers are mechanically interlocked (Figure 2.12). When functioning as a composite reinforcement, braid exhibits exceptional properties because it distributes loads very efficiently.

2.3.5.1 Types of Braided Structures

Braided textile structures are manufactured with mutual intertwining of yarns in a tubular form. There are three typical braid structures: diamond, regular, and hercules. Diamond structure is obtained when the yarns cross alternatively over and under the yarns of opposite direction. The repeat notation is 1/1. Regarding this way of notation, the regular braid structure has notation 2/2 and hercules 3/3. The braids are mostly produced in a regular structure. Generally braids are produced in a tubular form of biaxial yarn direction. By insertion of longitudinally oriented yarns (middle-end fiber) into the structure, the triaxial braids are obtained. Moreover, in the center of the tubular braid, additional fibers called axial fibers can be inserted. When the number of braiding fiber bundles is the same, the tubular braid increases the fiber volume fraction more than the flat braid. The main feature of the braid is the angle of intertwining that can vary between 10° and 80° and depends on the yarn fineness, the type of the structure (biaxial or triaxial), cover factor (tightness of the structure), and the volume ratio of the longitudinal yarns. Since the braids have tubular form, they are often replaced with the filament winding structures. In this respect, it has been proven that the braids can be competitive regarding the price. The braid is a flexible product and can be adjusted to various shapes. With the special device called mandrel, the braids can be shaped into various forms directly on the machine at the manufacturing stage.

Braided composites, once used for such applications as drive shafts, propeller blades, and sporting equipment, are becoming popular again in recent years partly due to the development of large, computer-controlled 2-D and 3-D braiders and partly due to the experience gained in using textile composites in the aerospace and automotive industries. Braiding has the potential to produce complex near-net shapes with fiber continuity at the edges and around holes and branches. However, unlike other quasi-laminar composites, the unit cell geometry of a braided composite is controlled by both the machine parameters and the component geometry.

2.3.5.2 Conventional 2-D Braided Structures

Braided structures may be classified as (1) 2-D braids produced on conventional maypole braiders and (2) 3-D through-thickness braids produced on specialized machinery. Three-dimensional braiding was popular in the 1980s for aerospace applications. However, in recent years, composites industry has been taking a fresh look at 2-D braids for developing affordable composite structures (Figure 2.13). For example, stitched 2-D braided performs are being used for stiffeners and stringers in aircraft structures.

2.3.5.3 Multiaxial Differentially Oriented Structures

Multiaxial differentially oriented structures (DOSs) either using Karl Mayer's warp-knitting-based method with variations in axial orientation of construction yarns or using LIBA's method of multiple weft-yarn stations give very interesting possibilities of producing technical textiles for a number of end-use applications. Karl Mayer's DOS incorporating thermoplastic yarns or split films as matrix material has been used to produce high-performance composites. This material is also suitable as substrate for coated products, and this technology allows incorporating nonwovens and other cellulose-based material for introducing bulk in these structures. Because the inlaid yarns in DOS are placed straight without any built-in crimp, the resultant

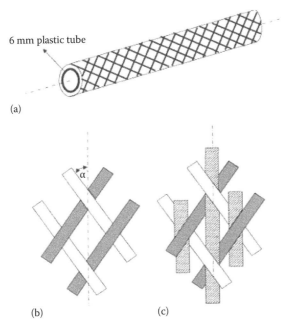

FIGURE 2.13
Triaxial braided structure for gears: (a) plastic core braided, (b) two-filament braids, and (c) multifilament braids.

stress distribution is an interesting factor in designing products for different applications where load-bearing aspect is important (Figure 2.14).

2.3.5.4 Biaxial Structures

Biaxial technical structures are formed as a combination of magazine weft insertion and a mislapping guide bar to obtain the lengthwise and width-wise reinforcements respectively.

2.3.5.5 Multiaxial Structures

Multiaxial multi-ply structures are fabrics bonded by a loop system, consisting of one or several yarn layers stretched in parallel. Said yarn layers may have different orientation and different yarn densities of single ends. Multiaxial multi-ply fabrics are used to reinforce different matrices. The combination of multidirectional fiber layers and matrices has proved capable of absorbing and distributing extraordinarily high strain forces. The typical feature of warp-knitted multiaxial multi-ply structures for substrates is the interlacing of single ends in line with the stitch courses and the associated almost smooth processing of fed yarns. These products are multi-ply structures with angles of 30° up to 60° and/or 90°/0°. A multiaxial multi-ply fabric

Monoaxial structures

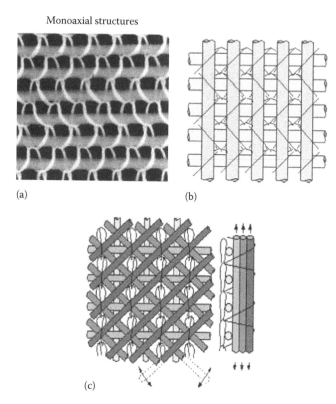

(a)   (b)

(c)

FIGURE 2.14
Braided structures: (a) monoaxial, (b) biaxial, and (c) multiaxial.

is demonstrating the two diagonal yarn sets in addition to the biaxial yarn sets created by the mislapping guide bar (warp or ST-yarn) and the magazine weft insertion system (weft).

2.3.5.6 Multiaxial Structures for Fiber-Reinforced Plastics

The typical feature of warp-knitted multiaxial multi-ply structures for fiber-reinforced plastics is the stitching-through principle, ensuring a uniform distribution of yarn ends and preventing gap formation. In addition, fibrous webs, films, foams, and other materials can be incorporated. Angle positions of −45° through 90° up to +45° and 0° are generally used, being infinitely variable. Multiaxial multi-ply structures are particularly suitable to fiber-reinforced plastics due to their noncrimped and parallel laid structures.

The special characteristics of such composites are the following:

- Low specific weight
- Adjustable stiffness between extremely stiff and extremely stretchable

- Resistance to corrosion and chemicals
- Highest mechanical load resistance

Applications:

- Rotor blades for wind power stations
- Molded parts for automotive, aircraft, and ship building
- Equipment for sports and leisure-time activities, for example, skis, snowboards, surfboards, sports boats

2.3.6 Nonwoven Structures

Nonwoven technology is a newer one regarding other technologies of fiber forming assemblies. A broader definition is that nonwoven is a flexible structure manufactured by bonding or interlocking of fibers, or both, accomplished by mechanical, chemical, thermal, or solvent means and combination of thereof. Nonwoven products make up more than 35% of industrial textile processing, and more than 90% of the nonwoven fabric is used for industrial textiles. The versatile and innovative nature of these nonwoven fabrics makes them an increasingly important raw material for industrial textiles. Nonwoven technology has made a breakthrough in the areas of isotropy, uniformity, hand feel, and thickness of the products. Nonwoven filter media are produced by forming a mat of fibers. The fiber diameter, orientation, packing density, and web weight all determine the filter media properties. The smaller the nonwoven web pore size, the finer the filtration efficiency. Nonwoven fabrics dominate nearly more than 90% of the filtration market: the reasons attributed are variability and economical manufacture. They can be easily adapted to all kinds of filtration jobs. The nonwovens generally are produced with polyolefin, polyester or nylon polymers, and fibers. Melt-blown media are one of the most versatile nonwovens for liquid filtration and the only one that can achieve submicron filtration. Melt-blown media have nominal ratings from 1 to 50 μm and, when calendered or laminated into composites, can have submicron and absolute ratings. Wet-laid media are generally produced on a paper machine and with cellulose, polymeric, or glass fibers. High-efficiency wet-laid media have equivalent efficiency as microfiltration (MF) membranes but with significantly higher dirt-holding capacity or life.

Fabrics made by nonwoven technology can be made up to five times more durable than conventional textile fabrics of the same weight. They can be designed to be extremely abrasion and heat resistant. Some fabrics can withstand extremely high temperatures, for example, mechanical bonded glass fibers can be used at operating temperatures up to 1000°F and silica materials can be used up to 2000°F. For acoustic insulation, nonwoven webs are used that weigh 50% less than any comparable material and provide the same or higher absorption values. Nonwoven webs have high barrier properties; they can filter almost anything ranging from macro- to nanoscale particle sizes. The nonwovens have a broad range of performance: from light materials for wadding and insulation where the fiber volume ratio equals only 2%–3%, to compact fabrics for reinforcement where the fiber volume ratio comes up to 80%. Multilayer combinations of woven and nonwoven fiber web can serve as noise absorption elements in a wide range of applications including acoustic ceilings, noise-reducing quilts, and noise-proof barriers.

Nonwovens are a class of fabric that are produced directly from fibers and in some cases directly from polymers, thereby obviating a number of intermediate processes such as spinning, winding, warping, and weaving/knitting. Hence nonwovens can be produced inexpensively for both single use and durable applications. Nonwovens are produced in two distinct steps:

1. Web formation: arrangement of fibers into a 2-D sheet
2. Consolidation: bonding the fibers together to create a nonwoven fabric

2.3.6.1 Web Formation Methods

Web formation may be classified into dry-laid, spun-laid, and wet-laid processes.

2.3.6.1.1 Dry-Laid Process

Dry textile fibers are carded, using a carding machine similar to the one used in the spinning industry, to arrange the fibers in a 2-D sheet with fiber orientations predominantly in the machine direction. The web is subsequently folded using a cross-lapping machine to increase the web thickness and to achieve transverse fiber orientation. In some cases, conventional carded and cross-lapped webs are combined to produce a web with bidirectional fiber orientation. Alternatively, an aerodynamic system is used for creating a web random fiber orientation.

2.3.6.1.2 Spun-Laid Process

This is a method of producing fabrics directly from polymer chips, hence eliminating the entire textile supply chain. Fibers are extruded from a spinneret similar to conventional melt spinning process. These fibers are attenuated (stretched) using high-velocity air streams before depositing on a conveyer in a random manner. The spun-laid process is the most commonly used method for producing both disposable and durable nonwovens for protective application. There are other related systems such as flash spinning, melt blowing, and electrospinning. Flash spinning involves extrusion of a polymer film dissolved in a solvent; subsequent evaporation of the solvent and mechanical stretching of the film result in a network of very fine fibers. These fibers are subsequently bonded to create a smooth, microporous textile structure used for protective applications. The melt-blowing process produces microfibers by attenuating the polymer jet, coming out of the spin-neret, using high-velocity air jet. Since the polymer is stretched in the molten state, extremely fine fibers can be produced. Because of the lack of molecular orientation, melt-blown fibers are weak and hence are generally used in conjunction with other types of nonwovens. For example, a composite non-woven consisting of melt-blown layer and a spun-bond layer is becoming popular for medical protective applications.

2.3.6.1.3 Wet-Laid Process

Developed from the traditional paper-making process, relatively short tex-tile and wood fibers are dispersed in large quantities of water before depos-iting on an inclined wire mesh. These materials find application in hospital drapes and filters.

2.3.6.2 Consolidation Processes

Fibrous webs can be consolidated using a number of techniques depending on the area density and the desired properties. They can be classified into mechanical, chemical, thermal, and stitch-bonding processes.

2.3.6.2.1 Mechanical Bonding

Needle punching and hydro-entanglement are two complementary mechan-ical processes. Relatively thick webs (150–1000 g/m^2) are felted with the aid of oscillating barbed needles. The hydro-entanglement process uses high-velocity water jets to consolidate relatively thin webs (<140 g/m^2). The result-ing spun-laced fabrics are highly drapable and hence popular for medical protective clothing.

2.3.6.2.2 Chemical Bonding

Fibers are bonded with a suitable adhesive and subsequently cured under heat. Saturation bonding is seldom used for protective applications, as this process results in a relatively stiff nonporous material. Spray and print bond-ing instead of saturation bonding improve the flexibility and permeability.

2.3.6.2.3 Thermal Bonding

Relatively thin webs are passed through a heated calender, resulting in par-tial melting and bonding of fibers. Thermal bonding is a high-speed process and hence commonly used in conjunction with spun laids.

2.3.6.2.4 Stitch Bonding

Cross-laid webs are stitched together with a relatively large number of nee-dles across the width. Alternatively, stitch bonding is also used to bond a series of noninterlaced thread systems.

FIGURE 2.15
Nonwoven fabric made out of polypropylene.

For making staple nonwovens, fibers are first spun and cut into staple form and then compressed into bales. These bales are then opened and scattered on a conveyor belt, and the fibers are spread in a uniform web by a wet-laid process or by carding. These nonwovens are bonded either thermally or by using resin. The spun-laid nonwovens are made in one continuous process (Figure 2.15). Fibers are spun and then directly dispersed into a web by deflectors or with air streams. Melt-blown nonwovens have extremely fine fiber diameters but are not strong fabrics. Spun laid is also bonded either thermally or by using resin. Both staple and spun-bonded nonwovens would have no mechanical resistance without the bonding step. Nonwoven fabric has gradually gained importance in various industrial applications along with medicine, personal care, hygiene, and household uses. They are used in interlinings and apparel, carpet backing, and underlay; needle punched felt for backing of PVC floor covering, home furnishing, and household products; medical, sanitary, and surgical applications; book cloths, industrial wiping cloths, filtration, shoe linings, automotive applications, laundry and carry bags in hospitality industry, etc.

2.3.7 Membranes

MF membranes are frequently used as a pretreatment for reverse osmosis systems. MF membranes are available in pleated cartridge or hollow fiber formats at a range of efficiencies. They can be run in normal or cross flow mode, and many hollow fiber formats are back-flushable. The development of back-flushing hollow fibers has made membrane-based water filtration an economic process. MF membranes are produced by a wide variety of processes. The most common process is to dissolve a polymer in a solvent and produce a liquid film on a support material. The solvent is removed by evaporation or dilution in a nonsolvent, and the polymer is precipitated forming a porous structure. The concentration of polymer in the solution and the rate of precipitation determine the degree of porosity and the pore size distribution. MF membranes are produced with efficiencies from 0.05 to 5 μm.

2.3.7.1 Industrial Textile Product Manufacturing Processes

There are several ways for manufacturing flexibility in production/manufacturing of industrial products that are shown in Figure 2.16.

The products of industrial textile, since the general definition of textile as an assembly of textile fibers into useful product is very broad, can be spread onto another segment of flexible industrial engineering. It utilizes high-performance fibers, where the bulk of the products use conventional polymer materials such as PE, PP, PA, viscose, cotton, jute, and even glass is surprising.

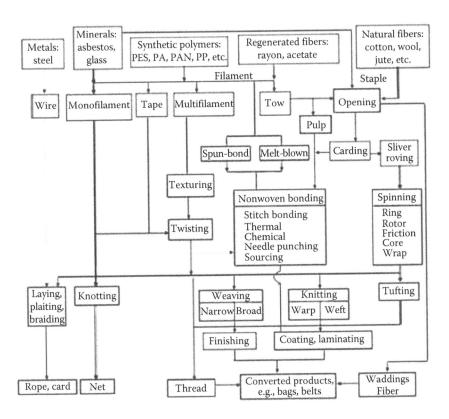

FIGURE 2.16
Industrial textile product manufacturing processes.

However, the properties and the structures of conventional fibers have been substantially modified compared to the ones used in everyday textile. The technology of industrial textile is not always confined to the products themselves or to the production technology, but it incorporates the "know-how" for the application of these products for a broad range of end users such as aviation, industry, medicine, defense, security, transport, construction, and agriculture. The fact that leads to confusion is that manufacturing of many products (e.g., manufacturing of metal wires covered with braids or fabrics) employs pure textile techniques.

Bibliography

Byrne, C., Technical textiles market—An overview, in *High Technology Fibers*, A. Horrocks, S. C. Anand (eds.), Woodhead Publishing Ltd., Cambridge, U.K., 2000, pp. 6–9.

Caleske, R., Focus on design: Carbon fibre epoxy violin captures critical acclaim, *High-Performance Composites*, 12(2), 14 (2004).

Carothers, W. H., Association polymerization and their properties of adipic anhydride, *Journal of the American Chemical Society*, 52, 3470–3471 (1930).

Demboski, G. and G. Bogoeva-Gaceva, Textile structures for technical textiles, *Bulletin of the Chemists and Technologist of Macedonia*, 24(1), 67–75 (2005).

Horrocks, A. R. and S. C. Anand, *Handbook of Technical Textiles*, Woodhead Publishing, Cambridge, U.K., 2000.

Matsuo, T., Overview and trend of technical textile technologies in Japan, *Nonwovens & Industrial Textiles*, 48(1), 46–48 (2001).

Ormerod, A. and W. S. Sondhelm, *Weaving—Technology and Operations*, Textile Institute, Manchester, England, 1999.

Piller, B., Integral knitted fabrics with increased moisture transfer, *Knitting Technique*, 9, 358–364 (1987).

Spencer, D. J., *Knitting Technology*, Pergamon Press, Oxford, U.K., 1986.

Turbak, A. F. and T. I. Vigo, Textile structure for technical textiles, in *High-Tech Fibrous Materials*, T. I. Vigo and A. F. Turbak (eds.), American Chemical Society, Washington, DC, 1991, pp. 1–15.

3

Medical Textiles

3.1 Market Scenario—Medical Textiles

Medical textiles are one of the most dynamically expanding sectors in the technical textile market. Growth rates are above average as a result of increases in consumption in developing countries in Asia and growth rates in the Western market. The prospects for medical textiles are rather better, especially for nonwoven materials and disposable medical textiles used in surgical rooms. Medical textile is defined as "fiber-based products and structures used in first aid or the clinical treatment of a wound or medical condition." Hygiene textiles are defined as those that "deal with the absorption of bodily waste products." The textile materials for medical and healthcare products range from simple gauze or bandage materials to scaffolds for tissue culturing and a large variety of prostheses for permanent body implants. Surgeons' wear, wound dressings, bandages, artificial ligaments, sutures, artificial liver/kidney/lungs, nappies, sanitary towels, vascular grafts/heart valves, artificial joints/bones, eye contact lenses, and artificial cornea are some of the examples of medical textiles. Textile materials used in the medical and applied healthcare and hygiene sectors are an important and growing part of the textile industry.

For medical textile segment alone, 1.54 million tons of textile materials were consumed, and their value is $5.4 billion, with 4% annual growth rate (AGR) until 2010. Market size was analyzed by the David Rigby Associates (DRA, 2011), which indicates a value $3.5–4 trillion out of which U.S. market alone is worth $1 trillion in case of healthcare market concern. In case of medical and hygiene textile sector as a concern, niche market dominated by large players. In 2004, medical consumer goods used in the global market were worth $220 billion out of which U.S. market alone is $87 billion and Europe market $65 billion. The total global scenario consumption was $73 billion out of which United States and Europe contributes 30% and 28%, respectively, in the case of hygiene products.

On the basis of DRA's research, more than 1.5 million tons of textile materials, with a value of $5.4 billion, were consumed worldwide in the manufacture of medical and hygiene products in 2000. The consumption of medical textiles worldwide was 1.5 million tons in 2000 and is growing at an annual rate of 4.6%. This is predicted to increase in volume terms by 4%–6% per annum to 2015 to reach 3.2 million tons with a value of $12.4 billion. This sector probably offers the greatest scope for the development of the most sophisticated and highest value textiles for niche applications. Technical textiles will find many different kinds of application with medical and hygiene products in the healthcare sector. The diversity of applications encountered in medical and healthcare products is quite remarkable, for example, simple bandages, biocompatible implants and tissues, antibacterial wound treatment material, prosthetics, and intelligent textiles. Each of these categories covers a broad range of applications, and the many end users with their disparate requirements create opportunities for all kinds of textile such as fibers, mono- and multifilament yarns, woven, knitted, nonwoven, braiding, and composite fabrics.

The medical textiles embrace all those textile materials used in health and hygiene applications in both the consumer and medical markets. Nonwovens account for a high part of the sector overall in terms of tons of fiber used. Also another feature of the medical textile market will be the growing proportion of composite materials used in wound management products. This will mean the combination of textiles with such materials as films, foam, and adhesives to form structures for the treatment of wound and healthcare products. The increased use of textiles in composite applications will provide major growth in fiber consumption in terms of volume. The increase and forecast in world consumption of medical textiles are presented in Figure 3.1.

3.1.1 World Trade of Medical Textiles

The market for medical textiles is being driven by a number of factors:

1. Population growth rates
2. Changes in demographics

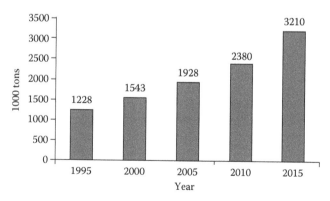

FIGURE 3.1
Forecast world consumption of medical textiles.

3. Changes in living standards
4. Attitude to health risks
5. Ongoing enhancement in product performance
6. Product usage and awareness
7. Increasing share of nonwovens

3.1.2 Global Consumption of Medical Textiles

3.1.2.1 Market Size: Hygiene Textiles

1. *Baby diapers:* European market—$3.6 billion; highest growth rates in Asia and Africa. Many product innovations and patents are enforced.
2. *Incontinence products:* 5%–7% suffer from urine incontinence; market growth 5% fueled by aging populations; $5.3 billion.
3. *Feminine hygiene products:* $9.7 billion; Europe—sanitary pads (54%), panty liners (23%), and tampons (23%); strong growth in Asia and Latin America.

3.1.2.2 Market Size: Wound Care Products

1. *Global market:* rapidly evolving technologies and intense competition; $4.5 billion.
2. *Conventional wound dressings:* clean and protect, absorb blood, for example, gauze-based; dominate the market.
3. *Advanced wound dressings:* Stimulate and promote healing; minimize scar formation; market untapped.

3.1.2.3 Market Size: Surgical Textiles

1. Gowns, drapes, caps, face masks, and shoe covers.
2. Market value of $1.7 billion in the United States and Europe.

3.1.2.4 Market Size: Nonwovens—Medical and Hygiene Textiles

1. Most disposable hygiene products.
2. A "significant proportion" of medical textiles.
3. Superior functionality and low cost.
4. Market value of $14.5 billion.

Other market sizes that involve the innovation and product developments are

1. *Developed markets:* AGR of 6%; new applications for nonwovens; large proportions of nonwovens used in products such as diapers.
2. *Emerging markets:* AGR of 8.5%; increased use of disposable products as per capita income rises.

Future growth will also depend on improvements in raw materials and technology.

1. Higher-performance nonwovens
2. New uses and markets
3. Lower costs

Technical textile consumption under Meditech is estimated at Rs. 1514 crore. Surgical dressing alone accounts for over 50% of the total technical textile consumption across Meditech segment. Surgical sutures account for around 21% of the total Meditech consumption followed by contact lenses and artificial implants with shares of around 12% and 8% respectively in the total consumption. Meditech products will prevent hospital-acquired infection and cross-infection, and provide savings in overall healthcare costs due to reduced cross-infections. Improved Meditech products will provide comfort and quicker healing.

3.1.3 Economic Aspects: Market Scenario

3.1.3.1 European Textile Sector

Technical textiles have seen their share of production grow considerably in Europe over the last 15 years, increasing constantly both in value and in volume. This market has increased from €65 to €85 billion from 1995 to 2005. About half of global technical textile production, 8.5 million tons, is consumed in Asia, followed by United States and Europe with 5.8 and 4.8 million tons, respectively. In Europe, four countries consume about half of the technical textiles in terms of value: Germany, France, the United Kingdom, and Italy. The technical textile industry in Germany represents 45% of the European textile industry, followed by France (30%), United Kingdom (30%), and Italy (12%). Europe is one of the world's leading exporters of textiles; according to EURATEX, the annual turnover of the textile industry in 2008 was over €203 billion, and the sector employed 2.3 million workers in more than 145,000 companies (mainly small and medium enterprises). However, over the last couple of decades, the European textile sector has dropped its production due to the globalization of the economy and the relocation of the European companies outside Europe. In the case of medical textiles, the market penetration is significantly lower (less than 1% of the total market share). This market segment has a very high potential, but nowadays the use of nanotechnology in this field is limited to nanosilver-enabled products (e.g., wound care products and other minor applications making use

FIGURE 3.2
Textile value chain in medical sector.

of antibacterial properties). The value chain for medical textiles is similar to that of sports/outdoor textiles; retailers deliver products for which the customer is willing to pay more than the sum of the cost activities in the value chain shown in Figure 3.2.

3.1.4 Asia

Asia's consumption of technical textile products was around 10.3 million tons in 2010 and is expected to rise to 14.8 million tons by 2015. This market is increasing at approximately 4%. As compared to world markets, a growth rate of 9.6% per annum is expected for nonwoven products in Asia. China is one of the important countries, which covers 50% of technical textiles from total Asian utilization growing at a rate of demand 10% per year. Nonwoven growth rate is 30% per year. The Chinese nonwoven industry has more than 500–600 nonwoven manufacturing companies. In China, growth rate of geo-nonwovens is about 15%, exceeding the growth of China's GDP. In the past 20 years, the Chinese government has invested in its infrastructure on a large scale. It is forecasted that China will become the largest nonwoven fabric market in the world by 2015 and will account for 42% of total demand. Other important Asian country is India. Currently, the consumption of technical textiles in India forms only 3% of the total world consumption; however, it is growing at a rate higher than most developed countries. A forecast by Goldman Sachs says that the Indian economy will overtake the economy of Europe and Japan by 2020 and that of United States by 2042. For the fiscal 2009–2010, India's economy grew by 7.4%, which is an upward revision from earlier estimates of 7.2%, and stated that the strong domestic consumption and the growing investments will put India's economy in the growth trajectory. India is expected to be one of the largest producers and consumers of technical textiles and nonwovens by 2020. India's share in the global technical textile market was around 6% in 2005 and 13.27% in 2012 and is expected to be 12%–15% by 2015: growing at a compound AGR of 10.4%. At present, the total nonwoven production in India is 90,000 tons and is expected to grow to 280,000 tons by 2014. The market size for technical textile and nonwoven in India in 2005–2006 was $6.7 billion, which has increased to $9 billion by 2007–2008. It is interesting to note that 93% of the present market is domestic and the remaining 7%–8% is for exports. The Indian market size of medical textiles was estimated to be INR 14.8 billion in 2003–2004 and had grown to INR 23.3 billion by 2007–2008. The market is expected to grow by 8% per annum after 2010–2016.

3.2 Textile Structures and Biomaterials in Healthcare

Medical textile is one of the most dynamic research fields of technical textiles, and its range of applications varies such as intrabody/extrabody, implantable/nonimplantable, and also textiles used in biological systems to estimate, treat, increase, or regenerate a tissue, organ, or the function of the body using plasters, dressings, bandages, pressure garments, etc. They also play a vital role in the manufacture of various implants, including the replacement of diseased or nonfunctioning blood vessels, segments of aorta, and other large arteries.

Textiles and textile fibers have long played a vital role in the medical and healthcare sector. Traditional products include bandages for covering wounds, sutures for stitching together the sides of open wounds to promote healing, substrates for plaster of paris casts, and incontinence products. Today, bioglass fibers are used in tissue engineering to create new bone structures, and textile scaffolds are being used to promote cell growth and build cell structures. Textile-based stents, small cylindrical tubes made from biocompatible materials, are helping to support and keep open veins and arteries. Textile stents can also be biodegradable over a predetermined period of time, thus avoiding the need to remove them surgically when they are no longer needed. Fibers are also being used in nerve regeneration techniques to repair injuries resulting from trauma or surgery. Bandages have themselves evolved into advanced dressings for wounds and burns, which enable antibiotic and other drugs to be delivered directly to the parts of the body where they are needed. Sutures have evolved from natural materials obtained from animals' intestines to advanced biodegradable or bioabsorbable materials that eliminate the need for further medical attention once stitching has taken place.

3.2.1 Material Selection and Fiber Types

Of the many different types of polymers, only a few can be made into useful textile fibers. A polymer must meet certain requirements before it can be successfully and efficiently converted into a fibrous product. Some of the most important of these requirements are

- Polymer chains should be linear, long, and flexible
- Side groups should be simple, small, or polar
- Polymers should be dissolvable or meltable for extrusion
- Chains should be capable of being oriented and crystallized

Common fiber-forming polymers include cellulosics (linen, cotton, rayon, acetate), proteins (wool, silk), polyamides, polyester (PET), olefins, vinyls, acrylics, polytetrafluoroethylene (PTFE), polyphenylene sulfide, aramids (Kevlar, Nomex), and polyurethanes (Lycra, Pellethane, Biomer). Each of these materials is unique in chemical structure and potential properties.

TABLE 3.1

Textile Materials Used in Human
Body Implants

Biodegradable Textile Fibers	Nonbiodegradable Textile Fibers
Cotton	Polyester
Viscose	Polypropylene
Polyamide	PTFE
Polyurethane	Carbon
Collagen	Nylon
Alginate	Polyethylene
Chitin	
Catgut	
Polylactic acid	
Polyglycolic acid	

For example, among the polyurethanes is an elastomeric material with high elongation and elastic recovery, whose properties nearly match those of elastic tissue fibers. This material when extruded into fiber, fibrillar, or fabric form derives its high elongation and elasticity from alternating patterns of crystalline hard units and noncrystalline soft units.

The reactivity of tissues in contact with fibrous structures varies among materials and is governed by both chemical and physical characteristics. Textile materials used in human body implants are given in Table 3.1. Absorbable materials typically excite greater tissue reaction, a result of the nature of the absorption process itself. Among the available materials, some are absorbed faster (e.g., polyglycolic acid (PGA), polyglactin acid) and others more slowly (e.g., polyglyconate). Semiabsorbable materials such as cotton and silk generally cause less reaction, although the tissue response may continue for an extended time. Nonabsorbable materials (e.g., nylon, polyester, and polypropylene) tend to be inert and to provoke the least reaction. To minimize tissue reaction, the use of catalysts and additives is carefully controlled in medical-grade products.

Implantable materials are used to repair the body whether it be closure or replacement surgery. The implantable medical textiles must have some characteristics according to their goals: biocompatibility, durability, impermeability or controlled permeability, flexibility, strength to the blood pressure and to the bacteria actions, positional stability, and stability in biological environments. Worldwide attempts have been made to engineer almost every human tissue.

3.2.2 Form of Textile Materials

Medical textile products are based on fabrics, of which there are four types: woven, knitted, braided, and nonwoven. The first three of these are made from yarns, whereas the fourth can be made directly from fibers, or even from polymers.

Fibers: Of the many types of polymers, only a few can be made into useful fibers that can then be converted into textile medical products. Polymers are extruded by wet, dry, or melt spinning and then processed to obtain the desired texture, shape, and size of fiber or filament. Through careful control of morphology, fibers can be manufactured with a range of mechanical properties. Tensile strength can vary from textile values (values needed for use in typical textile products such as apparel) of 2–6 g/d (gram/denier) up to industrial values (values typical of industrial products such as tire cords or belts) of 6–10 g/d. For high-performance applications, such as body armor or structural composites, novel spinning techniques can produce fibers with strengths approaching 30 g/d. Likewise, breaking extension can be varied over a broad range, from 10%–40% for textile to 1%–15% for industrial and 100%–500% for elastomeric fibers. Commonly used medical textile fibers are

- Polypropylene (PP)
- Polyester (PET)
- Polyethylene (UHMWPE—Dyneema Purity®)
- Nylon: polyamide—Supramid™
- PTFE
- Polyetheretherketone
- Polyetherketoneketone
- Silk
- Metallic wires (stainless steel, Nitinol, Pt–Ir, etc.)

Absorbable Fibers

- Polyglutamic acid
- Poly(lactic acid)
- 90:10 copolymer blend: PLGA—poly(lactic-*co*-glycolic acid)
- Polydioxanone/peroxydisulfate

Yarns: Fibers or filaments are converted into yarns by twisting or entangling processes that improve strength, abrasion resistance, and handling. Yarn properties depend on those of the fibers or filaments as well as on the angle of twist, modulus, and strength.

- *Monofilament yarn:* Single yarn, extruded; and measured by Mil or mm diameter.
- *Multifilament yarn:* Yarn consisting of many strands that can be plied or twisted together, multifilament materials has better conformity, softer

hand, and typically higher tenacity than monofilament. Measured in Denier or Tex.

- *Staple fiber:* Short lengths of multifilament yarn used for nonwovens and other custom applications.

Fabrics: Yarns are interlaced into fabrics by various mechanical processes—that is, weaving, knitting, and braiding. The three prevalent fabric structures used for medical implants or sutures are *woven*, in which two sets of yarns are interlaced at right angles; *knitted*, in which loops of yarn are intermeshed; and *braided*, in which three or more yarns cross one another in a diagonal pattern. Knitted fabrics can be either weft or warp knit, and braided products can include tubular structures, with or without a core, as well as ribbon. There are also numerous medical uses for nonwoven fabrics (wipes, sponges, dressings, gowns), made directly from fibers that are needle-felted, hydro-entangled, or bonded through a thermal, chemical, or adhesive process. Nonwovens may also be made directly from a polymer. For example, expanded PTFE (ePTFE) products such as sutures and arterial grafts and electrostatically spun polyurethane used as tubular structures are examples of medical applications of polymer-to-fabric nonwovens.

3.2.3 Medical Fabric Structures and Properties

The properties of medical fabrics depend on the characteristics of the constituent yarns or fibers and on the geometry of the formed structure: whether a fabric is woven, knitted, braided, or nonwoven, it will affect its behavior.

3.2.3.1 Woven Fabrics

Today's most advanced medical devices are benefiting from lightweight fabric and manufacturing flexibility of woven structures. With a variety of geometries possible through the engineering process, increasing numbers of manufacturers are looking to woven materials utilizing finer fabrics to meet the performance and functional requirements for a wide range of therapeutic indications, including cardiovascular and orthopedic applications. A variety of woven styles are possible, from single plain patterns to thicker, stronger, or shaped multidimensional weaves. Tapes and webbings are designed to meet specific needs for strength, porosity, morphology, and geometry and can be woven for particular width requirements. Woven structure also produces tapes and webbings at straight and varying widths and profiles (both across and within single biomedical textiles) depending on the design challenge. Woven tubes can also be manufactured with additional customization to meet any combination of functional requirements, including tapering or bifurcation to accommodate anatomic dimensions. Fine woven fabrics can be used to produce highly precise tubes, which enable sophisticated woven grafts for endovascular stent systems and other coronary intervention stent procedures. Common woven applications include cardiovascular grafts, heart valves, annuloplasty rings, orthopedic spacers, tethers, containments, ligament repair, tendon reinforcement, and fixation devices.

3.2.3.2 Knitted Fabrics

Today state-of-the-art equipment and techniques to deliver knitted solutions for implantable devices and other applications are available in textile industry. Knitting produces a structure with good stretch and high strength that requires a concentration of power in a thin form factor that can take advantage of the multitude of design options in permanent and absorbable materials to fulfill specific device capabilities. Knitting typically involves a higher number of individual fibers than most other biomedical textile engineering techniques, which allows for greater intricacy and performance capabilities in created structures. Knitting techniques include warp knitting and weft knitting (flat and circular knitting) and allow for many different configurations, including extra strength without increasing thickness, a flexible mesh with high conformability, or even flat structures with designed apertures to allow for cutting or other alterations without sacrificing edge integrity. Common knitted applications include surgical mesh, hernia repair, urogynecologic slings and prolapse devices, reconstructive and cosmetic surgery mesh, and other containment devices.

Compared with woven fabrics, weft-knitted structures are highly extensible, but they are also dimensionally unstable. Warp-knitted structures are extremely versatile and can be engineered with a variety of mechanical properties matching those of woven fabrics. The major advantage of knitted materials is their flexibility and inherent ability to resist unraveling when cut. A potential limitation of knitted fabrics is their high porosity.

3.2.3.3 Braided Fabrics

Today's most advanced medical procedures are pushing the boundaries of materials with demands for strong, flexible, and biocompatible textiles. For achieving these requirements, advanced technology and methods beyond traditional braiding processes will create customized structures with designed-in performance features to satisfy a variety of clinical applications in general surgery, soft tissue repair, and arthroscopic procedures. By intertwining three or more strands of biomaterial in highly complex and precise ways, to create flat or hollow structures with a high degree of strength but without a large surface area are required. Manufactured with a combination of different absorbable and permanent fibers, braids can provide a material for partial degradation over time or maintain a precise geometry for implantable replacement. Engineering customized biomedical textiles to possess softness, fatigue resistance, abrasion resistance, expandability, and compression. Braids can maintain a structural composition without sacrificing flexibility, and the result is a more natural, lifelike movement within the body than materials of metal, plastic, or ceramic. Common braided applications include device assembly and carriers, sutures, and sewing threads, tethers and component attachment, retrieval/deployment systems, tubes and tubing reinforcement, ports, sheaths, reinforcement devices, catheters, and tendon/ligament fixation.

Braided structures: Typically employed in cords and sutures, braided structures can be designed using several different patterns, either with or without a core. Because the yarns crisscross each other, braided materials are usually porous and may imbibe fluids within the interstitial spaces between yarns or filaments. To reduce their capillarity, braided materials are often treated with a biodegradable (polylactic acid [PLA]) or nonbiodegradable (Teflon) coating.

3.2.3.4 Nonwoven Fabrics

Nonwoven biotextiles have become the material of choice for many tissue engineering and regenerative medicine applications. With a superior surface area, high void volume, and excellent permeability, they are now increasingly used in a wide variety of restorative applications ranging from orthopedic reconstruction and wound management to cosmetic surgery. Composed of felt created by a carding and needle-punch process, nonwoven structures provide greater surface area than most other biomedical textiles to bring a different set of benefits to implantable. Most notably, the ability to encourage cellular in-growth through specific spacing, layer thickness, and material integrity allows for customized performance and controlled absorption profiles. Currently, most nonwoven scaffolds composed of absorbable biomaterials meant to facilitate regrowth before disintegration. Common nonwoven applications include scaffolds for tissue engineering, scaffolds for stem cell applications, dental and cosmetic applications, absorbable additives for cements and hydrogels, high surface area material for trauma applications, markers, pledgets, washers, spacers, and cuffs.

The properties of nonwoven fabrics are determined by those of the constituent polymer or fiber and by the bonding process. For instance, expanded PTFE products can be formed to meet varying porosity requirements. Because of the expanded nature of their microstructure, these materials compress easily and then expand a suture, for example, can expand to fill the needle hole made in a tissue allowing for tissue in-growth in applications such as arterial and patch grafts. Polyurethane-based nonwovens produce a product that resembles collagenous material in both structure and mechanical properties, particularly compliance (extension per unit pressure or stress). The porosity of both PTFE- and polyurethane-derived nonwovens can be effectively manipulated through control of the manufacturing processes.

3.2.3.5 Medical Fabric Manufacturing: Trends and Innovations

Medical fabric manufacture is the most important part of textile industry. Categories of medical textiles include

- Nonimplantable materials
- Implantable materials
- Extracorporeal devices
- Healthcare and hygiene products

The applications of different textile fibers for manufacturing various medical products are illustrated in Tables 3.2 through 3.4.

3.2.4 Medical Products

3.2.4.1 Vascular Grafts

Vascular grafting is the use of transplanted or prosthetic blood vessels in surgical injuries. PTFE and Dacron are some of the most commonly used

TABLE 3.2

Nonimplantable Materials Used in Medical Textiles

Textile Fibers	Textile Structure	Applications
Viscose, lyocell, plastics film	Woven, nonwoven	Base material
Alginate, chitosan, silk, lyocell	Woven, knitted, and nonwoven	Wound contact layer
Cotton, lyocell, elastomeric fiber	Woven, knitted	Compression bandages
Cotton, wood pulp, linters	Nonwoven	Wadding
Cotton, viscose, chitosan, alginate	Woven, knitted, and nonwoven	Gauze dressing
Cotton, viscose, lyocell	Nonwoven	Absorbent pads

TABLE 3.3

Implantable Materials Used in Medical Textiles

Textile Fibers	Textile Structure	Applications
Polyester fiber	Woven, knitted	Heart valves
PTFE fiber, polyester fiber	Woven, knitted	Vascular grafts
Chitin	Nonwoven	Artificial skin
PTFE fiber, silk, polyamide	Woven, braided	Artificial tendon
Silicone, polyacetyl fiber, polyethylene fiber	Filament form	Artificial joints
Collagen, catgut, polylactide fiber, polyglycolide fiber	Monofilament, braided	Biodegradable sutures

TABLE 3.4

Healthcare/Hygiene Textile Products

Textile Fiber	Fabric Structure	Application
Cotton	Woven	Sheets, pillowcases
Cotton, polyester fiber	Woven	Uniforms
Polyethylene fiber	Nonwoven	Outer layer
Viscose, lyocell	Woven, nonwoven	Clothes/wipes
Viscose	Nonwoven	Surgical caps
Cotton, polyester fiber, polypropylene	Woven, nonwoven	Surgical gowns

grafts. Grafts can be used for the aorta, femoral artery, or the forearm. Coronary artery bypass graft is used for people with occluded coronary arteries, and often the saphenous vein or left internal thoracic artery is used. This is 20 mm woven Dacron graft. This type and size of graft is frequently used for abdominal aortic aneurysm surgery. Dacron grafts are manufactured in either a woven or a knitted form, which is shown in Figure 3.3a. Woven grafts have smaller pores and do not leak as much blood. To reduce the blood loss, knitted grafts should be preclotted prior to insertion. They are less frequently used than woven grafts. Dacron grafts have recently been manufactured coated with protein (collagen/albumin) to reduce the blood loss and antibiotics to prevent graft infection. Vascular grafts can be classified as either biological or synthetic. There are two commonly used types of

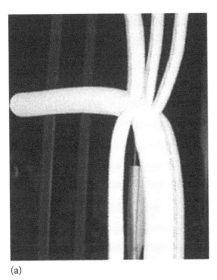

(a)

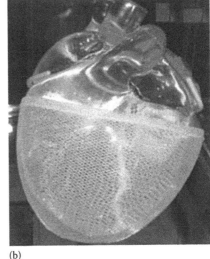

(b)

FIGURE 3.3
Medical devices. (a) Vascular grafts and (b) cardiac support device.

biological grafts. An autograft is one taken from another site in the patient. An allograft is one taken from another animal of the same species. Synthetic grafts are most commonly made from Dacron or PTFE, and its smooth surface is less thrombogenic than Dacron. Its smooth wall is prone to kinking as it passes around joints necessitating it to be externally supported. Dacron grafts are frequently used in aortic and aortoiliac surgery. Below the inguinal ligament, the results of all synthetic grafts are inferior to those obtained with the use of vein grafts.

3.2.4.2 Cardiac Support Device

The cardiac support device is a proprietary polyester mesh wrap implanted around the heart to provide support and reduce ventricular wall stress, which is shown in Figure 3.3b. It provides beneficial changes in cardiac structure associated with a reverse remodeling effect as defined by a reduction in left ventricular size and a change to a more elliptical shape and also an overall improvement in quality of life.

3.2.4.3 Embroidered Implants

Embroidered implants are stents, as used for the repair of abdominal aortic aneurysms. Hand embroidery has been used for decorating textiles for thousands of years; even automatic embroidery machinery has been in use for over 150 years. Surgical implants are developed within the last few years. Textile surgical implants have, until now, been constrained by the use of the traditional methods of knitting, weaving, or braiding to make fiber assemblies (Figure 3.4). Embroidery is the formation of stitches on a base cloth, it being possible to place the stitches in any position. The base cloth can be dissolved away after the stitching process has been carried out, and assuming appropriate design features have been incorporated, the structure holds securely together as a stable entity.

3.2.5 Implantable Medical Textiles

Medical textiles represent structures designed and accomplished for a medical application. The number of applications is diverse, ranging from a single thread suture to the complex composite structures for bone replacement and from the simple cleaning wipe to advanced barrier fabrics used in operating rooms. Textile materials and products, which have been engineered to meet particular needs, are suitable for any medical and surgical application where a combination of strength, flexibility, and sometimes moisture and air permeability is required. For developing medical textile products, keep in kind by microbionanotechnology and materials development in interdisciplinary fields (chemical, medical, pharmaceutical, and textile).

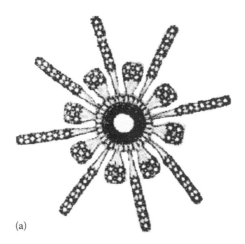

(a)

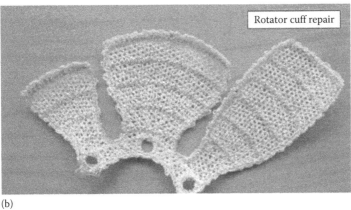

Rotator cuff repair

(b)

FIGURE 3.4
Embroidery technology for the development of (a) scaffolds and (b) rotator cuff.

A fiber for implants has been chosen for three important aspects: the replacement of fibrous fibers, strong interactions of tissue/implant, and the combination of fineness and strength. Besides the fibers or polymers employed in manufacturing the traditional medical textiles, there have been devised lately a series of new and innovative medical products with particular properties such as biocompatibility, biodegradability, biosafety, and bioabsorbability. The biofunctional medical textiles can be obtained by the additive treatments of the yarns by proofing with antiseptic and antimicrobial agents against contaminations and infections, textiles with aromatherapeutic effects, and virus- and microbe-proof textiles for increasing the immunity. The human body implantable devices are shown in Figure 3.5. There are two different kinds of biomaterials: biological (or naturals) and synthetic, but frequently, in tissue engineering, are used both synthetic

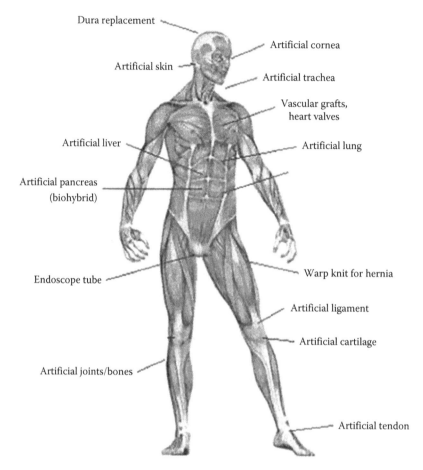

FIGURE 3.5
Implantable devices in human body.

components, for example, resorbable polymers, and biological components, either in the form of whole cells or cell-secreted extracellular matrix. Textile medical implants also used biodegradable and nonbiodegradable fibers.

3.2.6 Tissue Engineering

Tissue engineering uses the principles of biology, chemistry, physics, and engineering for repair, replacement, and enhancement of biological functions of diseases or ill human body parts through the manipulation of the cells in their extracellular microenvironment, a combination of cells and biomaterials. The scaffold cans be achieved in the form that we want with the help of techniques such as self-assembly, polymer hydrogel, nonwoven matrix, nanofibrous electrospun matrices, 3-D weaving, or any other textile technology-based techniques. The tissues can be obtained in vivo, by stimulating the body for autogeneration with in vivo biomaterials by attaching cells to a scaffold and reintroducing into the body. The cells can be gathered from the patient, a donor, or other species. These cells are then incorporated into a 3-D, biodegradable, polymeric support of the appropriate phenotype and growth factors are also added.

The cells are implanted into an artificial structure capable of supporting 3-D tissue growth. These structures are critical, both before implantation outside the body and within the body to provide a support structure that allows the cells to interact with their microenvironments. Most of these structures are designed to disintegrate at a specified point following implantation, once the cells have grafted or become integrated in the surrounding tissue. The main characteristics for scaffold design are as follows: (1) high porosity; (2) adequate pore size to facilitate cell seeding and diffusion; (3) functional with biomolecules; (4) rate of biodegradation should coincide with rate of tissue formation; (5) compatible; and (6) should resist biodegradation. They should allow cell attachment, migration, release, and maintenance of these cells, facilitating the spread of vital nutrients to the cells and exercise the mechanical and biological influences for behavior modification of inputted cells. When selecting materials for scaffolds, we need to consider important factors such as natural–synthetic, biodegradable–permanent, size, and reabsorption rate, inject ability, and manufacturing cost and process. Woven and nonwoven structures were introduced in in vitro development of various tissues (liver, skin, bone, cartilage, muscle). In clinical applications, implant complications are often associated with mechanical and structural incompatibility between implant and host tissues. A novelty in tissue engineering is a thin 3-D scaffold. Their use is indicated for ulceration. During culture, the cells proliferate and secrete proteins characteristic of human dermis resulting in a human dermis that contains the metabolic active cell and a dermal matrix.

The scaffold must be flexible and resistant to the people's moves, and for that, a suitable porosity and a 3-D form implant will be needed. Using the protein silk fibers will result in a biocompatible and biodegradable fabric (Figure 3.6).

Recent research in medical textiles made attempt for developing a polymeric extracellular matrix (biodegradable polymers such as polyglycolic acid [PGA] and PLA and nondegradable such as polyvinylidene fluoride), which can be used as biomaterial. Warp knitted are used for scaffolds that need elastic deformation and for organ reinforcement, for example, meniscus scaffold, silicone keratoprosthesis, and vascular grafts, because of the mechanical flexible properties and designing. Nonwoven are textile structures with shapes and scaly areas that are capable to make a preorientation of the cells in structures and with good water absorption.

An important characteristic for scaffolds in tissue engineering is the porosity that must offer enough space for cell to grow up and to be small

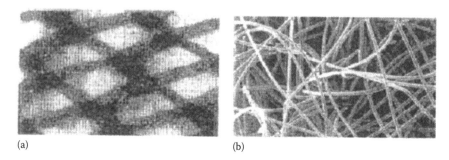

(a)   (b)

FIGURE 3.6
Scanning electron microscope image of (a) dermal matrix on the scaffold and (b) needle felt structure for scaffolds.

enough to maintain the stability. It is important to choose the right materials starting from different kinds of pure polymers or blended. Biodegradable or nonbiodegradable materials are used for a long or short life implant, but the degradation can be controlled through the chemical structure, process, fiber profile, and technique of materials achieving.

3.2.7 Biomedical Textiles

Biomedical textiles are fibrous structures designed for use in specific biological environments, where their performance depends on biocompatibility with cells and biological tissue or fluids. For example, in tissue engineering, developing new biocompatible fibrous scaffolds upon which new tissue is grown for implants. It combines skills in chemistry, fiber, and materials science with biological expertise to develop fit-for-function biomedical textiles and understand their interactions with biological systems. In biomedical research, researchers and expertise across the globe develop and evaluate new platform materials and devices for applications in

- Wound management
- Rehabilitation
- Injury prevention
- Tissue repair, replacement, and regeneration

Medical textiles are textile products and constructions for medical applications. They are used for first aid, clinical or hygienic purposes, and rehabilitation. Examples of their application include

- Protective and healthcare textiles
- Dressings, bandages, pressure garments, and prosthetics

- Hygiene products
- Antiseptic wound dressings

Examples of the application of biomedical textiles and biomaterials include

- Implantable materials and devices
- Biocompatible materials for regenerative medicine, such as tissue engineering
- Neural repairs

Research into medical and biomedical textiles covers several areas, including

- Injury prevention
- Fibrous scaffolds for tissue growth
- Biofeedback and medical devices
- Quality control instruments for medical textile production

Current research in biomedical textiles includes use of carbon nanotube yarns with fibroblast cells for analyzing biocompatibility nature on body implants.

3.2.8 Tissue Engineering: Cell Scaffolds

Tissue engineering supports structures or "scaffolds" that are artificial devices, designed to act as templates for attached cells and newly formed tissues. The scaffolds' 3-D porous structures encourage cell attachment, proliferation, and migration through an interconnected network of pores. New tissue forms gradually and can be implanted into the body. For a tissue engineering scaffold to work effectively, its structural and mechanical properties must be suitable for the type of tissue being grown. Using knitting, weaving, nonwoven, and electrospinning technologies, the researchers produce scaffolds of tubular or flat design. Image analysis and tensile testing reveal the structural and mechanical characteristics of the scaffolds produced. The scaffolds are subjected to testing a wide range of biocompatible, nonbiodegradable, and biodegradable polymers for scaffold fabrication.

Current research includes scaffolds for the regeneration of peripheral nerves and high-performance scaffolds for tissue engineering of spinal disks.

- *Fiber formation:* for example, carbon nanotubes, regenerated protein fibers, electrospinning, melt extrusion.
- *Fiber manipulation:* fibers to form structures with specific architectures and properties.

- *Fiber measurement and process modeling:* includes capability in metrology and instrument design, and the modeling of fibrous structures and fiber processing.

- *Biomedical:* using nanofibers and biocompatible polymers to build fibrous scaffolds for the purpose of growing cell cultures to regenerate human tissues.

- *Nanomaterials:* expertise in the production and manipulation of nanoscale fibers including the synthesis of carbon nanotubes, electrospinning, and processing of polymer nanofibers and the extrusion of nanocomposite fibers.

- *Surface, fiber, and protein chemistry:* combining advanced biopolymer chemistry in fiber surface science and applying to the textile, food futures, and health domains.

- *Instrumentation:* design of sensors, instruments, and fiber manipulation machinery and commercial prototyping.

- *Flexible electronics:* skills in embedding electronics into flexible structures with wireless interfacing to remote computers.

3.3 Advanced Wound Dressing: Structure and Properties

An ideal wound dressing must be able to maintain warmth and moisture, and also should provide many specific functions depending on wound type, injury or infection, healing scenario, age of the patient, and others. Specialized materials with determined functions can be included in dressing, extending it into multifunctional system made from natural or/and synthetic materials. Researches in wound care dressings are especially technical and technological as well as functional and effective oriented because of regular scientific inquiry into many new aspects of the wound-healing process and novel developments that continually enrich the knowledge base.

Surgical dressings include wound care products and bandages. Wound care products include wound contact layer/absorbent pad/base material/ nonadherent dressings/perforated films. Bandages include inelastic bandages/elastic bandages/light support bandages/orthopedic cushion bandages/plasters/waddings/guazes/lint. Wound healing is a dynamic process, and no single dressing is universally available for all types of wounds. Wound healing depends not only on medication but also on the use of a proper dressing technique and suitable dressing material. The prerequisites of wound dressings are ease of application, good padding characteristics, nonsticking nature to the wound and painlessness on removal, creation of an optimal environment for wound healing, softness, pliability, and high absorbency. Modern wound dressings are composed of absorbent layers held between the wound contact layer and a base material. The wound contact layer (primary dressing) is generally placed directly over the wound and covered with an absorbent pad and whole dressing retained with a base material. The wound contact layer has low adherency and can be easily removed without disturbing new tissue growth. The wound contact layer is made out of silk, polyamide, viscose, or polyethylene, with woven or nonwoven structure. The absorbent layer (pad) is of nonwoven type, made of cotton, viscose, acrylic, etc. Viscose helps to absorb the fluid while acrylic helps to maintain the thickness of pad even after absorbing the fluids. The base material is nonwoven or woven type made of viscose or is a plastic film.

Elastic bandages are cotton crepe bandages consisting of high twist yarn imparting the necessary elasticity and used in treating varicose veins. Inelastic bandages are medicated cloth bandages. These two bandages are grouped together as adhesive bandages, and they have a layer of adhesive impregnated on the cloth layer. Orthopedic cushions are made of cotton and synthetics. These bandages retain their cushioning effect in the moist atmosphere between skin and plaster. The plaster of paris bandages are made of cotton gauze material of leno weave cloth. The interlocking thread is impregnated in the plaster of paris solution and dried to get the bandages. Waddings are single-use cotton pieces in great demand in market; for clinical practices as well as domestic purpose, cotton rolls are preferred, and also sterile single-use cotton waddings are highly popular.

Advanced medical textiles are significantly developing area because of their major expansion in such fields like wound healing and controlled release, bandaging and pressure garments, implantable devices as well as medical devices, and development of new intelligent textile products. Present-day society is undergoing changes such as aging of the population, increase in life span of individuals especially in Europe and United States, and various situations and hazards of human activity and civilization including transport accidents, chemical materials, fire, cold, diseases, sports. Such factors stimulate the rapid movement of wound care product market with the requirement of novel technique and technologies to develop modern textile materials and polymers. Nowadays textile products are able to combine traditional textile characteristics with modern multifunctionality. The role of dressings in wound management is constantly evolving. Virtually new products are regularly being developed and approved. Today's worldwide industry reports estimate the wound care market to exceed $11.8 billion by 2009 and yearly growth for all products (devices for wound closure such as sutures and staples, dressing's adhesives) projected in excess of 7%–8% in 2014–2015.

3.3.1 Advanced Polymers and Textiles

Textiles include fibers, filaments, yarns, woven, knitted, and nonwoven materials made from natural and man-made materials. Table 3.5 shows the classification of fibers generally used in wound care, grouped into natural or man-made. Most important natural fibers are cotton, silk, and linen. Man-made synthetic polymers cover fibers manufactured from chemically synthesized polymers like polyester, polyamide, and polypropylene. Natural polymers include alginates and proteins.

3.3.2 Specialty of Textile Fibers

High surface area, absorbency, lightweight, and fineness in nature, and variety in product forms are advantageous properties of textile fibers, which are desirable when using them in wound dressing applications. The important and unique quality of medical textiles is biodegradability. Fibers used as wound care textiles are classified as biodegradable and nonbiodegradable. Biodegradable fibers are cotton, viscose, alginate, collagen, chitin, chitosan, and other ones that can be absorbed by the body in 2–3 months. Such synthetic fibers like polyamide, polyester, polypropylene, and PTFE that take more than 6 months to degrade are nonbiodegradable fibers.

TABLE 3.5

Classification of Textile Fibers Used in Wound Care Applications

Origin	Source	Fiber Type
Natural	Animal	Silk (spider, silkworm)
	Vegetable	Cotton seed
		Bast fibers (linen, hemp)
Man-made (synthetic)	Synthetic polymer	Polyester
		Polyamide
		Polypropylene
		Polyurethane
		Polytetrafluoroethylene
	Natural polymers	Regenerated cellulose
		Proteins (collagen, catgut, branan ferulated)
		Alginate
		Polyglycolic acids
		Chitin
		Chitosan
		Hyaluronan
	Other (nonfibrous material)	Carbon
		Metals (silver, iron, etc.)

TABLE 3.6

Various Textile Structures Used in Wound Care

Textile Structure	Technology/Processes
Sliver	Carding, drawing
Roving	Preparatory spinning
Spun yarn	Spinning
Filament yarn	Filament spinning
Woven fabric	Weaving
Nonwoven	Bonding/thermal/adhesive
Knitted fabric	Knitting
Braided	Braiding

3.3.3 Structure and Specific Qualities

Textile structures used for wound management are sliver, roving, yarn, woven, nonwoven, knitted, composite materials, etc., processed using various technologies (Table 3.6). In skin healing, mesh grafts are also used.

3.3.4 Drawback of Traditional Medical Dressing

Majority of traditional textile dressings is made of cotton bleached gauze. Its construction characterized by dimensional instability, flying edges, woolliness, and flat surface has undesirable features. New-generation dressing consists of a bleached cotton fabric of leno weave structure, onto which a layer of soft paraffin material is applied. The paraffin makes the dressing hydrophobic, so that wound secretion easily penetrates to the absorbing dressing that is placed on the paraffin dressing. Paraffin eliminates the problem of loose fibers getting caught in the wound as well as dressing is chemically neutral, so the agents shortening healing can be applied to it. Fibrin bandages are developed using blood-clotting chemicals and an enzyme to prevent excessive blood loss in injures like gunshots and automobile accidents.

3.3.5 Function of Modern Wound Dressing

The following structural, physical, and specific properties are expected from modern wound dressing products, such as stable and spatial structure, nontoxicity, barrier against microorganism, pain relieving, moisture, and liquid absorption.

Figure 3.7 shows the functions connected with the development of new applications of modern wound dressings. Specialized materials or/and additives with special functions can be introduced in advanced wound dressings. Such materials (additives) absorb offensive odors from bacteria-infected wounds, provide antibacterial properties (silver metals or their salts), include

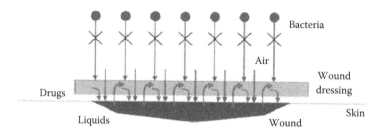

FIGURE 3.7
Function of wound dressing.

antiseptics, antibacterial constituents, antibiotics, zinc pastes used to sooth pain and relieve irritation, sugar pastes as deodorizing agent, provide honey therapy, etc.

3.3.6 Advanced Wound Dressing Products

Generally wound dressing products are classified into three groups: (1) passive products, (2) interactive products, and (3) bioactive products. Passive products are traditional dressings that provide cover over the wound, for example, gauze and tulle dressings. Interactive products are polymeric films and forms that are mostly transparent, permeable to water vapor and oxygen, nonpermeable to bacteria, for example, hydrogel, and foam dressings. Bioactive products are dressings that deliver substances active in wound healing, for example, alginates, chitosan. Medical substances into textile structure are developed by immobilization of medical products upon the polymer. Soft silicone mesh is a nonadherent, porous dressing with a wound contact layer consisting of a flexible polyamide net coated with silicone. The porous nature allows fluid to pass through the secondary dressing. The role of artificial skin substitutes and the treatment of chronic wounds are constantly evolving. A bioartificial skin has been produced from gelatin and D-glucan homopolysaccharides. Skin substitutes are very important in burn care and are used for early burn coverage, increasing survival and leading to a better recovery of function and appearance as well as for chronic wounds. Coated mesh fabric is used in skin substitutes for wound coverage.

The clinical performance of hydrogel dressings concludes the effectiveness for chronic wound treating. Researchers are focused on modifying the collagen glycosaminoglycan matrix through the incorporation of antibiotics. Bacterial cellulose is a natural polymer consisting of microfibrils containing glucan chains bound together by hydrogen bonds. Bacterial cellulose/chitosan wound dressings are innovative because of good antibacterial and barrier properties as well as good mechanical properties in wet state and high moisture-keeping properties. Such features make modified bacterial cellulose an excellent dressing material for treating various kinds of wounds and also burns and ulcers. This modified bacterial cellulose consists of microfibers with diameters in the order of tens of micrometer, which form a 3-D network. Besides that, it was estimated that chitosan has a favorable impaction on the mechanical properties of modified bacterial cellulose. High elongation at break indicates good elasticity, so such dressing fits the wound site well and therefore provides good protection against external infection. Bioactive material made from chitosan-modified bacterial cellulose provides optimal moisture conditions for rapid wound healing and stimulates wound healing without irritation or allergization. Such composite structures have applications in the management of burns, bedsores, skin ulcers, hard-to-heal wounds, as well as wounds requiring frequent dressing change. Bacterial cellulose modified with chitosan combines properties such as bioactivity, biocompatibility, and biodegradability of the two biopolymers creating an excellent dressing material greatly isolating the wound from environment and stimulating the healing of wound.

3.3.7 Absorption Characteristics: Textile Structures

Many nonwoven structures used for absorbent wound dressing exhibit anisotropic fluid transmission characteristics in-plane or in the transverse plane of the fabric structure. Fiber orientation distribution in such materials is the main factor that has major influence on anisotropic fluid transmission, so it can be manipulated to design desirable transport properties. The directional permeability and the anisotropy of permeability are determined by fabric porosity, fiber diameter, and fiber orientation distribution. Phenomenon of absorption is significant for various characteristics of the textile, such as skin comfort, static buildup, water repellence, shrinkage, and wrinkle recovery. The absorbency of alginate fiber can be improved by 120 times its own weight. Super absorbent fibers are made from super absorbent polymers, which absorb up to 50 times their own weight of water (conventional wood pulp and cotton filler absorbents absorb only six times their weight). Super absorbent fibers also present advantages such as high surface area, flexible handle, and availability to form soft product of different shapes to fit the surface of the wound. Compared with the powders, the super absorbent fibers absorb fluids much faster and to a very high level. Besides that, the fiber absorbing body fluid does not lose its fiber structure and returns to its original form.

3.3.8 Textile Architectures for Tissue Engineering

The purpose of tissue engineering is to culture viable human tissues outside the body. The novelties in the repair of large deep wounds over recent years have needed advanced skills in tissue engineering. Biomaterials into cells and tissues if to reconstruct and repair the living organisms extended the field of tissue engineering. Manufactured structures made from natural or synthetic materials provide stabile scaffolds and allow tissues to regenerate the damaged as well as missing organs. Three-dimensional textile structures are widely used as scaffolds in tissue engineering applications. Such textile scaffolds demonstrate the ability to provide necessary conditions for cell maintenance. The scaffold should have a certain shape, porosity, and volume. Transplanted scaffolds holding a 3-D cell culture should copy the cartilage characteristics. Looking for regeneration of injured cartilage, the scaffold material should disappear while real cartilage is healing the wound. Collagen and hyaluronans that are extracted from human or animal tissues could be used as natural scaffolds for the assistance of healing process, and such scaffolds subsequently disintegrate and are absorbed by the body. Similar properties of synthetic biodegradable materials have also been developed, but their assessment in vivo and in vitro practices are still limited. PGA, PLA, and also their derivatives are successful materials used for these applications. Tissue-engineered skin is indicated for wounds that are difficult to heal, such as diabetic foot ulcers. In this case, the scaffold is a crocheted mesh manufactured from multifilament yarn on which human dermal fibroblasts are seeded. The result is a dermal tissue containing metabolically active cells and a dermal matrix. Tissue engineering provides viable alternatives to autograft skin that may be used in various clinical scenarios: from covering raw areas as in burns to stimulating the healing in the chronic wounds. Such products could be single layered or bilayered. The potential of embroidery technique was investigated for the development of textile scaffold structures for tissue engineering. An experimental study about the influence of in-growing tissue on the mechanics of the thereby-formed vital–avital composite was fulfilled.

The use of alginate in textile scaffolds that may be knitted, woven, nonwoven, braided, embroidered, or combined has certain specialized functions. Flexibility provides versatility, and so alginate fiber systems are ideal for encouraging cells to reconstruct the tissue structure in 3-D. Scaffold structure and porosity are key properties that will guide the formation of new tissue. There is a need for structural biocompatibility of the scaffold and the host tissue. Three-dimensional embroidered textile architecture that combines different kinds of pores and holes and also stiff elements is designed for effective wound treatment. Novelties made in genetic technology have increased growth factor availability and their potential therapeutic role in wound healing and care. Engineered growth factors capable of proving speed and safe healing in all types of injuries including wound-related leg amputation and the like are actively being developed and improved.

3.3.9 Nanotechnology for Advanced Wound Management

Nanotechnology has acquired tremendous impulse in the last decade. Nanofibers are very attracted due to their unique properties: high surface area to volume ratio, film thinness, nanoscale fiber diameter, porosity of structure, lighter weight. Nanofibers are porous, and the distribution of pore size could be of wide range, so they can be considered as engineered scaffolds with broad application in the field of tissue engineering. There is a high potential for nanofibers to be the carrier of various drugs to the specific sites. If to incorporate drugs into the nanofiber matrix, drug must be encapsulated into the nanofibrous structure. There are a number of materials, like metals, metal oxides at nanoscale, biological materials such as enzymes, drugs, etc., that can add functionality to nanofibers. Such value-added nanofibers can be used effectively in tissue engineering and biomaterials, drug delivery, protective clothing, etc. Natural tissue can be weakened or lost by injury, disease, etc., so the artificial supports are required to heal wounds and to repair damaged tissues. Nanofiber scaffolds having enough mechanical and biological stability can be very important as a degradable implant. Wound dressings composed of electrospun polyurethane nanofibrous membrane and silk fibroin nanofibers are developed. Such materials are characterized by a range of pore size distribution, high surface area-to-volume ratio, and high porosity, which are proper qualities for cell growth and proliferation.

Among the antimicrobial agents, silver has long been known to have strong antimicrobial activities, so antibacterial disinfection and finishing techniques are developed for many types of textiles using treatment with nanosized silver. The prospective evaluation results in the effect of nanocrystalline silver dressing for ulcers that completely healed after 1–9 weeks of treatment; however, further studies with a larger number of patients are required. Silver-impregnated textiles are used as wound dressings for infected wounds, also for wounds at high risk of infection. Antimicrobial yarns can be produced from cotton, linen, silk, wool, polyester, nylon, and their blends having nanosilver particles. Electron microscopic studies indicated that the yarns contained nanosilver particles mostly below or about 10 nm size with silver content 0.4%–0.9% by weight. Such treated yarns showed effective antimicrobial activity against various bacteria, fungi, etc. Silver-containing antimicrobials have been incorporated into wound care devices as safe and effective means in improved healing. It is known that silver in contact with wound enters it and becomes absorbed by undesirable bacteria and fungi, so silver ions kill microbes resulting in the treatment of infected wound.

Functionalized nanofibers are nanofibers with specific foreign materials for adding special functionalities and capabilities to nanofibers, and so their application possibilities are unlimited. By improving mechanical stability of biodegradable nanofibrous structures and finding novel ways to incorporate functional materials, it would make functionalized nanofibers as potential candidates for highly efficient biomaterials. Currently, the research has started investigating the interaction between cell and nanofiber matrices.

3.4 Natural and Biopolymer Finishes for Medical Textiles

New-generation medical textiles are important growing field with great expansion in wound management products. Virtually new products are coming, but also well-known materials with significantly improved properties using advanced technologies and new methods are in the center of research, which is highly technical, technological, functional, and effective oriented. The key qualities of textile fibers and dressings as wound care products include that they are bacteriostatic, antiviral, fungistatic, nontoxic, high absorbent, nonallergic, breathable, hemostatic, biocompatible, and manipulatable to incorporate medications and also have reasonable mechanical properties. Today advantages over traditional materials have products modified or blended with also based on alginate, chitin/chitosan, collagen, branan ferulate, and carbon fibers. Textile structures used for modern wound dressings are of large variety: sliver, yarn, woven, nonwoven, knitted, crochet, braided, embroidered, and composite materials. Wound care also applies to materials like hydrogels, matrix (tissue engineering), films, hydrocolloids, and foams. Specialized additives with special functions can be introduced in advanced wound dressings with the aim to absorb odors, provide strong antibacterial properties, smooth pain, and relieve irritation. It may have unique properties as high surface area-to-volume ratio, film thinness, nanoscale fiber diameter, porosity, and lightweight; nanofibers are used in wound care.

New potential is open to textile fibers that could well themselves possess biofunctions as well as have reasonable mechanical properties or possibility to carry medications. Alginates have many advantages over other traditional dressings. Alginate-based products form a gel on the absorption of wound exudates to prevent wound from drying out on the contrary to traditional cotton and viscose fibers, which can entrap in the wound developing discomfort during dressing removal. It confirms that the alginate fibers are nontoxic, noncarcinogenic, nonallergic, hemostatic, biocompatible, of reasonable strength, capable of being sterilized, manipulatable to incorporate medications, and easy processable.

3.4.1 Wound Healing

The skin is considered the largest organ of the body and has many different functions. The epidermis or outer layer is made of mostly dead cells with a protein called keratin. This makes the layer waterproof and is responsible for protection against the environment. The dermis or middle layer is made up of living cells. It also has blood vessels and nerves that run through it and is primarily responsible for structure and support. The subcutaneous fat layer is primarily responsible for insulation and shock absorbency. Cells on the surface of the skin are constantly being replaced by regeneration from below with the top layers sloughing off. The repair of an epithelial wound is merely a scaling up of this normal process. Science of wound healing is recorded as "three healing gestures"; it describes the three gestures as washing the wound, making plasters, and bandaging the wound. Wounds are generally classified as wounds without tissue loss (e.g., in surgery), and wounds with tissue loss, such as burn wounds, wounds caused as a result of trauma, abrasions or as secondary events in chronic ailments, for example, venous stasis, diabetic ulcers or pressure sores, and iatrogenic wounds such as skin graft donor sites and dermabrasions. Wound healing process may be divided into four continuous phases: (a) homeostasis, (b) inflammation, (c) proliferation, and (d) maturation or remodeling.

3.4.2 Wound Dressing Products

Wound dressings are generally classified as (a) passive products, (b) interactive products, and (c) bioactive products. Traditional dressings like gauze and tulle dressings that account for the largest market segment are passive products. Interactive products comprise polymeric films and forms, which are mostly transparent, permeable to water vapor and oxygen, but impermeable to bacteria. These films are recommended for low-exuding wounds. Bioactive dressing is that which delivers substances active in wound healing, either by delivery of bioactive compounds or using dressings constructed from materials having endogenous activity. These materials include proteoglycans, collagen, noncollagenous proteins, alginates, or chitosan.

3.4.3 Alginate

Wound dressing based on alginic material is well known, in literature as well as from commercial point of view, in wound management. Calcium alginates being a natural hemostatic, alginate-based dressings are indicated for bleeding wounds. The gel-forming property of alginate helps in removing the dressing without much trauma and reduces the pain experienced by the patient during dressing changes. It provides a moist environment that leads to rapid granulation and reepithelialization. In a controlled clinical trial, significant number of patients dressed with calcium alginate was completely healed at day 10 compared with the members of the paraffin gauze group. Calcium alginate dressings provide a significant improvement in healing split skin graft donor sites. In another study with burn patients, calcium alginate significantly reduced the pain severity and was favored by the nursing personnel because of its ease of care. The combined use of calcium sodium alginate and a bio-occlusive membrane dressing in the management of split-thickness skin graft donor sites eliminated the pain and the problem of seroma formation and leakage seen routinely with the use of a bio-occlusive dressing alone. Calcium alginate fibers can be used to produce yarns and fabrics for medical applications, as drug carriers for wound healing.

3.4.4 Chitin

Chitin is a valuable natural polymer indicating excellent bioactive properties. Chitin products are antibacterial, antiviral, antifungal, nontoxic, and nonallergic. Three-dimensional chitin fiber products with qualities such as soft handle, breathability, absorbency, smoothness, and nonchemical additives are the ideal dressings with wound healing properties. The novel method for making a chitin-based fibrous dressing material uses a nonanimal source, microfungal mycelia, as the raw material, and the resulting microfungal fibers are different from the normal spun ones. A method of chitin separating from the bodies of dead honeybees has been developed with the goal to prepare soluble derivatives useful for the manufacture of novel textile dressing materials. Preliminary research of modified honeybee chitin has been carried out, and soluble mixed polyesters of chitin with bioactive properties have been obtained. Chitosan is a partially deacetylated form of chitin.

3.4.5 Chitosan

Chitosan is a natural biopolymer, biocompatible, biodegradable, and nontoxic, and is able to be used as gels, films, fibers, beads, support matrices, and in blends as well. Chitin/chitosan fibers and chitosan derivatives process excellent antibacterial properties and wound healing. The key properties of chitin and chitosan fibers as biomedical products also include that they are hemostatic and fungistatic. Chitosan is used in a broad range of wound healing, drug delivery, and tissue engineering applications. Recent studies in medical textiles have resulted in progress in the modification of traditional materials that are widely used as wound care products. Further, it also possesses other biological activities and affects macrophage function that helps in faster wound healing. It also has an aptitude to stimulate cell proliferation and histoarchitectural tissue organization. The biological properties including bacteriostatic and fungistatic properties are particularly useful for wound treatment. Like alginate material, there is also a number of references on chitosan in wound treatment. Flexible, thin, transparent, novel chitosan–alginate polyelectrolyte complex membranes caused an accelerated healing of incision wounds in a rat model compared with conventional gauze dressing.

Faster wound healing was observed when a variety of chitosan-based skin graft material was tested in guinea pigs and rabbits. A dressing with an optimal combination of chitosan, alginate, and polyethylene glycol containing a synergistic combination of an antibiotic and an analgesic was studied on human subjects with chronic nonhealing ulcers. Chitosan provides a nonprotein matrix for 3-D tissue growth and activates macrophages for tumoricidal activity. It stimulates cell proliferation and histoarchitectural tissue organization. Chitosan is a hemostat, which helps in natural blood clotting and blocks nerve endings reducing pain. Chitosan will gradually depolymerize to release *N*-acetyl-*b*-D-glucosamine, which initiates fibroblast proliferation and helps in ordered collagen deposition and stimulates increased level of natural hyaluronic acid synthesis at the wound site. It helps in faster wound healing and scar prevention.

3.4.6 Alginate Filaments

Alginate filaments coated by chitosan are developed for advanced wound dressings. Cotton fabric surface modified by chitosan absorbs antibiotic molecules from aqueous solution. The quantity of absorption depends on the degree of modification of the samples. With the higher degree of modification, higher amount of antibiotic can be bonded by the textile. Such cotton textile finishing enables to achieve therapeutic new-generation dressings for the protection of surgical wounds against infections.

3.4.7 Branan Ferulate

Branan ferulate is a carbohydrate polymer extracted from corn bran that may infiltrate the biological activities in the body and so accelerate the wound healing process. Hyaluronan is able to interact with various biomolecules. Of course, because of solubility, rapid resorption, and short tissue residence time, the direct use of hyaluronan in wound care has limitations, but the hyaluronan derivatives of different solubility and being of different forms including fibers, membranes, sponges, microspheres, have potential states hyaluronan and chondroitin sulfate chemically modified and made into hydrogel films with wound healing application as biointeractive dressings. Collagen, gelatin, casein, zein, and elastin are proteins widely used in the production of medical textiles. Collagen has been used for sutures for many years. It has controlled the biodegradation rate and is also biocompatible and highly pure.

3.4.8 Collagen Materials

Collagen materials have great potential in scaffolds for tissue culture and wound healing. Advanced product in the generation of biologic dressings is a bilayered composite, a collagen sponge supporting live human allogeneic skin cells. Hybrid scaffolds for tissue repair have been produced from collagen and chitin. Novel use for carbohydrates in textiles is modification of textiles with cyclodextrins. They can trap body odor compounds by inclusion complexation or can be used to release perfumes as well or to deliver pharmaceuticals/cosmetics on skin contact.

Specific potential in the designing of advanced biomaterials for various applications gives carbon fibers that have been used in the reconstruction of soft and hard tissue injuries. In recent years, high attention has been given to the materials that are bioactive. Such quality along with biocompatibility, biosafety, and high absorbency is very desirable in drug delivery systems for surgical implants as well as in scaffolds for tissue regeneration. Bioactive fibers manufactured from modified man-made fibers with introduced antibacterial additive in spinning process provide protection against cross transmission of diseases and infections. The data present an overview that reflects on important developments in advanced wound management materials from fibers to finished products and technique as well that are likely to refresh and enlarge the concept. Wound care textiles and products are recognized, understood, and evaluated. The information indicate that the significant properties of advanced fibers and dressings for wound management include that they are bacteriostatic, fungistatic, hemostatic, nonallergic, nontoxic, high absorbent, biocompatible, breathable, manipulatable to incorporate additives, of reasonable mechanical qualities. The advantage is that the materials are able to be used as gels, films, sponges, foams, beads, fibers, support matrices, and in blends as well. The main structures used for modern wound care are sliver, rover, yarn, woven, nonwoven, knitted fabrics, crochet, braided, embroidered, and composite materials. Specialized additives providing special functions can be introduced in advanced wound dressings. Up-to-date overview of latest innovations in the field of wound care materials confirms that modern techniques like tissue engineering and nanoapplications have a great impact on advanced wound structures.

3.5 Healthcare Products in the Hospital Environment

The scope of medical textiles embraces all textile materials used in health and hygiene applications in both consumer and medical markets. Depending on the nature of application, many medical products are disposable and made out of nonwoven fabrics. Textile products are used in medical and healthcare sectors in various forms. The complexity of applications has increased with research and developments in the area of medical textiles. The surgical gown, operating room garments, and drapes require special antibacterial properties combined with the wearer's comfort. Other major uses of medical textiles are incontinence diapers, sanitary napkins, and baby diapers. Wound dressing, bandages, and swabs are also widely used conventional medical textiles. Wearing of suitable dresses in hospitals and health clubs by the doctor and supporting staff has been accorded a very high priority by the hospital administration in order to protect them from getting infected by the germs and microorganisms and also from spreading the diseases to other patients. Due to the increase in awareness and concern about the healthcare textiles, it has become the need of the hour to develop hospital textiles with functional properties like antimicrobial, odor resistance, and comfort characteristics. Textiles are also being used as sutures, orthopedic implants, vascular grafts, artificial ligaments, artificial tendons, heart valves, and even as artificial skins. Recent advances in medical textiles to be used as extracorporeal devices are also significant; these include artificial kidney, artificial liver, and mechanical lungs. New materials are finding specialized applications like antimicrobial and antifungal fibers and additives used in barrier fabrics, abdominal postoperative binders, applications in neurodermatitis treatment, and various other wound management and surgical treatments.

3.5.1 Healthcare Textiles

Healthcare textiles comprise surgical clothing (gowns, caps, masks, uniforms, etc.), surgical covers (drapes, covers, etc.), and beddings (sheets, blankets, pillow cases, etc.). Healthcare textiles can be disposable or nondisposable. All over the world, disposable healthcare textiles are replacing nondisposables due to ease of use and hygiene, infection-free nature, and also being cost-effective by eliminating laundering. For the disposable healthcare items, polypropylene spun bond is most popular due to its low cost. However, in Western countries, spun bond–melt blown–spun bond and spun lace are more popular because of their inherent advantages in terms of absorption, breathability, etc. The growth of healthcare textiles is expected to be at a very high rate over the years with increase in awareness about advantages of its usage coupled with cost-effectiveness vis-à-vis reusable healthcare textiles. The growth of the healthcare textiles is linked to the growth of healthcare sector, which is growing at around 13%–16% every year. However, nonwoven disposal is expected to register higher growth as it would be penetrating into the share of reusable medical textiles. It is estimated that disposable healthcare textiles would increase from Rs. 11 crore in 2003–2004 to Rs. 120 crore in 2007–2008.

3.5.2 Medical Implants and Devices

Medical implants and devices cover items like cardiovascular implants (vascular grafts, heart valves, etc.), soft tissue implants (artificial tendon, artificial skin, artificial ligament, artificial cornea, etc.), orthopedic implants (artificial joints), and extracorporeal devices (artificial kidney, artificial liver, mechanical lung, artificial heart, etc.).

3.5.2.1 Vascular Grafts

Vascular grafts are used to treat hindrances to blood flow caused by vascular and other diseases. A vascular graft replaces the damaged artery or creates a new artery in order to increase blood flow. The vascular grafts are sterile and are for single patient use only. They are of following types: polyester grafts—used to repair thoracic and abdominal occluded arteries,

Dacron grafts—for aortic surgeries, and PTFE grafts—to repair occluded arteries and veins in the hands and feet and for dialysis treatment of chronic renal failure patients. The prerequisites of good vascular grafts are biocompatibility, nonfraying properties, flexibility, durability, resistance to sterilization, bacteria resistance, and nonthrombogenicity.

3.5.2.2 Heart Valves

The heart valves assist cardiothoracic surgeons in treating valvular diseases. The heart valves are of two types: mechanical valves and tissue valves. Mechanical valves are used for younger patients and require periodical checkups, and after a particular period, the patients need to be operated a second time. Mechanical valves are made of titanium, around which is a knitted fabric to be stitched to the original tissue called as sewing ring. The sewing ring of the caged-disk type of prostheses uses a silicone-rubber insert under a knitted composite PTFE and polypropylene fiber cloth.

3.5.2.3 Artificial Tendon (Mesh)

The composite meshes made up of polyester, polypropylene, and polyester/carbon fiber are used for repairing hernia. The utilization of mesh grafts in humans for hernia operations is based on the fact that during the absorption period, a neomembrane is formed at the site where the mesh has been implanted. The mesh graft prevents recurrence of hernia and hence has an advantage over the tissue repair technique practiced for a long time in India.

3.5.2.4 Artificial Joints

The artificial joints are made of stainless steel, chromium cobalt, titanium, or some other inert material. The textile material present in the joints is ultra-high-molecular-weight HDPE. Artificial joints are covered under BIS No. IS: 5810. The imports of artificial joints have been from Germany, France, Switzerland, United States, etc.

3.5.2.5 Artificial Kidney

Artificial kidney consists of a semipermeable membrane, on one side of which blood passes while a special dialysate solution is passed along the other. The artificial kidney is made of polyacetate and polysulfone in equal proportions.

3.5.2.6 Cardiovascular

Cardiovascular device has unique uses for knitted and woven fabrics, as well as felt scaffolds. To aid not only in support and repair, but also in the regrowth of natural cells within the body for recovery from conditions affecting the heart and circulation, high-density, nonstretch biomaterials have become increasingly important to device design. It includes cardiovascular grafts, heart valve repair, annuloplasty rings, containments, tethers, and pledgets.

3.5.2.7 General Surgery

From sutures to slings, developing high-quality biotextiles for applications in general surgery is varied and deep. Designing high-strength and low creep sutures to aid in ease of use for surgeons, or surgical meshes for organ containment within the abdomen, an intimate knowledge of advanced textile engineering combined with anatomic applicability offers a proven track record of success in device development. It includes sutures, suture fasteners, sewing thread, surgical mesh, hernia repair, urogynecologic slings, prolapse devices, and catheters.

3.5.2.8 Orthopedics

Orthopedic is based in extensive successful applications across a variety of procedures, repair and replacement devices, and regrowth systems that require strong but supple materials not only in the defect, but also in the surrounding soft tissue areas where organ architecture and function are affected. Textiles from woven tapes to knitted meshes continue to help build next-generation orthopedic devices. It includes sutures, knee, shoulder, and small joint arthroscopic procedural assistance, knee, shoulder, and spinal implants, soft tissue repair, spinal/cervical disk repair containment sleeves and meshes, orthopedic reconstruction via cellular regrowth, and cuffs.

3.5.2.9 Bariatric

As obesity threatens to create more health problems than ever among an increasingly heavy population, bariatric devices and procedural instrumentation will demand a higher standard of development innovation to satisfy performance requirements and mimic natural body materials. Biomedical solutions (BMS) fiber expertise extends to tissue knowledge and anatomic applicability for containment devices and other tools for weight-loss surgery and maintenance.

3.5.2.10 Scaffolds: Tissue Growth

Scaffold technology is designed for the high surface area and multilayer requirements of tissue engineering and regenerative medicine. Cellular ingrowth within these scaffolds is enabled by a high void volume and open architecture, both of which support cellular proliferation. The base polymer will then degrade and is replaced by natural tissue. These applications demand an advanced engineering approach to ensure fabric integrity and profile. Now an increasingly wide variety of applications from wound healing to stem cell development rely on medical textile production.

3.5.2.11 Cosmetic/Reconstructive Surgery

Cosmetic surgery demands high-quality repair and replacement capabilities for stressful procedures and reconstructive efforts. It takes advantage of the lifelike properties of fiber to meet these demands. From high surface area to measured density can satisfy a host of procedural requirements, from minimally invasive procedures to fabric tissue support. It includes reconstructive and cosmetic surgery mesh, absorbable additives for cement and hydrogel void fillers, and pledgets.

3.5.2.12 Dental Implants

Research and discovery of new, useful, and innovative applications in biomedical structures are growing every day. Applications like dental implants and procedural components benefit from the variety of precise geometries and lifelike texture of the textiles that help build them. Versatility in shape, size, and durability profiles, implants, and other tools for every device challenges.

3.5.3 Other Applications: Healthcare Textiles

3.5.3.1 Pressure-Relieving Mattress

The mattresses provided in most of the hospitals are made of hard polyurethane foam, covered by waterproof-coated fabric, over which a simple single-layered cotton bedspread is used, which makes the patient highly uncomfortable due to the heat generated and strain on the contact areas. To overcome the earlier-said problem, a pressure-relieving mattress was developed along with multilayered functional bed cover to reduce the magnitude, direction, and/or duration of pressure and temperature, thereby avoiding excessive tissue distortion on vulnerable parts of the body. The interface pressure and temperature between body and the mattress were measured and analyzed using the parameters such as Deformation Index and Pressure gradient. The pressure-relieving mattress developed reduces the interface pressure by 30%–60% and the heat generated by around 3°C.

Most of the hospitals are using tough mattress, covered with waterproof-coated fabric, over which simple single-layered cotton bedspread is used, which makes the patient highly uncomfortable due to the strain and heat-generated on the contact areas. One of the common manifestations of chronic disease and disability is the abnormal loading of skin and other

surface tissues to large mechanical forces, which develops into a significant injury that damages tissues through the entire thickness of the body
wall. Excess compression causes damage of the blood vessels, leading to
bedsores of different degrees with unbearable pain. When the pressure on
any part of the body increases beyond 33 mm mercury level, blood circulation is arrested, which leads to bedsore development. There is also compel-
ling evidence that factors in addition to pressure are contributors and must
also be considered when attempting to fully understand the pressure-sore
phenomenon. Studies have implicated factors such as shear stress, impact
loading of tissue, elevated temperature and humidity, age, nutritional sta-
tus, general health, activity level, deformity, posture and postural change,
body stature, and psychological deficits. Increased friction and shear, poor
nutrition, disease, and pressure aggravate compromised skin. Even the
chemical irritation of frequent washings with soaps can cause irritation.
The adhesiveness of moist skin to bed linens is estimated to increase the
risk of ulceration fivefold.

Pressure: Bedsores form where the weight of the person's body presses the
skin against the firm surface of the bed. The pressure that causes bedsores
does not have to be very intense. Pressure of less than 25% the pressure of a
normal mattress can lead to bedsores. Complete muscle necrosis was demonstrated at 100 mmHg for 6 h, and pressures of 70 mmHg for 2 h resulted in
pathologic changes within muscle and that lower pressures of 35 mmHg for
4 h resulted in no changes.

Temperature: The metabolic heat generated by the body must be transferred through the bed linen and the failure of which leads to increase in
interface temperature between the body and mattress. Elevated body temperature raises the metabolic activity of tissues by 10% for every 1°C of
temperature increase, concurrently increasing the need for oxygen and an
energy source at the cellulose level.

Shearing and friction: Shearing and friction cause skin to stretch and blood
vessels to kink, which can impair blood circulation in the skin. In a person
confined to bed, shearing and friction can occur when the person is dragged
or slid across the bed sheets. Shear stress occurs when a force is applied
in the plane of the skin surface. Friction occurs when there is displacement
between the skin and the supporting surface.

Moisture: Wetness from perspiration, urine, or feces can make the skin too
soft and more likely to be injured by pressure. Moisture from sweating or
incontinence will hydrate the skin, dissolve the molecular collagen cross-
links of the dermis, and soften the stratum corneum. Another result of skin
hydration is the rapid increase in the epidermal friction coefficient, which
promotes adhesion of the skin to the support surface and increase shear, easy
sloughing, and ulceration.

Research studies show a prevalence of pressure ulcers in 11% of the hos-
pitalized population and in 20% of nursing home residents at any given
time. For patients in nursing homes, the prevalence of pressure sores ranges

from 7% to 35%, resulting in a fourfold increase in mortality. In spinal cord-injured patients, pressure ulcer incidence is as high as 42%–85% in some centers. The principal approach in the past to the challenge of maintaining healthy skin and avoiding breakdown has been prevention. For example, patients restricted to bed rest, a subject population at high risk of pressure ulcer formation, will be turned frequently by the nursing staff to relieve prolonged pressure. Mattresses designed to cyclically change the distribution of pressure have been developed. Prevention measures also involve the application of interface surfaces (mattresses, cushions, liners) and frequent pressure reliefs to avoid sustained pressures in one position.

3.5.3.1.1 Development of Mattress with Air Circulation Device

A mattress was developed using soft polyurethane foam with a thickness of 10 cm, provided with horizontal and vertical drill holes connected to an air circulation device to give enough air circulation through the mattress. Vertical and horizontal holes were drilled to enhance air circulation inside the foam mattress. Slots were cut on the surface of the foam to accommodate a continuous air tube, and the air tube is provided with small drill holes on its surface, through which mild air circulation is maintained throughout the length and width of the mattress. Both the ends of air tubes are connected to an air-circulating device THAT has two outlets, one for air outlet and another for air suction. This foam part of the mattress acts as the firm bottom support for the mattress, over which needle-punched nonwoven fiber web made of hollow polyester fiber is laid one over the other for a thickness of about 20 cm. The foam and fiber-filled mattress were covered by means of a knitted fabric cover, made of lyocell fiber and microencapsulated with phase-changing material.

3.5.3.1.2 Design Features of Mattress Cover Material

The design features of the cover material warrants sufficient elastic property, which plays an important role in the ability of the mattress to deform and envelope around the body. For example, a fluid-filled support surface, such as a water bed, would not envelop as water does, because the membrane cover containing the water does not have enough elastic properties. A mattress cover under high tension (overstretched or very tight bed covers) may cause locally high peak pressures. Hence the mattress cover material is an important element of the support surface, which affects the loading in the plane of the skin and the microclimate at the interface between the subject and the mattress. The choice of cover material significantly influences these factors. Relatively inelastic covers tend to produce high interface stresses at the skin surface. Primarily, weight of the body is taken by both deformation of the support structure, for example, foam top surface and the elastic stretching of the cover material. If the cover material is relatively inelastic, then more of the load is taken by cover material in the plane of the skin, and this tends to have a shearing effect on the skin. This form of loading can significantly contribute to the development of pressure ulcers. Cover material with relatively homogeneous elastic properties, hence, two-way stretch can be used to minimize the stretching effect. Hence, lyocell knitted fabric, which is capable of stretching in both length- and width-wise directions, is selected as the cover material for the mattress.

A pressure-relieving mattress was developed along with multilayered functional bed cover to reduce the magnitude, direction, and/or duration of pressure, thereby avoiding excessive tissue distortion on vulnerable parts of the body. From the interface pressure measurement on the hospital mattress and the mattress developed, it can be observed that the reduction in pressure ranges from 30% to more than 60%. This reduction in pressure is due to the highly soft nature of the hollow fiber-filled part of the mattress that deforms to fit to the shape of the body, thereby increasing the area of contact between the body and the mattress leading to pressure distribution to more area. The pressure-relieving mattress with air circulation system reduces the heat generated between the body and the top surface of the mattress by around 3°C and keeps the body comfortable. The deformation index, which is a measure of the difference between the peak and average pressures, is reduced by 23%–33% in the mattress developed, when compared to the hospital mattress.

3.5.4 Current Research in Healthcare Textiles

Medical textiles always look for new products, new uses for existing products, and better substitutes. Healthcare textiles have a well-established market in the developed countries where people are conscious of the risks posed to the healthcare workers, especially from blood-borne diseases. Massive growth in the population in developing countries and rising standard of living have helped in creating a vast potential for healthcare textiles. Hospital textiles such as bedding, clothing, surgical gowns, and hospital cloths are expected to fulfill comfort and hygienic properties such as moisture management, thermal conductivity, breathability, antimicrobial activity, and odor resistance. Since the activated carbon is efficient in adsorbing odorous volatile microorganisms, thereby reducing the odor and growth of microorganisms, it is used in hospital textiles for the reduction of microbial growth and adsorption of wound odor.

Bamboo charcoal is a nongraphite form of activated carbon made from the pieces of bamboo plants and then converted into powder form. Fibers from bamboo charcoal can be produced in many kinds, such as single, multifilament, and staple fiber. It can also be spun with pure cashmere, cotton, and others. There are two main ways to produce bamboo charcoal fiber: the first way is to add nanobamboo charcoal powder during the process of spinning in the spinning solution; the second is to add the established bamboo charcoal composite polymer master batch in the stage of synthesizing fiber. Bamboo charcoal viscose fiber can be produced from natural plant cellulose pulp by adding bamboo charcoal micropowder milk dissolved by the solvent and then spinning the solution by extrusion and solidification. Polyester-based bamboo charcoal fiber is produced in a similar way from polyester master batch with bamboo charcoal content about 50%. Bamboo charcoal has countless small cavities when compared with wood charcoal, which gives about three times more mineral constituent and four times better absorption rate. Bamboo charcoal has a surface area of 300 m/g, which is 10 times more than wood charcoal of 30 m/g. The unique properties of bamboo charcoal include uniform composition, high porosity, antibacterial and antifungal properties, breathability, thermal regulation, odor control, absorption and emission of far infrared energy, preventing static electricity buildup, and good wash durability. These fabrics can absorb and disperse sweat fast, making them feel dry and comfortable. They also do not stick to the skin on hot summer days. Bamboo charcoal fabrics absorb and decompose benzene, phenol, methyl alcohol, and other harmful substances. As the bamboo charcoal nanoparticles are embedded in the fiber rather than simply coated on the surface, these fabrics are washable without diminished effectiveness of the charcoal powder's special qualities, even after 50 washes. Cotton, bamboo/cotton, and bamboo charcoal bed linens were analyzed for their suitability as hospital textiles by applying antimicrobial and blood repellent finish, and it was found that antimicrobial activity, blood repellency, and odor resistance are higher for bamboo charcoal fabrics than 100% bamboo/cotton union fabrics and 100% cotton fabrics. The other constituent of the union fabric selected for producing hospital textiles is lyocell, which is a high-performance, solvent-spun 100% cellulosic fiber, ideal for use in many woven and nonwoven applications. The advantages of lyocell include more softness, gentle to the skin, high absorbency, excellent water management property, cool and dry to touch, strong retardation to bacterial growth, and excellent wet strength.

3.5.5 Textile-Based Medical Products Used in Hospitals

1. *Face mask:* A surgical face mask is an important medical device used to protect both surgical patients and operating room personnel from the transfer of microorganisms, body fluids, and particulate material.

2. *Surgical gown:* Surgical gowns are worn by the doctors and nurses in the operating theater to address the dual function of preventing transfer of microorganism and body fluids from the operating staff to the patient and also from patient to staff. Different textile structures used in surgical gown are single-use type, made up by nonwoven techniques and reusable category and normally developed through weaving. International specifications for surgical gown (ASTM F 2407)—this specification establishes requirements for the performance, documentation, and labeling for surgical gowns used in healthcare facilities. European EN 13795 aims to establish

requirements for surgical gown, surgical drape, and clean air suits used as medical devices for patients, clinical staff, and equipment. There are four special tests that must be performed in order to evaluate the performance of surgical gown: (1) spray impact penetration test, (2) hydrostatic head test, (3) resistance to synthetic blood, and (4) viral penetration resistance.

3. *Surgical drape:* Surgical drapes are impervious to liquid strike through which bacterial transfer and subsequent contamination of the surgical site can be reduced. These drapes feature an absorbent, impervious material throughout the drape to incorporate the true critical zone in the entire sterile field.

4. *Hospital bed linen:* The bottom sheet wraps around the mattress deeper than regular fitted sheets and has two-way stretch so the corners would not slip out from underneath on the bed. The set also comes with a top sheet, a pillowcase, and a thermal blanket, which is perfect for homecare hospital-style beds.

5. *Baby diaper*—Diaper is used for wrapping the newly born or young children, who have not developed the fixed routine for making water or latrine. Diapers retain the liquid for about 2 diaper hours. Diapers can play an important role in the prevention of contamination and the reduction of infection. Disposable diapers are more effective in urine/feces containment than cloth. Increased contamination of surfaces in day care settings has been linked to infectious diarrhea outbreaks.

6. *Incontinence product:* The incontinence products are described as "step-in" underwear or protective underwear, goods described commonly as "belted briefs," a two-piece system that consists of a belt that is fastened around the waist of an individual and used with a diaper folded through the crotch.

7. *Sanitary napkin:* It is an absorbent item worn by a woman while menstruating, recovering from surgery, for absorption of blood or liquid.

8. *Gauze bandage:* It is a simple woven strip of material, or a woven strip of material with a Telfa absorbent barrier to prevent adhering to wounds.

9. *Plaster of paris bandage:* It is made up of leno gauze base fabric with high-purity gypsum in special high-tech formulation, which gives a strong, rapid setting and durable cast.

10. *Crepe bandage:* This bandage is also known as an ACE bandage, elastic wrap, or compression bandage or elastic bandage. It is a "stretchable bandage used to create localized pressure." Elastic bandages are commonly used to treat muscle sprains and strains by reducing the flow of blood to a particular area by the application of even stable pressure that can restrict swelling at the place of injury.

11. *Nonwoven gauze bandage:* Nonwoven gauze is of mesh structure (similar to woven cloth) made from viscose rayon, polyester blend fabric with minimum 65% viscose content.

12. *Elastic adhesive bandage:* It consists of a woven fabric, elastic in warp direction and coated with the adhesive mass containing zinc oxide. The adhesive mass may be porous or permeable to air and water vapor.

13. *Cellulose wadding:* Cellulose wadding consists of compressed sheets of felted fibers, consisting almost entirely of cellulose. The fibers are bleached good white.

14. *Vascular graft:* Arteries are the blood vessels, which carry oxygenated blood throughout the body. These arteries sometimes get damaged and fail to do the work properly. Artificial grafts are developed to act as blood vessels, and they are called as vascular grafts.

15. *Sutures:* Sutures are the simplest example of a textile biomedical device. Sutures are used for wound closure to close cuts and incisions and thus prevent infection and are an integral part of all operations. In fact, no surgery can be performed without the sutures. Absorbable sutures are ideal for wounds inside the body as they dissolve and get absorbed into the body after the operation. Absorbable sutures are of two types: natural and synthetic. Synthetic absorbable sutures are made up of PGA, which are absorbed into the body within 20–90 days, and natural absorbable sutures are made up of mucosa of sheep intestine. Nonabsorbable sutures that are made up of nylon, polypropylene, silk, polyester, and PTFE are not absorbed into the body and need to be removed by the surgeon. Nonabsorbable sutures are generally used in external applications where they are easily accessible, removal is easy, and prolonged high strength is required. Nonabsorbable sutures are used for serious and complex wounds, where the need is that stitches should not dissolve fast to give the wound a chance to heal while preventing wound reopening and scar tissue formation.

3.6 Evaluation and Testing of Medical Textiles

Textiles have always played a major role in human hygiene. Textiles provide a barrier against all kinds of germs or serve as bandage or plaster. Modern surgery uses textile "stents" and "scaffolds" degradable or not by the human body. The microbiological testing is equipped to assess antibacterial and antifungal properties of textiles. The test performs on biological degradability, barrier properties against microorganisms, particles, laser rays in operation theaters, and the microbiological hygiene of medical textiles. Textiles used in operation theaters have to be conforming to the requirements of the European Medical Devices Directive 2007/47/EC (modified 93/42/EEC). This conformity depends on the type of medical device and more in particular on the class to which it belongs. There are indeed four different classes of medical devices, the most severe one being class 4 for those medical devices including the highest risk of contamination. Textiles used in operation theaters belong to class 1 (low risk) *of the medical devices* and are conform to if they comply with different parts of standard EN 13795. This standard has been written in the spirit of the European directive and aims at preventing the transfer of infections between the medical staff and the patient during surgical operations and other invasive interventions. Moreover, the standard wants to guarantee the same safety and performance levels for all products, disposable or reusable articles alike, and this during the entire product life.

EN 13795 deals with (woven or nonwoven) surgical drapes, gowns, and clean air suits, used as medical devices for patients, hospital staff, and equipment. The standard defines the requirements and corresponding test methods. The standard was modified for the last time in 2009 and includes three parts: EN 13795-1:2002+A1:2009, EN 13795-2:2004+A1:2009, EN 13795-3:2006 + A1:2009. These minimum values account for the circumstances of the surgical operation. Therefore, they distinguish between (1) products for standard performances and products for higher performances and (2) the critical and less critical zones of the product.

- The critical zone of a product has a greater chance to be involved in the transfer of infection carriers on the operation site or on the invasive zone or vice versa. For example, the front piece and the sleeves of surgical gowns are in the immediate proximity of the operation site.
- A product for standard performances or high performances is defined in the function of the exposure to biological or other fluids, to mechanical pressure, or to the duration of surgical operation.

Part 2 of EN 13795 describes the following tests:

- Tensile strength—dry and wet—EN 29073-3:1992
- Burst strength—dry and wet—EN 13938-1
- Resistance to fluid penetration—EN 20811
- Cleanliness—microbial—EN 1174 (replaced by EN ISO 11737)
- Cleanliness—particulate matter—ISO 9073-10
- Linting—ISO 9073-10
- Resistance to microbial penetration—wet—ISO 22610
- Resistance to microbial penetration—dry—ISO 22612 (Figures 3.8 through 3.10)

FIGURE 3.8
ISO 22610: Resistance to microbial penetration—wet.

FIGURE 3.9
ISO 22612: Resistance to microbial penetration—dry.

3.6.1 Protection of the Patient versus Medical Staff

Standard EN 13795 is aimed at the protection of the patient; this determines the direction in which the sample is brought into contact with the contaminating agent during the test. If the manufacturer also claims the protection of the medical staff, the surgical gown will no longer be considered as a medical device but as a personal protection equipment. In this case, the product has to comply with the corresponding directive 89/686/EEC (protective clothing) and standard EN 14126: Protective clothing—performance requirements and tests methods for protective clothing against infective agents.

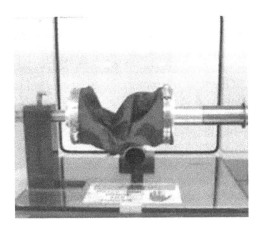

FIGURE 3.10
ISO 9073-10: Linting test of textile fabric.

Standard EN 13795 mentions some interesting additional properties such as "liquid control" or adhesion properties. The following test methods are to measure the comfort properties of the products:

- Skin model—ISO 11092—EN 31092
- Thermal manikin—EN ISO 15831—ASTM F2370
- Air permeability—ISO 9237
- WP resistance to water—ISO 811

Besides tests that have to prove the conformity of the product, the European directive introduces the notion of risk analysis by proposing the following tests:

- Nontoxicity of the medical device—ISO 10993
- Measurement of the electrical risk—ISO 2878 (BS2050) and EN 1149
- Protection against laser beams
  1. ISO 11810-1: primary ignition and penetration
  2. ISO 11810-2: secondary ignition

3.6.2 Microbiology Tests for Textiles

Microbial biotechnology is one of the branches of applied science with diverse applications. Microorganisms are present everywhere in soil, water, and air, and they need nutrition for growth, depending on the substrate available for the growth of microorganisms, the durability, and the aesthetic value of material. The microorganisms play a vital role in deterioration of any material, which is susceptible to attack. The routinely used textile material or which is coming in contact with the body is more prone to attack by microorganisms. This is because during perspiration or when in contact with the body, the organisms present on the body gets transferred on to cloth or related material. If such textile material is nonresistant to attack by microorganisms, then it can lead to deterioration and hence change in the physical properties. Such material not only is of poor quality but also can cause some skin irritation or allergies. Hence, it is utmost important to incorporate some form of antimicrobial agent in the textile material to get good quality, longer durability, and better aesthetic value. Many textile manufacturers and research laboratories are currently engaged in developing various antibacterial and antifungal finishes for textile products. The textile finish needs to have properties such as skin friendly and resistant to washings, for its effective usage along with its antibacterial or antifungal effect. The following types of tests are recommended at microbiology/biotechnology tests for textile fabrics, which are given as follows:

- Antibacterial activity assessment of textile materials—parallel streak method (AATCC 147)
- Antibacterial finishes on textile materials (AATCC 100)
- Antifungal activity, assessment on textile material, mildew, and rot resistance of textiles (AATCC 30)
- Standard practice for determining resistance of synthetic polymeric materials to fungi (ASTM G 21)
- Determining the antimicrobial activity of immobilized antimicrobial agents under dynamic contact conditions (ASTM E2149)
- Testing for antibacterial activity and efficacy on textile products (JIS L 1902)
- Bioburden testing of textiles, food material, medical textiles, and other pharmaceutical products as per USP, BP, IP
- Identification of specific microorganisms from given material as per specified IS
- Sterility testing of treated material as per USP
- Microbial limit tests as per USP
- Detection of mildew/rot proofness (MIL-STD-810 F Method 508.5.1-12/01/2000)
- Determination of resistance of geotextiles to microbial attack by soil burial test (DIN EN 12225)
- Evaluation of bacterial filtration efficacy of medical textile (ASTM F 2101)
- Determination of resistance of medical textiles to penetration by synthetic blood (ES 21–92)

- Bacteriological analysis of water as per WHO/APHA standards
- Insect pest deterrents on textiles (AATCC 28)
- Method for testing cotton cordages for resistance to attack by micro-organisms (IS 1386)
- Method for testing cotton fabric for resistance to attack by microorganism (IS 1389)
- Assessment of antimicrobial activity of carpets (qualitative and quantitative; AATCC 174)

Some other tests for quality assessment of healthcare products are given in Table 3.7.

Major organizations have published guidelines for healthcare workers:

- Centers for Disease Control and Prevention
- Association of Pre-Operative Registered Nurses
- Occupational Safety and Health Administration
- Operating Room Nurses Association of Canada
- Association for the Advancement of Medical Instrumentation (AAMI)

AAMI classification system: There are four levels of barrier performance, level 4 being the highest protection available, which is given in Table 3.8.

3.6.3 Special Test Methods and Their Importance

Bacterial filtration efficiency %—ASTM F 2101: This test method measures the percent efficiency at which the face mask filters bacteria passing through the mask. The maximum filtration efficiency that can be determined by this method is 99.9%.

Splash resistance mmHg—ASTM f1862: This test method is used to evaluate the resistance of medical face masks to penetration by the impact of a small volume (2 mL) of a high-velocity stream of synthetic blood. Medical face mask pass/fail determinations are based on visual detection of synthetic blood penetration.

Spray impact penetration test—AATCC 42: A volume of synthetic blood is allowed to spray against the taut surface of a test specimen backed by a weighed blotter. The blotter is then reweighed to determine water penetration, and the specimen is classified accordingly.

Water resistance—hydrostatic pressure test: One surface of the test specimen is subjected to a hydrostatic pressure, increasing at constant rate, until three points of leakage on its surface. The water may be applied from above or below the test specimen.

TABLE 3.7

Some Other Tests for Quality Assessment of Healthcare Products

Products	Basic Tests Requirements
Face mask	Air permeability
	Bacterial filtration efficiency %
	Splash resistance
Surgical gown	Resistance of liquid penetration
	Resistance to microorganisms
	Breathable
	Flexible
	Water vapor transmission rate
	Tensile strength
	Air permeability
	Stiffness
	Flammability
	Bursting strength
	Thermal resistance
Surgical drape	Drape
	Air permeability
	Weight per unit area
	Breaking strength and elongation of textile fabrics
	Flammability
	Thermal resistance
	Antibacterial activity assessment (qualitative)
	Linting test
Hospital bed linen	Weight/square meter
	Tensile strength
	Tear strength
	Antibacterial activity assessment (qualitative)
	Antibacterial activity assessment (quantitative)
Baby diaper	Soft
	High absorbency
	Protection against leakage
	Should not rewet
	Comfortable
	Speed of absorption
	Rewet
	Absorbent capacity
	Absorbent retention
	Fit and comfort
Incontinence product	Volume of leaked liquid
	Absorption rate
	Wicking rate
	Wettability
	Permeability

TABLE 3.7 (continued)

Some Other Tests for Quality Assessment of Healthcare Products

Products	Basic Tests Requirements
Sanitary napkin	Absorbency
	pH
	Disposability
Gauze bandage	Yarn count (tex)
	Threads/10 cm
	Coloring matter
	Surface active substances
	Sulfated ash their soluble substances
	Water-soluble substances
	Loss on drying
	Foreign matter
	Viable microorganism prior to sterility (cfu/g)
	pH
	Absorbency
Plaster of paris bandage	Threads per unit length
	Weight of the fabric (GSM)
	% Calcium sulfate
	Tensile strength (kg/cm^2)
	Compressive strength (kg/cm^2)
	Setting time (min)
	Alkalinity
Crepe bandage	Yarn count
	Warp yarn twist
	Threads/10 cm
	Stretchability
	Breaking load
	pH
Nonwoven gauze bandage	Fluorescence
	Sterility
	Foreign matter
	Absorbency
	Weight (GSM)
	Length and width
Sutures	Diameter
	Tensile strength
	Bending stiffness
	Surface roughness
	Knot pull strength
	Knot security

(*continued*)

TABLE 3.7 (continued)

Some Other Tests for Quality Assessment of Healthcare Products

Products	Basic Tests Requirements
Circular knitted bandages	Thickness
	Air permeability
	Weight per unit area
	Breaking strength and elongation of textile fabrics
	Breaking fabrics
	Skin irritation
	Antibacterial activity assessment (qualitative)
	Antibacterial activity assessment (quantitative)
	Thermal resistance
Combined wound dressings	Wound contact layer
	Middle absorbing layer
	Base material
	Skin irritation test
	Air permeability
	Weight per unit area
	Breaking strength and elongation of textile fabrics
	Absorbency
	Antibacterial activity assessment (qualitative)
	Antibacterial activity assessment (quantitative)
	Thermal resistance
Elastic adhesive bandage	Threads per 10 cm
	Weight of the fabric (GSM)
	Weight of adhesive mass
	Zinc oxide content in adhesive mass
	Adhesive strength g/2.5 cm
	Moisture vapor permeability
	Regain length
Cellulose wadding	Weight per unit area (GSM)
	Sulfated ash
	Loss on drying
	Absorbency
	Chloroform soluble substances
Vascular graft	Porosity
	Bursting strength
	Tensile strength
	Water permeability
	Biocompatibility

TABLE 3.8

AAMI Test Method of Fabric Barrier
Performance Level

Level	Test	Result
1	ATCC 42, water impact (WI)	≤4.5 g
2	AATCC 42, WI	1.0 g
	AATCC 127, hydro head (HH)	≥20 cm
3	AATCC 42, WI	≤1.0 g
	AATCC 127, HH	≥50 cm
4	ASTM F1671, gowns	Pass
	ASTM F1670, drapes	Pass

Note: Level 1, least protective; level 4, most protective. AAMI (Association for the Advancement of Medical Instrumentation).

Resistance to synthetic blood—ASTM F 1670: A specimen is subjected to body fluid stimulant (synthetic blood) for a specified time and pressure. Visual observation is made to determine when penetration occurs. Any evidence of synthetic blood penetration constitutes failure. Results are reported as pass or fail.

Resistance to penetration by blood-borne pathogens—ASTM F 1671: This test method is used to measure the resistance of materials used in protective clothing to penetration by blood-borne pathogens using a surrogate microbe under conditions of continuous liquid contact. Protective clothing material pass/fail determinations are based on detection of viral penetration.

Liquid strike through time (simulated urine): This test method measures the strike through time, that is, the time taken for a known volume simulated urine applied to the surface of a test piece of nonwoven cover stock, which is in contact with an underlying absorbent pad, to pass through nonwoven.

Diaper rewet: The purpose of the test is to examine the ability of diaper cover stock to resist the transport back onto the skin of a liquid, which has already penetrated the cover stock.

Absorbency: The sanitary napkins shall absorb 30 mL of colored water or oxalated sheep or goat blood or test fluid when lowed onto the center of the napkin (at the rate of 15 mL/min), and it shall not stain through/leak through at the bottom or sides of the sanitary napkin.

pH value: The sanitary napkin shall be free from acids and alkali and the pH of the absorbent material shall be 6–8.5.

Disposability: A disposable sanitary napkin with the covering removed shall be immersed in 15 L of water and stirred. The pad shall disintegrate in the water in not more than 5 min.

3.7 Future Medical Textiles

The global market in 2007 for medical textiles was about $8 billion; every year this niche market becomes more relevant, and its importance will increase even more in the future. The increase in population over 60 years and consequently the increase in doctor visits are the main problem that new technologies will have to cope in order to assure a good medical service. However, the incorporation of nanotechnology in textiles may boost the so-called telemedicine (use of sensors and telecommunications in textiles to transfer medical information with the aim of consulting and remote medical measures or examinations). Health and hygiene textile materials have a wide variety of uses; however, nanoenabled developments remain in the early stages of development. The products range from simple gauze or bandage materials to scaffolds for tissue culturing and a large variety of prostheses for permanent body implants.

3.7.1 Current Medical Textile Research

The literature of the past decade, including patents, provides a broad overview of current research activities as well as of some of the problems and concerns related to implantable medical textiles. Among the more intensively studied product groups are surgical sutures, vascular grafts, and artificial ligaments and tendons. In fact, several factors such as suitable temperature and humidity, presence of dust, soil, spilled food and drink stains, skin dead cells, sweat and oil secretions of skin gland, and also finishing materials on the textile surfaces can make textile optimal enrichment cultures for a rapid multiplication of microorganisms. Regarding too rapid improvement of hygienic living standard, controlling of aforementioned terrible effect is necessary. Therefore, many researchers have focused on the antibacterial modification of textiles. Recently, using natural material has been preferred for textile modification because of possible harmful or toxic effects of many chemical antimicrobial agents. Application of inorganic nanoparticles and their nanocomposites would be a good alternative, and consequently, they can open up a new opportunity for antimicrobial and multifunctional modification of textiles. This review chapter is concerned with properties and application methods of inorganic nanostructured materials with good potential of antimicrobial activity and their nanocomposites for textile modification. Inorganic and metal-based nanostructured materials have created a new interesting field in all sciences for the continuous investigations due to their undeniably unique properties. Their applications have already led to the development of new practical productions.

Surgical sutures: For surgical sutures, the predominant areas of concern are strength, capillarity, sliding and positioning of knots, knot security, and handling characteristics. The recent focus of suture research has been on improving the structure of the braids (two recently proposed products are spiral- and lattice-braided materials), reducing the difference in the elongation properties between the core and the sheath yarns, using finer-denier filaments in the sheath yarns, and improving knot security and performance by exposing a two-throw square knot to laser-beam energy.

Vascular grafts: Regarding vascular grafts, the lack of healing, compliance, and suture-line patency continue to be concerns, especially in small-caliber (<6 mm diameter) grafts. Three important efforts that highlight global developmental activities in this area are (a) use of semiabsorbable structures, with absorbable components, woven or knitted, in the inner tube wall; (b) use of spray technology in conjunction with elastomeric polymers to produce collagen-like fiber structures with biomechanical compliant properties; and (c) the incorporation of elastomeric components in the weft threads of woven prostheses. Using this technique, woven grafts of 4–6 mm diameter could be produced, with transverse compliance comparable to that of canine and other similarly sized arteries.

Ligaments and tendons: In the area of artificial ligaments and tendons, desirable properties include high strength, high elasticity, low abrasion, low creep, and low stiffness. Current research endeavors are examining the use of ultra-high-strength fibers (e.g., spectra from AlliedSignal), threads containing layers of both absorbable inelastic and nonabsorbable elastic fibers, and coatings with biocompatible polymers to reduce abrasion and restrict escape of abraded particles from within the structure.

3.7.2 Nanotechnology in Medical Textiles

Textile materials continue to serve an important function in the development of a range of medical and surgical products. The introduction of new materials, the improvement in production techniques and fiber properties, and the use of more accurate and comprehensive testing have all had significant influence on advancing fibers and fabrics for medical applications. As more is understood about medical textiles, there is every reason to believe that a host of valuable and innovative products will emerge. Nanotechnology-related textiles can play an important role in the medical sector. Currently, woven and nonwoven antibacterial fabrics are the most used applications of nanotechnology in the medical textile segment, being used to prevent infection or deodorize medical clothing, wound dressing, and bedding. The numbers of fields where nanotechnology-related textiles are finding applications are growing and include the following:

- *Surgical, with surgical drapes* for the aseptic techniques used in everyday wound dressing, catheter changing, and the like, to reduce the chances of contamination and cross-infection.
- *Medical,* 3-D textiles to prevent and reduce contact irritations and wound infections.

- *Prostheses*, with fibers that are able to facilitate the bonding of the implant to the living bone, or with resorbable guidance devices for the regeneration of peripheral nerves.
- *Dental*, with textile that release medical active gases for therapeutic applications, or with multicomponent nanofilament for dental care applications.
- *Garments*, with lightweight, flexible, lead-free x-ray shielding aprons, or clothing incorporating electronic functions to monitor biological parameters or improve the quality of life.
- *Drug delivery*, with drug-loaded fibers for the delivery and the controlled release of therapeutic agents.
- *Fabrics surface-functionalized* and utilized for tissue engineering.
- *Nonwoven nanofiber filters* used in a variety of medical equipment, such as respiratory equipment and transfusion/dialysis machines.
- *Hygiene*, with composite nonwovens with improved liquid-absorbing features for nappies, sanitary napkins, adult incontinence pads, panty liners, etc.

In the medical/healthcare sector, the nanomaterials principally utilized are silver nanoparticles, for their recognized antibacterial activity.

3.7.3 Antibacterial Textiles

The applications in the medical sector cover the range of antibacterial textiles with a broad spectrum of antimicrobial activity and the absence of drug resistance, capable to prevent mite sensitization in dermatitis, antibacterial wound dressings, patient dresses, bed linens, or reusable surgical gloves and masks. But they can be extended also to protective face masks and suits against biohazards or to toothbrushes. The need to sanitize clothing, and many other everyday items, has resulted in the extension of the antibacterial war to many other objects such as sports clothing, domestic and automotive interior textiles, and toys. The antibacterial activity of these textiles is utilized, in particular, to produce antiodor clothes for the sport/outdoor and furniture sectors. In addition, possible uses of antibacterial textiles are considered for the household products such as kitchen clothes, sponges, or towels.

3.7.4 Antimicrobial Wound Dressings

This is a very important application of nanotechnology in the medical field. Wound dressings are manufactured by means of a bilayer of silver-coated, high-density polyethylene mesh with a rayon-adsorptive polyester core. The dressing delivers nanocrystalline silver from a nonadherent, nonabrasive surface. In vitro studies have shown that the sustained release of this ionized nanocrystalline silver maintains an effective antibacterial and fungicidal activity. In addition, nanocrystalline silver dressings have been clinically tested in a variety of patients with burn wounds, ulcers, and other nonhealing wounds facilitating wound care by adequate debridement, and bacterial and moisture balance. Wound dressings have also been developed, which combine an electrospun polyurethane nanofibrous membrane and silk fibroin nanofibers. These electrospun materials are characterized by a wide range of pore size distribution, high porosity, and high surface area-to-volume ratio, which are favorable parameters for cell attachment, growth, and proliferation. The porous structure is particularly important for fluid exudation from the wound, avoiding wound desiccation, and impairing exogenous microorganism infection.

3.7.5 Antiadhesive Wound Dressings

Textile wound dressings such as plasters or bandages find wide range of uses in medical applications to cover wounds until the healing process can protect the wound against external environmental attack. Traditional wound dressings generally adhere to the healing wound, causing a new injury on removal, and thereby interrupting the healing process. The close control over fiber architecture offered by embroidery is also of potential interest for highly loaded structures, enabling fibers to be placed in the position and with orientations necessary to optimize strength and stiffness locally. The textile surface of these wound dressings is also of importance for comfort and prevention of mechanical irritation. Innovative wound dressings with antiadhesive properties to the healing wound have been obtained by coating the common viscose bandages with silica nanosol modified with long-chain alkyltrialkoxysilanes. An additional, not secondary, feature of the earlier innovative wound dressings is their ability in water uptake. Good absorption properties for the wound exudates are of great help to the healing process and are of special value for bedridden patients with chronic wounds.

3.7.6 Product Examples

- *Acticoat*™ (Smith & Nephew plc, United Kingdom): Smith & Nephew has created a fast-acting, bacteria-destroying wound dressing. It contains safe bactericidal concentrations of silver with patented nanocrystalline technology.
- *Face masks* (Nanbabies® Face Masks, United States): It works against all types of bacteria and viruses, even killing antibiotic-resistant strains as well as all fungal infections. The nanocrystalline silver particles used remain active up to 100 washes.

- *Nanocyclic towel* (NanoCyclic, Inc., United States): Super absorbing and antibacterial cloth. It absorbs water and repels germs.
- *NanoMask* (Emergency Filtration Products, United States): It is the first protective face mask in the world to utilize nanoparticle-enhanced filters to address potentially harmful airborne contaminants.
- *Eco-fabric* (Green Yarn, United States), which is antimicrobial, anti-static, and has other health benefits.
- *NanoPro Wrist Supporter; NanoPro Elbow Supporter; NanoPro Back Supporter* (Vital Age, United States): It helps to increase microcirculation, in the elbow and in the lower back areas to help relieve tired muscles. It utilizes an exclusive ceramic compound utilized in the production of all NanoPro products.
- *Nanover™ wet wipes* (GNS Nanogist Co. Ltd, United States): Safe to use for children's toys, soft like cotton, protect babies' frail skin. Low-irritative natural ingredients protect and moisturize your skin, and prevent skin trouble. Cleans hands and around lips.

3.7.7 Medical Smart Textiles

Textiles provide an excellent substrate for integration with electronic devices, including sensing, monitoring, and information-processing tools, able to react to the conditions and stimuli, like the mechanical, thermal, chemical, and electrical, and transmitted by the wearer. The healthcare field could take advantage of these smart textiles to provide for a patient's extended monitoring during a long rehabilitation period. In this field, a new class of electrically conductive material, called Quantum Tunneling Composite (QTC), is being produced by the U.K.-based Peratech. QTC has the unique ability to smoothly change from an electrical insulator to a metal-like conductor when placed under a pressure; these features will find a number of medical applications including blood pressure control, respiratory monitoring, and sensing in prosthetic socket.

3.7.8 Medical Industry Key Players

In the medical industry, the key players in the textile sector are the following:

Johnson & Johnson (United States; www.JNJ.com): Global pharmaceutical and medical devices and consumer packaged manufacturer. Among its consumer products are bandages, Johnson's baby products, Neutrogena skin and beauty products, clean & clear facial wash, and Acuvue contact lenses.

Baxter Healthcare Corporation (United States; www.baxter.com): Healthcare company with headquarters in Deerfield, IL. The company is a subsidiary of Baxter International Inc. Baxter develops, manufactures, and markets products to treat hemophilia, kidney disease, immune disorders, and other chronic and acute medical conditions.

Beiersdorf AG (Germany; www.beiersdorf.com): Manufacturer of personal care products. The total revenue in 2009 was € 5748 billion, and among its different brands are Elastoplast, Eucerin, Labello, Liposan, Nivea, and Marlies Möller. Beiersdorf already sells products that include nanotechnology, including sunscreen and deodorants, and is looking to prove an effective technology for both sunscreen and antiaging products to use with textiles.

3M Company (United States; www.3M.com): Minnesota Mining and Manufacturing Company. Among the products they manufacture include adhesives, abrasives, laminates, passive fire protection, dental products, electronic materials, electronic circuits, and optical films. The company sells an Aldara skin cancer cream that is improved by nanotechnology. Using nanotechnology, it is also improving the ScotchGard protector for use in textiles and clothes.

Smith & Nephew (United Kingdom; www.smith-nephew.com): Smith & Nephew is a medical company specializing in orthopedic reconstruction, orthopedic trauma, and clinical therapies, and endoscopy and advanced trauma management. They are involved in business like orthopedic reconstruction, trauma endoscopy, and wound management. Smith & Nephew has developed the brand Acticoat, a wound management that contains nanocrystalline silver.

Bibliography

Aibibu, D., S. Houis, M. Sri Harwoko, and Th. Gries, *Textile Scaffolds for Tissue Engineering—Near Future or Just Vision*, Woodhead Textiles Series 75, Woodhead Publishing, Cambridge, U.K., 2010.

American Red Cross, Fibrin bandages include natural clotting agent—US Army and Navy tackle bleeding in different ways, *Medical Textiles*, 11, 116 (1999).

Bagherzadeh, R., M. Montazer, M. Latifi, M. Sheikhzadeh, and M. Sattari, Evaluation of comfort properties of polyester knitted spacer fabrics finished with water repellent and antimicrobial agents, *Fibers and Polymers*, 8(4), 386–392 (2007).

Biomedical Structures LLC, Warwick, RI, www.bmsri.com/structures-overview.com (accessed May 25, 2013).

Bower, C. K., J. E. Parker, A. Z. Higgins, M. E. Oest, J. T. Wilson, B. A. Valentine, M. K. Bothwell, and J. McGuire, Protein antimicrobial barriers to bacterial adhesion: In vitro and in vivo evaluation of nisin-treated implantable materials, *Colloids and Surfaces*, 25, 81–90 (2002).

Centexbel Ghent, Zwijnaarde, www.centexbel.be/evaluation-of-medical-textiles (accessed May 25, 2013).

Chen, X., G. Wells, and D. M. Woods, Production of yarns and fabrics from alginate fibres for medical applications, in *Medical Textiles*, S. C. Anand (ed.), Woodhead Publishing, Cambridge, U.K., 2001, pp. 20–29.

Chen, X., G. Wells, and D. M. Woods, Production of yarns and fabrics from alginate fibres for medical applications, *Proceedings of International Conference on Medical Textiles*, 1999 August 24–25, Leeds, U.K., Woodhead Publishing, Cambridge, U.K., 1999, pp. 20–29.

Ciechanska, D., Multifunctional bacterial cellulose/chitosan composite materials for medical applications, *Fibres and Textiles in Eastern Europe*, 12(48), 69–72 (2004).

Cohen, I. K., *A Brief History of Wound Healing*, 1st edn., Oxford Clinical Communications, Inc., Yardley, PA, 1998.

CSIRO, Australia, www.csiro.au/en/Outcomes/Health/TissueEngineering.aspx (accessed May 25, 2013).

David Rigby Associates (DRA), May 5, 2011, http://www.davidrigbyassociates.co.uk (accessed May 25, 2013).

Dunn, R. L., J. W. Gibson, B. H. Perkins, J. M. Goodson, and L. B. Laufe, Fibrous delivery systems for antimicrobial agents, *ASC Polymeric Material Science and Engineering*, 51, 28–31 (1984).

Eaglstein, W. H., Moist wound healing with occlusive dressings: A clinical focus, *Dermatologic Surgery*, 27, 175–181 (2001).

Eming, S. A., H. Smola, and T. Krieg, Treatment of chronic wounds: State of the art and future concepts, *Cells Tissues Organs*, 172, 105–117 (2002).

Gao, Y. and R. Cranston, Recent advanced in antimicrobial treatments of textiles, *Textile Research Journal*, 78(1), 60–72 (2008).

Gebelein, C. G. and C. E. Carraher, *Biotechnology and Bioactive Polymers*, Plenum Press, New York, 1994.

Graham, K., H. Schreuder-Gibson, and M. Gogins, Incorporation of electrospun nanofibers into functional structures, *Proceedings of the International Nonwoven Technical Conference*, September 15–18, Baltimore, MA, 2003.

Guidoin, R., D. Marceau, J. Couture et al., Collagen coatings as biological sealants for textile arterial prostheses, *Biomaterials*, 10(3), 156–165 (1989).

Gulrajani, M. L., Nano finishes, *Indian Journal of Fibre and Textile Research*, 31, 187–201 (2006).

Gupta, B. S., Medical textile structures: An overview, *Medical Plastics and Biomaterials Magazine*, 5(1), 16–30 (1998).

Gupta, B. S. and V. A. Kasyanov, Biomechanics of the human common carotid artery and design of novel hybrid textile compliant vascular grafts, *Journal of Biomedical Materials Research*, 34, 341–349 (1997).

Hasebe, Y., K. Kuwahara, and S. Tokunaga, Chitosan hybrid deodorant agent for finishing textiles, *AATCC Review*, 1(11), 23–28 (2001).

Hellotrade International, India, www.hellotrade.com/biomedical-structures/product.html (accessed May 25, 2013).

Kamaruk, E., J. Mayer, M. During et al., Embroidery technology for medical textiles, *Proceedings of the International Conference Medical Textiles*, 1999 August 24–25, Leeds, U.K., Woodhead Publishing, Cambridge, U.K., 1999, pp. 200–206.

Kamaruk, E., G. Raeber, J. Mayer, B. Wagner, B. Bischoff, and R. Ferrario, Structural and mechanical aspects of embroidered scaffolds for tissue engineering, *Proceedings of the 6th World Biomaterials Congress*, Kamuela, HI, 2000.

Kandhavadivu, P., C. Vigneswaran, T. Ramachandran, and B. Geethamanohari, Development of polyester-based bamboo charcoal and lyocel-blended union fabrics for healthcare and hygienic textiles, *Journal of Industrial Textiles*, 41(2), 142–159 (2011).

Karamuk, E., J. Mayer, M. During, B. Wagner, B. Bischoff, R. Ferrario, M. Billia, R. Seidl, R. Panizzon, and E. Wintermantel, Embroidery technology for medical textiles, medical textiles, *Proceeding of the 2nd International Conference*, Bolton, U.K., 1999.

Kenawy, E. R. and Y. R. A. Fattah, Antimicrobial properties of modified and electrospun poly(vinyl phenol), *Macromolecular Bioscience*, 2(6), 261–266 (2002).

Khil, M. S., D. I. Cha, H. Y. Kim, I. S. Kim, and N. Bhattarai, Electrospun nanofibrous polyurethane membrane as wound dressing, *Journal of Biomedical Materials Research Part B: Applied Biomaterials*, 67, 675–679 (2003).

Khor, E. and L. Y. Lim, Implantable applications of chitin and chitosan, *Biomaterials*, 24, 2339–2349 (2003).

Knill, C. J., J. F. Kennedy, J. Mistry et al., Alginate fibres modified with unhydrolysed and hydrolysed chitosans for wound dressings, *Carbohydrate Polymers*, 55, 65–76 (2004).

Lala, N. L., R. Ramaseshan, L. Bojun, S. Sundarrajan, R. S. Barhate, J. L. Ying, and S. Ramakrishna, Fabrication of nanofibers with antimicrobial functionality used as filters: Protection against bacterial contaminants, *Biotechnology and Bioengineering*, 97(6), 1357–1369 (2007).

Le, Y., S. C. Anand, and A. R. Horrocs, Using alginate fibre as a drug carrier for wound healing, in *Medical Textiles 96*, S. C. Anand (ed.), Woodhead Publishing, Cambridge, U.K., 1997, pp. 21–26.

Lee, H. J., S. Y. Yeo, and S. H. Jeong, Antibacterial effect of nanosized silver colloidal solution on textile fabrics, *Journal of Materials Science*, 38, 2199–2204 (2003).

Lim, S. K., S. K. Lee, S. H. Hwang, and H. Kim, Photo catalytic deposition of silver nanoparticles onto organic/inorganic composite nanofibers, *Macromolecular Materials and Engineering*, 291, 1265–1270 (2006).

Macken, C., Bioactive fibers—Benefits to mankind, *Chemical Fibers International*, 53(1), 39–41 (2003).

Mao, L. and L. Murphy, Durable freshness for textiles, *AATCC Review*, 1, 28–31 (2001).

Matsuda, K., S. Suzuki, N. Isshiki et al., A bilayer artificial skin capable of sustained-release of an antibiotic, *British Journal of Plastic Surgery*, 44, 142–146 (1991).

Merhi, Y., R. Roy, R. Guidoin et al., Cellular reactions to polyester arterial prostheses impregnated with cross-linked albumin: In vivo studies in mice, *Biomaterials*, 10(1), 56–58 (1989).

Min, B. M., G. Lee, S. H. Kim, Y. S. Nam, T. S. Lee, and W. H. Park, Electrospinning of silk fibroin nanofibers and its effect on the adhesion and spreading of normal human keratinocytes and fibroblasts in vitro, *Biomaterials*, 25, 1289–1297 (2004).

Minami, S., Y. Okamoto, and S. Tanioka, Effects of chitosan on wound healing, in *Carbohydrates and Carbohydrate Polymer*, M. Yalpani (ed.), ALT Press, Chicago, IL, 1992.

Miraftab, M., Q. Quiao, J. F. Kennedy, S. C. Anand, and G. Collyer, Advanced materials for wound dressings: Biofunctional mixed carbohydrate polymers, *Proceedings of the International Conference Medical Textiles*, 1999 August 24–25, Leeds, U.K., Woodhead Publishing, Cambridge, U.K., 1999, pp. 164–172.

Muri, J. M. and P. J. Brown, *Alginate Fibres-Biodegradable and Sustainable Fibres*, Woodhead Publishing, Cambridge, U.K., 2005, pp. 89–109.

Pachene, J. M. and J. Kohn, Biodegradable polymers, in *Principles of Tissue Engineering*, R. P. Lanza, R. Langer, and J. Vacanti (eds.), Elsevier Science, New York, 1997.

Paul, W. and C. P. Sharma, Chitosan and alginate wound dressings: A short review, *Trends in Biomaterials and Artificial Organs*, 18(1), 18–23 (2004).

Petrulyte, S. Advanced textile materials and biopolymers in wound management, *Danish Medical Bulletin*, 55(1), 72–77 (2008).

Rajendran, S. and S. C. Anand, Development in medical textiles, *Textile Progress*, 10–13 (2002).

Rajendran, S. and S. C. Anand, Contribution of textiles to medical and healthcare products and developing innovative medical devices, *Indian Journal of Fibre and Textile Research*, 31, 215–229 (2006).

Rigby, A. J., S. C. Anand, and A. R. Horrocks, Textile materials for medical and health-care applications, *Journal of Textile Institute*, 88(Part 3), 83–93 (1997).

Russell, S. J. and N. Mao, Anisotropic fluid transmission in nonwoven wound dressings, *Proceedings of the International Conference Medical Textiles*, 1999 August 24–25, Leeds, U.K., Woodhead Publishing, Cambridge, U.K., 1999, pp. 156–163.

Scarlet, R., Deliu, R., and Manea, L. R., Implantable medical textiles: Characterization and applications, *7th International Conference—TEXSCI 2010*, September 6–8, Liberec, Czech Republic.

Shalaby, S. W., Antimicrobial fabrics, U.S. Patent Application, 6, 780–799 (2004).

Smith, M., Fibrous scaffolds for tissue culturing, *Proceedings of the International Conference Medical Textiles*, 1999 August 24–25, Leeds, U.K., Woodhead Publishing, Cambridge, U.K., 1999, pp. 173–179.

Smith, L. A. and P. X. Maa, Nano-fibrous scaffolds for tissue engineering, *Colloids and Surfaces*, 39, 125–131 (2004).

Studer, H., Antimicrobial protection for polyolefin fibers, *Chemical Fibers International*, 47(5), 373–374 (1997).

Tamura, H., Y. Tsuruta, and S. Tokura, Preparation of chitosan-coated alginate filament, *Materials Science and Engineering*, 20(1–2), 143–147 (2002).

Tavis, M. J., J. W. Thornton, R. H. Bartlett, J. C. Roth, and E. A. Woodroof, A new composite skin prosthesis, *Burns*, 7, 123–130 (1980).

THETA Reports, Market analysis, in *Advanced Wound Care Biologics: World Market Analysis*, PJB Medical Publications, Inc., New York, 2002, pp. 37–46.

Thiry, M. C. and S. Writer, Antimicrobials take the field, *AATCC Review*, 1(11), 11–17 (2001).

Vigo, T., Antibacterial fiber treatment and disinfection, *Textile Research Journal*, 51, 454–465 (1981).

Wang, J., J. Smith, W. Babidge, and G. Maddern, Silver dressings versus other dressings for chronic wounds in a community care setting, *Journal of Wound Care*, 16, 352–356 (2007).

Williams, S. K., T. Carter, P. K. Park et al., Formation of a multilayer cellular lining on a polyurethane vascular graft following endothelial cell seeding, *Journal of Biomedical Materials*, 26(1), 103–117 (1992).

Wintermantel, E., J. Mayer, J. Blum, K. L. Eckert, P. Luscher, and M. Mathey, Tissue engineering scaffolds using superstructures, *Biomaterials*, 17, 83–91 (1996).

Wong, Y. W. H., C. W. M. Yuen, M. Y. S. Leung, S. K. A. Ku, and H. L. I. Lam, Selected applications of nanotechnology in textiles, *AUTEX Research Journal*, 6(1), 1–8 (2006).

www.bmsri.com/structures-overview.com (accessed May 25, 2013). Bordenave, L., J. Caix, B. Basse-Cathalinat et al., Experimental evaluation of a gelatin-coated polyester graft used as an arterial substitute, *Biomaterials*, 10(3), 235–242 (1989).

www.centexbel.be/evaluation-of-medical-textiles (accessed May 25, 2013). Chapman, K., High-tech fabrics for smart garments, *AATCC Review*, 2(9), 11–15 (2002).

www.csiro.au/en/Outcomes/Health/TissueEngineering.aspx (accessed May 25, 2013). Dastjerdi, R., M. R. M. Mojtahedi, A. M. Shoshtari, and A. Khosroshahi, Investigating the production and properties of Ag/TiO$_2$/PP antibacterial nano-composites filament yarns, *Journal of the Textile Institute*, 101, 204–213 (2010).

www.hellotrade.com/biomedical-structures/product.html (accessed May 25, 2013). Hussain, M. M. and S. S. Ramkumar, Functionalized nanofibers for advanced applications, *Indian Journal of Fibre and Textile Research*, 31(3), 41–51 (2006).

Yang, J. M., H. T. Lin, T. H. Wu, and C. C. Chen, Wettability and antibacterial assessment of chitosan containing radiation-induced nonwoven fabric of polypropylene-g-acrylic acid, *Journal of Applied Polymer Science*, 90, 1331–1339 (2003).

Younsook, S., Y. Dong, and M. Kyunghye, Antimicrobial finishing of polypropylene nonwoven fabric by treatment with chitosan oligomer, *Journal of Applied Polymer Science*, 74, 2911–2916 (1999).

4

Finishing of Industrial Textiles

4.1 Introduction

Textile finishing is a general term applied to various processes that the textile materials undergo after pretreatments, dyeing, or printing for final adornments to enhance their attractiveness as well as comfort and durability. Chemical processing of textiles involves three stages such as pretreatment, coloration, and finishing. Finishing is the final step in the fabric manufacturing process, which will add value to the fabric. Finish can be either chemicals that change the fabric's aesthetic and/or physical properties or changes in texture or surface characteristics brought about by physically manipulating the fabric with mechanical devices. Finishing is understood to involve all those processes that cannot be classified under dyeing and printing. Besides washing, this includes desizing, steaming, setting, and the production of various effects such as calendering and raising as well as treatments to improve serviceability properties such as water-repellent, shrink-resistant, wrinkle-resistant, and soil-release finishes. Most finishes are applied to fabrics such as woven, knitted, and nonwovens.

It is evident that the application of naturally occurring products could produce quite dramatic effects and new fabric properties. The life of a fabric could be extended, or vastly improved properties, such as water repellency or increased strength or durability, could be achieved with simple or quite complicated finishing treatments. Nowadays, many of these finishes are carried out for very specific purposes in that they give special properties for specific end uses, and in most cases, the cost of processing is extremely competitive. Most of the end uses are industrial, where the life of the product is short term. Finishing treatments are basically meant to give the textile material certain desirable properties like

- Softness
- Luster
- Drape
- Dimensional stability

- Crease recovery
- Antistatic
- Nonslip
- Soil release
- Water repellency
- Flame-retardancy
- Mildew proofing

The finishing stage plays a fundamental role in the excellence of the commercial results of textiles, which strictly depend on market requirements that are becoming increasingly stringent and unpredictable, permitting very short response times for textile manufacturers.

4.2 Types of Finishing

The types of finishes required and their methods of application depend upon the nature of fibrous substrate and their arrangement in the yarn or fabric. Finishing is commonly divided into two categories: chemical and mechanical. In chemical finishing, water is used as the medium for applying the chemicals. Heat is used to drive off the water and to activate the chemicals. The chemical finishes are subject to intensive and high-quality research and are undergoing continuous changes to make these even more effectual, long lasting, easy to apply, and, last but not the least, cost-effective.

Mechanical finishing is considered as a dry operation even though moisture and chemicals are often needed to successfully process the fabric. Mechanical finishing is the process that alters the hand, look, and performance of the textiles. The mechanical treatments are mainly, though not exclusively, applicable to the natural fibers and have not changed much with the passage of time.

4.2.1 Mechanical Finishing of Industrial Textiles

Mechanical finishing refers to any process performed on yarn or fabric to improve the look, performance, or "hand" of the clothing. Fabric luster, smoothness, softness, residual shrinkage, and hand are some of the properties that can be enhanced by mechanical finishing. Mechanical finishing processes normally performed on a prepared and dyed fabric in order to improve its suitability for the desired end use, the aftertreatment being provided purely by mechanical means.

Industrial textile is a textile product manufactured for nonaesthetic purposes, where function is the primary criterion. In order to impart the required functional properties to the textile material, it is customary to subject the material to different types of physical and chemical treatments.

Industrial textiles cover several categories such as filtration textiles, geotextiles, medical textiles, reinforcement fabrics in hoses and belts, and ropes. Dimensional stability is the essential property required in many industrial textiles applications such as filter fabric, home textiles, geotextiles, roofing fabrics, wound care dressings, and fabric reinforcements. Hand is the desirable property needed in some of the industrial textiles such as medical gowns, surgical masks, and home textiles. Filter fabric requires smooth surface that facilitates easy release of dust cake from fabric surface without clinging. Singeing and raising processes are performed in filter fabric to enhance the dust release and dust-holding capacity respectively. The application of mechanically finished fabric in industrial textiles is enormous, and it covers wide range of sectors.

4.2.1.1 Heat Setting

Heat setting is a process done to make the fabric dimensionally stable, particularly synthetic fabrics. The main purpose of heat setting is to improve structure homogenization and the elimination of internal stresses within the fiber during manufacture resulting in reduced shrinkage, improved dimensional stability, and reduced creasing propensity in textile fabrics. It is a vital process in the pretreatment of man-made fibers, which is also of importance in subsequent stages as intermediate setting or postsetting. The effect of heat setting is greater in fibers with an increasingly hydrophobic character and is carried out on polyolefin, polyester, polyurethane (PU), polyacrylonitrile, polyamide, triacetate, acetate, and viscose fibers.

Stenters are widely used for stretching, drying, heat setting, and finishing of fabrics. The speed ranges from 10 to 45 m/min with a maximum setting time in the setting zone 30 s at the temperature ranging from 175°C to 250°C depending upon the thickness and type of the material. The application of heat in heat setting can be done by hot air, on a pin stenter at 220°C for 20–30 s for polyester goods and at a lower temperature range of 190°C–225°C for 15–20 s for polyamides. Acrylics may be heat set partially at 170°C–190°C for 15–60 s to reduce formation of running creases, but higher temperature should be avoided to prevent yellowing.

4.2.1.2 Calendering

Calendering is the mechanical finishing process to modify the surface of a fabric by the action of heat and pressure. It is carried out by compressing the fabric in between two or more rolls under controlled conditions of time, temperature, and pressure in order to alter its handle, surface texture, and appearance. Modern calenders commonly have two or three bowls in a

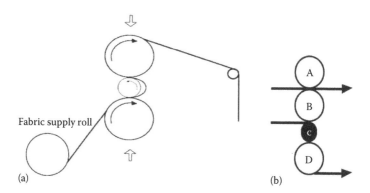

FIGURE 4.1
(a) Calendering process. (b) Bowl arrangement options on calenders (A, B, D, resilient rollers; C, heatable metal rollers).

vertical arrangement. The alteration in properties depends on the ability of the fabric to be mechanically changed.

Calendering is the important finishing operation for most types of woven fabric (except wool), narrow fabrics, etc. Calendering process influences the surface appearance, pore density, smoothness, luster/matt effects, and handle of the fabric. The number of passes, composition of the rolls, temperature controls, moisture control, and pressure can vary to produce the desired effect (Figure 4.1).

Calendering process alters the fabric properties as follows:

1. Reduced fabric thickness
2. Increased fabric luster
3. Increased fabric cover
4. Smooth surface feel
5. Reduced air porosity
6. Reduced yarn slippage

The type of calender used depends on the type of cloth to be run and the desired effect to be produced on the fabric. Calenders can be classified as embossing calenders, friction calenders, swissing calenders, chase calenders, and compaction calenders. The difference between them is the number of calender rolls and the drive system.

4.2.1.2.1 Friction Calendering

Friction calenders apply a friction force to the face of the fabric. The speed differentials ranging from 5% to 100% create friction on the fabric, so it is essential to have a strong fabric to withstand strains. This process brings about fabrics with a high degree of luster on the face side, and the final effect achieved is similar to ironing.

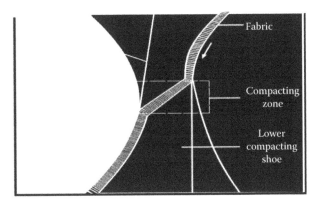

FIGURE 4.2
Friction calendering.

Friction calenders consist of highly polished chromium-plated steel bowl which runs at a higher circumferential speed than the fabric itself so that the fabric receives a lower or higher degree of glazing depending on the preselected advance speed (friction) of the calender bowl. Adequate fabric density is a substantial prerequisite for imparting a glazed effect on the fabric surface (Figure 4.2).

High calender roll nip pressure combined with high temperature applied to a fabric impregnated with starch, on repeated frictioning, produces a quite transparent fabric in which all the internal spaces are eliminated. Friction calendering plays a dominant role in the development of tracing cloth, book cloth, and some specialty fabrics.

4.2.1.2.2 Swissing

Swissing is a process in which a cold calender produces a smooth flat fabric. However, if the steel bowl of the calender is heated, then in addition to smoothness, the calender produces a lustrous surface (Figure 4.3).

Five-bowl calender system normally consists of a bottom bowl made up of cast iron, a second bowl of compressed cotton felt or paper, a third bowl of

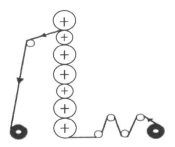

FIGURE 4.3
Swissing calender.

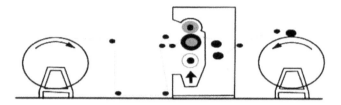

FIGURE 4.4
Schreinering calendering process.

hollow iron and fitted with steam heat apparatus. The fourth bowl is made of compressed cotton, and fifth of cast iron. The fabric is simply passed through for "swissing," that is, for the production of an ordinary plain finish.

4.2.1.2.3 Schreinering

"Schreinering" is a form of embossing whereby fine lines engraved on a metal roller are transferred to the fabric by passage through the nip between the heated engraved roller and a filled bowl. The pattern roll has anywhere from 250 to 350 lines/in., etched at 26° from the vertical. These lines are lightly embossed into the fabric and, being regular, reflect light so as to give the surface a high luster. This operation gives a silk-like brilliance to cotton fabrics (Figure 4.4).

Multipurpose Schreiner calender with exchangeable rollers include the two-roller embossing silk finishing and chintz calender, the three-roller silk finish and similar mercerizing calender. The position of the grooves in the fluting is important as the warp and weft twist need to have uniform direction so the fluting does not run at an angle through the fiber lay and cut into the fibers.

4.2.1.2.4 Moiré Effect

Moiré effects are produced by interference from two linear systems. The characteristic moiré patterns may be produced by an engraved roller in a manner analogous to the schreinering process, but the result will show a warp way repeat at a pitch equal to the circumference of the engraved bowl. A truly random moiré effect is obtained by passing two lengths of the same fabric face-to-face through the calender. Each fabric generates a moiré pattern on the other as a result of the inevitable small variations in yarn spacing.

Moiring (watering) aims at the production of wave-shaped moiré effects, which occurs due to partial even printing of weft ribs on viscose and silk fabrics. Real moiré without repeat of figures is produced by calendering with ribbed rollers and possibly following outside passage of an irregularly per-forated edge, whereby weft threads are shifted from their position and the moiré effect is increased. False moiré is produced on the embossing calender using moiré embossing.

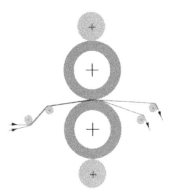

FIGURE 4.5
Embossing.

4.2.1.2.5 Embossing

Embossing is the impressing of patterns on normally smooth textile surfaces, produced under high pressure and at high temperature on the embossing calender, schreiner calender, etc., between a heated embossing cylinder and a mating or elastic counter roll (Figure 4.5).

Generally bleached, dyed, or printed fabrics of even wetting capacity, singed, possibly with preliminary finish but chemically unmodified are suitable for embossing. For embossing in particular, a variable adjustment of the working pressure is necessary. The working pressure is dependent upon the quality of the fabric and also the embossing design because of the different pressing areas given by each design.

4.2.1.3 Sueding or Emerizing

Sueding or emerizing is a process in which fabric at open width is passed over one or more rotating emery-covered rollers to produce a suede-like finish. Sueding operation is often carried out before the raising process to reduce the friction between the fibers making up the cloth and consequently to facilitate the extraction of the fiber end. Sueding is performed on the surfaces of woven and knitted fabrics and even laminated fabrics. The sueding process is normally carried out on both sides of the fabric and modifies the appearance and the final hand of the cloth. The major change in the fabric after emerizing is the production of a very low pile, that is, short fibers protruding from the fabric surface. The handle will vary according to the type of fiber, the fiber linear density, and the sueding intensity on the fabric. The fabrics made up of microfibers should be emerized prior to dyeing (Figure 4.6).

Emerizing machines can be classified into multiroller machine and single-roller machine. In the multiroller emerizing machine, the fabric is tensioned

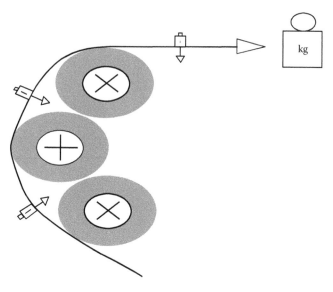

FIGURE 4.6
Emerizing.

over four to seven rollers with emery papered rollers that turn against or with the direction of the fabric. The single-roller emerizing machine is less productive compared to multiroller emerizing machine, with a typical operating speed of about 7.5 mpm on microfiber fabrics.

4.2.1.4 Raising or Napping

Raising is the process of imparting a surface effect that gives a brushed or napped appearance to the fabric. Raising operation creates a different feel and a velvety material surface on fabrics by loosening a large number of individual fibers from the fabric and subsequent raising (velour raising) and napping (nap raising) in order to create a dense raised surface. Raised fabrics that are produced from staple-fiber spun yarn fabrics, such as woolens, worsteds, and velours consist of pulling out a layer of fibers from the structure of a fabric to form a pile. Raised fabrics from filament yarns consist of stretching of loops in the fabric structure by the raising action. Raised loop fabrics are used for nightwear or bed sheets. Raising process also produces more fullness and softer handle in the fabric (Figure 4.7).

A typical modern roller raising machine is fitted with 24 rollers: 12 nap and 12 counter nap rollers, mounted alternately. Each set of rollers is independently driven; their relative speeds, together with that of the cloth, govern the raising effect. The ends of the needles protruding from the rollers are 45° hooks. The fuzzy surface is created by pulling the fiber end out of the yarns

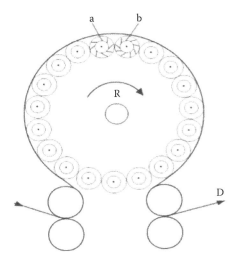

FIGURE 4.7
Raising. R, reel drum; D, direction of goods; a, counterpile roller; b, pile roller.

by means of metal needles provided with hooks shelled into the rollers that scrape the fabric surface.

4.2.1.5 Singeing

Singeing should be carried out with all fabric qualities made of spun yarns if this is made necessary by requirements such as desired wear properties or subsequent processing conditions. The object of singeing is to remove projecting fibers from the surface of the fabric so as to give it a smoother, cleaner appearance, which is essential both for subsequent processing in textile finishing and for desired serviceability properties. Knitted or woven fabrics made from all types of fibrous materials or their blends can be singed, with singeing having special significance in cotton finishing.

Singeing is often carried out on cotton fabrics, or fabrics with cotton blends, and results in increased wettability (better dyeing characteristics, improved reflection, no "frosty" appearance), a smoother surface (better clarity in printing), improved visibility of the fabric structure, less pilling, and decreased contamination through removal of fluff and lint.

Singeing machines for open-width knit-goods differ from those for wovens mainly by the guiding and transport of the fabric. The curled selvedges have to be opened as far as possible by particularly suitable devices for uncurling. Spreading and centering are effected by specially equipped slatted guide rollers. Counter-rotating driven scroll rolls are placed directly in front of the burner to keep the selvedges open when the flame meets the fabric. Osthoff® singeing machines are equipped with driven slatted guide rollers and singeing rollers to minimize the tension during the thermal stress when singeing.

4.2.1.6 Perforating and Slitting

The nonwoven bonded fabrics produced are too stiff and are, therefore, unsuitable for clothing. This is because the individual fibers are not free to move in relation to one another, as are threads in woven or knitted fabrics. Perforating and slitting are two methods practiced to improve the drape of nonwoven bonded fabrics.

Perforating process not only punches holes but also reinforces as a result of cross-linking and condensation of the bonding agent. Perforated webs made of synthetic fibers to produce nonwoven bonded fabrics are strong and yet supple enough for use as building and insulation materials.

Slitting originally developed to improve the softness and drape of films was used for interlinings, in particular for adhesive fixable interlinings. The optimum cut length and distance between the slits to get maximum softness and fall without serious reduction of strength can be calculated. The effect of slitting allows greatest flexibility at right angles to the direction of the slit. The slitting is accomplished by a roller with small blades mounted on it, for example, in an offset arrangement 1.7 mm apart, making slits of a maximum length of 6.5 mm. Rotary knives with spreaders can be fitted to the roller, thus making an interrupted cutting edge. Polyethylene or polyamide film shaped by splitting or embossing and stretching makes good air-permeable bonding layers for laminating nonwoven bonded fabrics.

4.2.1.7 Sanforizing

Sanforizing is a process in which fabric is stuffed together along its length (warp-way) on a machine resembling a felt calender and setting it in this state, so that when it is subsequently washed, no further shrinkage will occur (Figure 4.8).

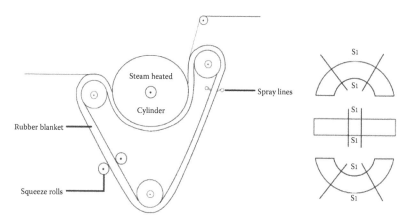

FIGURE 4.8
Sanforizing head.

The goods are extended to full width on a short frame at a higher speed before reaching the shrinking device. The shrinkage takes place immediately after passing the hot shoe at a temperature of approx. 180°C. The flexural strength of the fiber is decreased, the physical cross-linking is broken, and the textile material is more easily compressed. Drying takes place immediately after the subsequent felt calender. The shrinking capacity of this machine is max. 10%. The degree of shrinkage is determined by the thickness of the felt. The thicker the felt, the higher the shrinkage capacity. The degree of shrinkage is regulated by altering the shrinking shoe.

4.2.1.8 Decatizing

A finishing process is mainly used to improve the characteristics of wool and wool blend piece goods with regard to appearance, shape, handle, luster, and smoothness, thereby producing a fabric in the desired finished state. Decatizing is accomplished by steaming under pressure, that is, by subjecting the well-packed or firmly wound fabric to a steam treatment.

4.2.1.9 High-Pressure Water Jet Softening

Nonwoven fabric is subjected to high-pressure water jet softening having characteristics at least a 25% reduction in fabric bending modulus. High-pressure water jet softening is a new process developed to soften autogenously point-bonded nonwoven fabric in which the fabric is impinged with a water jet. The fluid jet employed will be a high-energy jet of the type obtained by ejecting highly pressurized fluids through appropriate nozzles or orifices. The jet impingement may be employed, simultaneously or sequentially, in conjunction with other fabric treatments tending to effect or enhance fabric softening.

Bending modulus is used as a measure of fabric softness. Hand, drape, and bending moduli are determined by analyzing the load-extension curve. Bending modulus is determined as an average of fabric faceup and facedown machine and transverse direction measurements.

4.2.2 Chemical Finishing of Industrial Textiles

Function is often the most important parameter to determine the value of industrial textiles. The industrial textiles are used for automotive, construction, filtration, medical, agricultural, and other applications. Chemical finishing has always been an important component of textile processing, but in recent years, the trend to "high-tech" products has increased the interest and use of chemical finishes. They usually have been treated with particular finishing and chemical processes for desired performance and durability. For instance, they can be more stable against mechanical, chemical, photochemical, or thermal influences; more stable under ultraviolet (UV) rays; more repellent against water, oil, and soil; and offering antistatic and electromagnetic protective properties. Other considerations include hand, appearance, hygiene, and laundering, preferred by end users. Medical clothing can be treated with a special finish based upon a microstructured titanium dioxide to diminish bacterial growth.

Increasing car ownership in Asia will, for example, generate enormous opportunities for increased use of industrial textile products, for example, door linings, boot linings, filters for air, fuel and oil, thermal/sound insulation, seat covers and linings, floor mats, tire cords, air bags, and seat belts. The growing industrial textile sector in Asia and elsewhere has attracted further development on finishing and coating technologies. In general, the technical performance of textile materials can be enhanced by the use of appropriate chemical finishes, to modify either the bulk or surface properties of the fibers. Such chemical finishes must be durable under the conditions of the end use, and the manufacturers of chemical finishes are devoting research and development facilities to service the growing industrial textile sector.

Industrial textiles for outdoor applications such as awnings, tents, and flags, especially manufactured from polyester filaments, benefit from possessing self-cleaning properties. Thus when exposed to rain, any dirt particulates and other soils can be simply removed, retaining a pristine-like appearance. One approach to obtaining a self-cleaning finish is Mincor TX-PES® from BASF (Germany). Mincor TX-PES® is applied in combination with a resin and is based upon silicate nanoparticles that are embedded in a fluorocarbon. As a result, the textile is provided with a nanostructured surface from which dirt particles are removed easily by rolling water droplets, either through rain or by hosing. Mincor TX-PES® exhibits very good oleophobicity, good soil repellency, good self-cleaning properties, and good soil release on washing. Important outlets for vector protection finishes are battle dress uniforms, tents, and apparel for outdoors in order to protect soldiers and travelers against mosquitoes, flies, ticks, lice, and fleas and diseases such as malaria, leishmaniasis, and Lyme disease.

Freshness-and-hygiene systems for garments worn under strenuous physical conditions, such as sports apparel and protective clothing worn in hot climates by military personnel, are increasingly being appreciated by consumers. The Silverplus® finish (Rudolf Chemie, Germany) is based upon a microstructured titanium dioxide. This acts as a carrier for the antibacterial agent, silver chloride, which under moist conditions (e.g., perspiration) releases antibacterial silver ions. The microstructure of Silverplus® provides an anchorage with durable effects to washing and dry cleaning coupled with physical bonding in the interfiber capillary spaces. Silverplus® can be applied via exhaust or pad application methods without any binder. In wastewater treatment plants, Silverplus® is converted into completely insoluble silver sulfide, which has no negative influence on biological wastewater treatment processes. For example, John Heathcoat & Co offers antibacterial wear treated with the Silverplus® technology to hospital staffers. The silver control of the hospital apparel restricts the growth of bacteria to reduce the risk of infection and cross-contamination. A novel expansion of the Silverplus® approach is the Ruco-Membrane PU10 Silverplus system® (Rudolf Chemie). This produces an optimum final dispersion of Silverplus in the membrane that offers optimum breathability and high hydrostatic head performance.

Coating is another approach to improve the performance of industrial textiles. Nanocoating layers on textiles, for example, are becoming increasingly important because of greater demands for more functional properties and surface characteristics. Nanocoatings include coating disperses enriched with nanoparticles or coating layers in the nanometric scale. Very thin coating layer be utilized in industrial textiles to provide

- Application of transparent conductive layers for films/foils
- Application of SiO_2 barrier layers
- Chemicals for low surface roughness, for example, silicone
- Coating for electro-conductive/optical transparency applications
- Application of thin photo-active layers, for example, TiO_2, ZnO
- Application of electroluminescent/thermoluminescent layers

Coating technologies based upon slot die, ultrasonic spray, micro roller, case knife roller, and two roller systems can all be used to provide numerous functional performance enhancements. Sol–gel coating offers higher abrasion resistance. Sol–gel coating treatments appear to be a promising area for exploitation in industrial textiles because thin inorganic metal oxide coatings can be generated with good adhesion and optical transparency, making them invisible to the naked eye. The sol–gel process is a wet-chemical technique and is also known as chemical solution deposition. Fabrication of materials is conducted with a chemical solution (or sol) that acts as the precursor for an integrated network (or gel) of either discrete particles or network polymers. The most common types of sol–gel coatings are based upon the deposition of colloidal suspensions leaving a 3-D structure based upon silicon oxide. Such treatments can be water-based and can be applied by pad-dry-cure techniques. Organic–inorganic hybrid silicon-based sols have been shown to be very effective in increasing the abrasion resistance of woven cotton and silk fabrics. In many industrial applications where industrial textiles are utilized in end uses in which abrasion resistance is important, for example, beltings, sol–gel coatings could offer a longer product service life.

4.2.2.1 Flame-Retardant Finishes

Fire safety is a significant cause of property damage and of death. Standards are therefore set for electrical appliances, textiles upholstery, and many other materials to minimize these losses. Textiles consist of highly ignitable materials and are the primary source of ignition. They contribute to rapid fire spread; however, reduction of ignitability can be obtained by

1. Use of inorganic materials (asbestos, glass, etc.)
2. Through chemical treatment with flame-retardant ([FR] chemicals)
3. Through modification of the polymer

Flame retardants are chemicals that are applied to fabrics to inhibit or suppress the combustion process. They interfere with combustion at various stages of the process, for example, during heating, decomposition, ignition of flame spread. The different types of flame retardants are as follows:

- Brominated flame retardants
- Chlorinated flame retardants
- Phosphorous-containing flame retardants (phosphate ester such as tri-phenyl phosphate)
- Nitrogen-containing flame retardants (i.e., melamines)
- Inorganic flame retardants

It is also confirmed that flame retardants based on aluminum trioxide, ammonium polyphosphates, and red phosphorous are less problematic in the environment (Figure 4.9).

One of the most preferred processes of applying FR on cotton is the "precondensate"/NH_3 process. This is an application of one of several phosphonium "precondensates," after which the fabric is cured with ammonia and

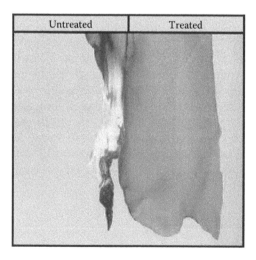

| Untreated | Treated |

FIGURE 4.9
Flame-retardant textiles.

then oxidized with hydrogen peroxide. Precondensate is the designation for a tetrakis-hydroxymethyl phosphonium salt prereacted with urea or another nitrogenous material. The amount of anhydrous sodium acetate is approximately 4% of the amount of precondensate used. Some precondensates are formulated along with the sodium acetate. Softeners are also added along with precondensates. The pH of the pad bath should be approximately 5.0. The amount of flame retardant required depends primarily on fabric type, application conditions, and test criteria to be met. Application of FR to fabric can be accomplished with conventional padding, padding with multiple dips and nips, followed by 30–60 s dwell gives good results. A critical factor in the successful application of precondensate/NH_3 flame retardant is control of fabric moisture before ammoniation. Generally, moisture levels between 10% and 20% give good results.

Flame retardancy is an essential requirement for many industrial textile products such as healthcare settings, intravenous pumps, hospital beds, hospital curtains, carpets, floor coverings, tarpaulins, automotive fabrics, and package fabrics. Cotton carpets can be given an FR finish using THP derivatives while titanium or zirconium salts are used for wool carpets. Aluminum hydrate is added to the back coatings in order to increase flame retardancy. The finish is applied by slop padding or spraying carpet materials with solutions or dispersions containing FR materials. FR finishing increases the static charging of the coverings. The flame retardancy is an essential requirement for materials to be used for construction purposes. These requirements are mainly concerned with various aspects relating to the safety of persons and goods, which also includes safety in the event of fire. Textile products affected by this directive consist exclusively of materials used for interior decoration firmly bonded to various supporting materials that cannot be removed under normal conditions of use and can therefore be compared with actual construction materials. This applies particularly to wall and ceiling coverings as well as floor coverings.

4.2.2.2 Waterproof and Water-Repellent Finishes

Waterproof fabrics have higher penetration resistance to water under much higher hydrostatic pressure than are water-repellent fabrics. Waterproof fabrics have fewer open pores and are less permeable to the passage of air and water vapor. Air permeability is greatly reduced in the case of finishes produced with waterproof coatings.

Textile fabrics used for raincoats, umbrellas, and tarpaulins have to be coated with chemicals to impart water-resistance property to the fabric. The degree of waterproofing is dependent on the fabric construction. The type of hydrophobic treatment chosen affects the quality of the fabric and its durability during washing or chemical cleaning. A fabric's resistance to water will depend on the nature of the fiber surface, the porosity of the fabric, and the dynamic force behind the impacting water spray. Finish is

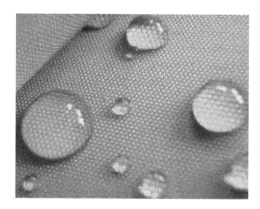

FIGURE 4.10
Waterproof textiles.

applied by filling the fabric pores with a film-forming compound or applied to individual fibers or fabrics of compounds that repel water and have a high surface tension (Figure 4.10).

Most waterproofs and tents are nowadays made of nylon (or to some extent polyester). Nylon fabrics are very light and strong, packing away into a small bundle. They can be coated in a variety of ways to render them either totally waterproof or breathable-waterproof. The various waterproof agents used on textiles are

- Wax
- Polyvinyl chloride (PVC)
- Neoprene
- PUs
- Silicone elastomer

The waterproof finishes can be applied on fabric by any one of the following methods:

1. Spray application
2. Application by infusion
3. Dipping process
4. Exhaust methods
5. Padding processes
6. Solvent impregnation

Water-repellent fabrics have open pores and are permeable to air and water vapor. A fabric is made water-repellent by depositing a hydrophobic material on the fiber's surface; however, waterproofing requires filling the pores as well. Water-repellent fabrics will permit the passage of liquid water once hydrostatic pressure is high. For fabrics to be water repellent, the surface free energy of the fiber's surface must be lowered to about 24–30 mN/m. Pure water has a surface tension of 72 mN/m, so these values are sufficient for water repellency. There are different ways that low-energy surfaces can be applied to textiles. The water-repellent finishes can be applied on fabric by any one of the following methods.

4.2.2.2.1 Deposition of Water-Repellent Agents on the Fiber by Purely Mechanical Means

Paraffin repellents were some of the earliest water repellents applied by mechanical means. These finishes can be applied by both exhaustion and padding. The lack of durability to laundering and dry cleaning and their low air and vapor permeability limit the use of paraffin-based repellents.

4.2.2.2.2 Chemical Reaction between Hydrophobic Substances and the Fiber

Compounds formed by reacting stearic acid and formaldehyde with melamine constitute another class of water-repellent materials. The hydrophobic character of the stearic acid groups provides the water repellency, while the remaining N-methylol groups can react with cellulose or with each other (cross-linking) to generate permanent effects. Advantages of the stearic acid–melamine repellents include increased durability to laundering and a full hand imparted to treated fabrics.

4.2.2.2.3 Formation of Water-Repellent Films on the Fiber

Polydimethylsiloxanes form a flexible surface film over textile fibers, which imparts a soft handle whereas polymethylhydrogen siloxanes polymerize to leave a hard brittle surface film with a harsh handle.

Goretex® is the waterproof breathable membrane, made of microporous polytetrafluoroethylene (PTFE) plastic, which will not allow cold rainwater droplets in but allows warm water vapor to pass out through the tiny pores. It can be laminated (i.e., stuck on) to a variety of fabrics or can also be sewn in as a separate drop lining between a nonwaterproof outer and inner. Microporous membranes need to be kept clean, because the pores can clog up and prevent the water vapor passing through—leading to condensation. Sympatex® is the water- and windproof breathable membrane that is nonporous polyester, absorbs warm water vapor inside, and passes it through to the cooler outside of the membrane.

4.2.2.3 Oil- and Stain-Repellent Finishes

Oil repellency is the capacity to withstand wetting and penetration by oily liquids. Oil is not absorbed, but is repelled in droplets. Fluorochemical repellents are unique in that they confer both oil and water repellency to fabrics.

The ability of fluorochemicals to repel oils is related to their low surface energy that depends on the structure of the fluorocarbon segment, the nonfluorinated segment of the molecule, the orientation of the fluorocarbon tail, and the distribution and amount of fluorocarbon moiety on fibers. Fluorocarbon polymers are used in the form of emulsion, sometimes in padding, sometimes in the exhaustion method.

Oil repellency is tested by placing a drop of oil on the fabric and observing whether the drop resides on top of the fabric or whether it penetrates. A homologous series of hydrocarbons decreasing in surface tension is used to rate the fabric's oil repellency. The hydrocarbon with the lowest surface tension to remain on top and not penetrate is indicative of the fabric's repellency (Figure 4.11).

Stain is a collective term for the local discoloration of a textile, which is usually undesirable. Stains are caused by chemical bonding between a fiber and soil or items such as fruit juice, wine, inks, coagulated blood, heavy metal salts, lubricating oils, and rust. In contrast to such easily removed soils, stains must either be decolorized or destroyed with the aid of chemically active compounds. In the last few years, rapid rise in the usage of fluorochemicals to impart stain repellency or stain-release properties to apparel goods have been noticed. Stain release fluorochemical finishes permit oil and water stains to penetrate the fabric; however, when the fabric is laundered, the stains are easily removed. Among the existing textile chemicals, fluorochemicals have the unique property to provide fabrics a low surface energy film with both high oil and water repellency properties to resist penetration of oil- and water-based stains (polar and nonpolar liquids). Fluorochemical hybrid finishes, also called "dual effect," which contain hydrophilic groups, have

FIGURE 4.11
Oil- and stain-proof textiles.

also been developed and improved successfully over the most recent years to impart both stain repellency and stain release properties.

4.2.2.4 Antimicrobial Finishes

Antimicrobial finishes are particularly important for industrial fabrics that are exposed to cold weather conditions. Fabrics used for awnings, wind screens, tents, tarpaulins, ropes, frost protection fabrics, and the like need protection from rotting and mildew. Home furnishing textiles such as carpets, shower curtains, bath mats, floor mats, mattress ticking, and upholstery also require antimicrobial finishes. Fabrics and clothing used in places where there might be danger of infection from pathogens can benefit from antimicrobial finishing. The application of antimicrobial finishes to prevent unpleasant odors on underwear, socks, and athletic wear is an important market need.

Microbes are the tiniest creatures that cannot be seen with the naked eye. They include a variety of microorganisms like bacteria, fungi, algae, and viruses. Bacteria are unicellular organisms that grow very rapidly under warmth and moisture. Some specific types of bacteria are pathogenic and cause cross-infection. Fungi, molds, or mildew are complex organisms with slow growth rate. They stain the fabric and deteriorate the performance properties of the fabrics. Dust mites are eight-legged creatures and occupy the household textiles such as blankets, bed linen, pillows, mattresses, and carpets. The dust mites feed on human skin cells, and liberated waste products can cause allergic reactions and respiratory disorders.

Antimicrobials are defined as the agents that either kill microorganisms or simply inhibit their growth. The antimicrobial agent works either by the slow release of the active ingredient or by surface contact with the microbes. The general name "antimicrobial finish" is given to all types of finishing agents that kill off fungi and bacteria or inhibit growth and thus have disinfecting properties. The method of application of antimicrobial finish on textiles can be classified into three categories:

1. Addition of an antimicrobial agent in the polymer before extrusion
2. Modifications involving grafting or other chemical reactions
3. Posttreatment of the fiber or the fabric during the finishing stage

The efficacy of the antimicrobial finish will depend on various factors, such as its chemical nature, method of application, and durability. Antimicrobial finishes are produced in general with products based on phenol derivatives, organic or inorganic heavy-metal compounds that split off formaldehyde and quaternary compounds. Cationic surfactants function as effective antimicrobial finishes. Polyhexamethylene biguanide hydrochloride (PHMB) claimed to possess a low mammalian toxicity and broad spectrum of antimicrobial activity. PHMB is particularly suitable for cotton and cellulosic textiles and can be applied to blends of cotton with polyester and nylon.

The antibacterial agents can be applied on fabric by any one of the following methods such as pad, spray, foam, or exhaust. Metallic silver combined with zeolite and dispersed in the polymer before extrusion spinning provides a polyester fiber that can be intimately blended with cotton to produce a durable antimicrobial composite. The blend is reported to have excellent antibacterial and antifungal properties. The bacteriostatic action of quaternary ammonium compounds on textiles is well known. Quaternary silicones like 3-trimethoxy-silylpropyldimethyloctadecyl ammonium chloride have been used for a number of years as a durable odor preventive on socks. Magnesium hydroperoxy acetate is another environmentally friendly compound, which can be fixed to cotton to impart antimicrobial properties. Actigard® finishes from Clariant are used in carpets to combat action of bacteria, house dust mites, and mold fungi.

4.2.2.5 Antistatic Finishes

Static electricity is most commonly generated when different materials are rubbed together. The material that loses the electrons becomes positively charged and the other material, negatively charged. Materials differ in their propensity to lose some of their electrons when in contact with another material. Textile materials that make up clothing are poor conductors of electricity when they are dry and they build up large charges that eventually discharge to a lower potential. Clothing layers will develop an opposite charge as they rub against each other. When the layers are separated, one will retain its positive charge and the other its negative charge. If these charges are great enough, sparks will fly from one layer to the other. Charged clothing can induce a charge on the body.

Footwear can have a great effect on the dissipation of static charge from the body. If one is wearing insulated footwear, the electricity cannot pass from the body to the ground. As a result, any electric charge which is built upon the body stays on there, and, under the proper conditions, is dissipated with sparks.

In order to reduce electrical charge in clothing and its induced charge on the body, it is necessary to make the fibers as conductive as possible. There are several ways to do this. One is to apply an antistatic agent to the fiber surface, a second is to disperse conducting or antistatic particles within the fiber, and a third is to incorporate metallic or highly conductive fiber into the yarns.

The majority of chemical antistatic agents work by attracting moisture to the fiber surface and so increase the fiber's conductive properties. Recently, polymer grafting techniques have used high-energy radiation or chemical techniques to graft ionic functional groups to polyester or polyamide fibers to give improved and permanent moisture regain properties. Another method of making fibers conductive is to embed fine particles of carbon black into polyester fibers that have been modified to have lower melting point surfaces. These are called "carbon epitropic fibers." About 2% carbon black by mass is needed in 100% polyester fabrics to give optimum antistatic properties. Antistatic agents are effective on practically all types of fibers, especially polyamide, polypropylene, polyethylene, and polyacrylonitrile.

4.2.2.6 Ultraviolet Protection Finish

Light is the prime energy source and essential element for life. Human body generates vitamin D when it is exposed to small doses of UV radiation. But overdose of UV radiation causes acute, chronic reactions and erythematic damages (skin reddening) or other diseases such as skin cancer. Ozone is a very effective UV absorber. So its decrease has led to the increased UV radiation reaching the earth's surface and has thus enlarged the risks of the negative effects of sunlight. UV radiation constitutes to only 5% of the incident light on earth (visible light 50% and IR radiation 45%). Even though its proportion is less, it has the highest quantum energy. This energy of UV radiation is of the order of magnitude of the organic molecules' bond energy. So it will have a detrimental effect on human skin (Figures 4.12 and 4.13).

When light falls on a textile surface, a part is reflected depending on the surface characteristics of the fabric, a part is absorbed, and a part is transmitted through the fabric, which falls on the skin. The transmitted part consists of the light diffused by the fibers and the light going through the pores unaltered, which depends on the cover factor and the thickness of the fabric. The absorbed component depends on the nature of the fiber and the nature of the additives, construction of the yarn and the fabric, as well as the presence of the dye, optical brighteners, or any finish.

Sun-protective factor or UV-protective factor is used to measure how much a textile protects human skin from UV rays. It is the ratio of the length of time of solar radiation exposure required for the skin to show

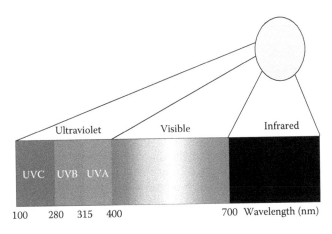

FIGURE 4.12
UV rays.

FIGURE 4.13
UV-protective textiles.

redness (erythema) with and without protection (Fung 2002). The SPF is calculated as follows:

$$SPF = \frac{MED\,protected\,skin}{MED\,unprotected\,skin}$$

where MED is the minimal erythemal dose or the quantity of radiant energy needed to produce the first detectable reddening of skin 22 ± 2 h after exposure.

UV absorbers are inorganic or organic compounds that absorb UV rays efficiently and convert the energy into relatively harmless thermal energy without itself undergoing any appreciable irreversible chemical change or inducing any chemical change in the polymeric host molecule. The important groups of UV absorbers are

- 2-hydroxybenzophenones
- 2-hydroxyphenyl benzotriazoles
- 2-hydroxyphenyl-s-triazines

Clariant® has commercially introduced a new finish technology that makes it possible to substantially increase the ultraviolet protection factor (UPF) of light garments under the name Rayosan®. The Rayosan® process produces an increased absorption of UV rays by the treated textiles without any impairment of their other properties such as appearance, handle, or bright ability. Ciba® TINOSORB® comprehensive UV absorbers are suitable to be applied via the laundry washing process to clothing. By their use, they help convert cotton fabrics into effective UV shields. TINOFAST® CEL is a reactive UV absorber that can markedly and durably improve the UPF of cotton, viscose, lyocell, polyamide, and their various blends, especially if these are dyed or printed in pale shades and provided they are close textured, that is, have low porosity.

The incidence of skin cancer has been rising for years worldwide. Elevated exposure to UV radiation is considered one of the main factors causing neoplasms of the skin. Too much exposure to UV radiation can result in skin damage such as sunburn, premature skin aging, allergies, and skin cancer. Thus for effective UV protection, a thin, only slightly porous fabric treated with the appropriate UV absorber is to be chosen, since it can offer a very high SPF while being cool to wear at the same time.

4.2.2.7 Other Finishes

4.2.2.7.1 Rotting/Mildew-Resistant Finish

Emulsion free metal salts consisting of fatty acid esters of dihydroxydichlorodiphenylmethane and a benzimide azole derivative are applied on fabric to impart mildew-resistant finish. Ciba® FUNGITEX ROP can be used for fungicidal finish of cotton, linen, and polyester/cotton fabrics for tents, tarpaulins, and rucksacks.

4.2.2.7.2 Thermal Insulation

Aluminum foil coated under high vacuum conditions: the foil is glued on in high vacuum, so that a bond between the surface of the fabric and the foil is guaranteed at nearly every point, for technical applications (mainly in use with Kevlar or preox/p-aramid) to insulate thermal radiation. Also in use are outside of suits and clothes made from Kevlar (melting ovens and blast-furnace plants).

4.2.2.7.3 Slip-Resistant/Antislip Finishes

These finishes are used to avoid the shifting of crossing warp and weft threads in fabrics containing a low number of yarns or woven in open constructions to prevent the formation of holes. Fabrics for technical applications (e.g., geotextiles, glass fiber wallpaper) are often woven with such thin constructions that there is a great risk of yarn slippage as they are delivered from the weaving machine. Aqueous-based coating using acrylic/PU compounds is used for adhesion of yarns successfully to prevent slippage under stress. Suitable slip-resistant finishes include film-forming polymer dispersions or silicic acid hydrosols. Ciba® DICRYLAN AHS gives good adhesion on nearly all substrates, especially on polyester and polyamide.

4.2.3 Plasma Treatment for Industrial Textiles

The modification of bulk and surface properties of textile materials can represent a promising approach for meeting technical and economical requirements. Due to the cost involved in the design and development of new fibers, fiber researchers now focus on imparting desired aesthetic and functional properties on existing fibers by fiber surface modification.

Conventional fiber modification methods include various thermal, mechanical, and chemical treatments. In the last decade, fiber modification is also performed with cold plasma in order to increase the dye uptake of fibers and to impart specific functionality to the fibers. The plasma is an ionized gas with equal density of positive and negative charges, which exist over an extremely wide range of temperature and pressure. The reactive species of plasma, resulting from ionization, fragmentation, and excitation processes, are high enough to dissociate a wide variety of chemical bonds, resulting in a significant number of simultaneous recombination mechanisms. Plasma is partially ionized gas, composed of highly excited atomic, molecular, ionic, and radical species with free electrons and photons.

Plasma surface treatments show distinct advantages, because they are able to modify the surface properties of inert materials, sometimes with environment-friendly devices. The advantage of such plasma treatments is that the modification turns out to be restricted in the uppermost layers of the substrate, thus not affecting the overall desirable bulk properties. Plasma includes less water usage and energy consumption, with very small fiber damage, then making plasma process very attractive.

Polymeric surfaces are usually not wettable and adhesion is poor. After plasma treatment, the surface energy increases, and wettability and adhesion enhancements are produced. Depending on the gas employed, plasma treatment may render the surface very hydrophilic, oleophobic, or hydrophobic (Table 4.1). The hydrophilic feature can be controlled very well. Oleo- and hydrophobic features are readily achieved in plasmas containing fluorine.

TABLE 4.1

Calendering

Properties to Impart	Material/Fabric	Type of Plasma Used
Enhance mechanical properties: softening	Cotton and other cellulose-based polymers	Oxygen
Electrical properties: antistatic finish	Rayon	Chloromethyl dimethylsilane
Wetting: improvement of surface wetting	Synthetic polymers (PA, PE, PP, PET PTFE)	O_2, air, NH_3
Hydrophobicity	Cotton, cotton/PET	Siloxan- or perfluorocarbon
Oleophobicity	Cotton/polyester	By means of grafting of perfluoroacrylate
Dyeing and Printing: 1. Improvement of capillarity 2. Improved dyeing	Wool and cotton Polyester Nylon	Treatment in oxygen Treatment in $SiCl_4$ Treatment in Ar
UV protection	Dyed cotton/polyester	Hexamethyldisiloxane (HMDSO)
Flame retardant	PAN, rayon, cotton	Phosphorus containing monomers in plasma

Plasma technology holds tremendous potential to develop processes that can limit the environmental impact of textile processing and contribute toward sustainable development. Physical etching of textile substrate can be used to create nanosized peaks on the surface of the fiber. This coupled with nanolayering of a hydrophobic fluorocarbon compound can be used to create the famous lotus effect on textile that makes surfaces of hydrophilic fibers effectively hydrophobic while still leaving it breathable. Plasma treatment, however, does have certain drawbacks. The treatment tends to produce harmful gases such as ozone and nitrogen oxides during operation. This happens due to the formation of free radicals and nascent oxygen during the treatment, which react with atmospheric gases to form harmful by-products. Some companies are now making plasma systems that employ an inert gas such as argon as the main plasma gas rather than atmospheric air. An inert gas does not react with contaminations to produce hazardous air pollutants.

4.2.3.1 Magnetron Sputtering of Textiles

A new flexible coating technology called magnetic sputtering can be used to coat virtually any material. The functionalization of textile materials to suit for specific end-use applications using metal oxide or polymer or composite coatings can carried out by means of sputtering. The sputtering process removes atomized material from a solid by energetic bombardment of its surface layers with ions or neutral particles. The magnetron sputtering method is realized at quite deep vacuum (about 5×10^{-5} mmHg) and allows coating the substrates by thin films of copper, aluminum, titanium, brass, silver, stainless steel, bronze, and other metals and their alloys. In comparison with other deposition methods, a most important advantage of sputtering is that even the highest melting point materials are easily sputtered. Sputtered films typically have better adhesion on the substrate than evaporated films. The applications of textile materials metalized by plasma sputtering method are as follows:

- Producing elements of special working cloth for people working in hot workshops as well as for firefighters. Sewing bactericidal burn-curing bed sheets and napkins.
- Manufacture of cloth elements protecting from detrimental effects of radio telephones, radio frequency and microwave frequency equipment, and cloth for attendants working in the medical offices for physical therapy.
- Manufacture of materials resistant to accumulation of static electricity: curtains for "clean rooms," operating rooms, and special cloth for personnel working in the electronic industry.
- Manufacture of bactericidal filters.

- Protection of computer systems and electronic control systems from false signals, manufacturing protective devices to prevent the information leakage.
- Use for military equipment to reduce the visibility by night-viewing devices of infrared and radio ranges.

4.2.4 Nanofinishes for Industrial Textiles

Nanotechnology encompasses a wide range of technologies concerned with structures and processes on the nanometer scale. Nanotechnology deals with the science and technology at dimensions of roughly 1–100 nm, although 100 nm presently is the practically attainable dimension for textile products and applications. The nanofinishes are applied on fabrics intended for industrial applications to achieve the desired attributes, such as fabric softness, durability, and breathability, and in imparting functional characteristics, such as stain repellency, water repellency, fire retardancy, and antimicrobial and antistatic requirements. The size of the elements in the nanofinishing treatments range on a 1–100 nm scale and are arranged in an orderly fashion, introducing novel properties to the textile material meant for specific end-use applications. Fabrics consisting of durable and wearable fibers consume less coating compound in nanofinishing compared to conventional finishes due to the ordered structure.

The application of nanotechnology in textile material property enhancement is expected to become a trillion dollar business in the next decade, with impressive technological, economical, and ecological advantages. Worldwide, many countries are funding hefty amounts annually to boost the research and development in the area of nanotechnology in addition to the millions of dollars invested by private industries.

Nano-Tex™, founded in 1998, has been one of the leaders in nanotreatments designed specifically for textiles. Nano-Tex treatments are applied to a fabric in a "bath." As the fabric goes through the bath, nanoparticles come in contact with the fibers of the fabric. When the fabric is cured or heated—the nanoparticles spread out evenly and bond to the fibers. Treatments are permanent and do not jeopardize the aesthetic characteristics or mechanical properties of the fabric. Treatments can be applied to a number of fibers including cotton, polyester, silk, and wool. A variety of enhancing characteristics can be imparted to the fabric through the application of special treatments. Nanotreated fabrics can be spill resistant, stain proof, wrinkle resistant, and static proof.

Wilhelm Barthlott (Bonn Institute) discovered that the lotus plant, admired for the resplendence of its flowers and leaves, owed the property of self-cleaning to the high density of minute surface protrusions. These protrusions catch deposits of soil preventing them from sticking. Resists Spills® (Nano-Tex™) was one of the first nanotreatments offered by Nano-Tex. It can be applied to cotton, polyester, wool, silk, or rayon. Stain release was designed to mimic the natural characteristics of lotus leaves.

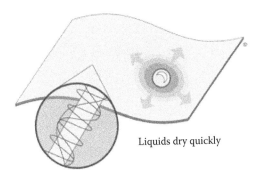

Liquids dry quickly

FIGURE 4.14
Coolest Comfort®.

The surface of lotus leaves is hydrophobic. In a rain shower, water droplets roll off the leaf's surface carrying away contaminants. A leaf's surface is also rough, decreasing the surface's ability to soak up water. Like a leaf's surface, treatments have been developed to make the fabric ultra-hydrophobic. Self-cleaning fibers might eventually replace conventional fluorochemical-based finishes currently used to provide water repellency (Figure 4.14).

A Coolest Comfort® (Nano-tex™) treatment imparts superior wicking properties to a previously hydrophobic synthetic. The treated fabric pulls away perspiration from the body allowing the wearer to stay dry and comfortable. Coolest Comfort is now being applied to resin-treated cotton. Resins are applied to cotton fabric to make them wrinkle free. Unfortunately, the resin treatment blocks cotton's natural ability to absorb moisture.

Resists Static® (Nano-tex™) is the first permanent antistatic treatment for synthetic fibers. Not only does it repel static, but the treatment also repels statically attractive substances such as dog hair, lint, and dust. Resists static can be applied to a variety of fabric constructions including rough-textured fleece and slick suit linings (Figure 4.15).

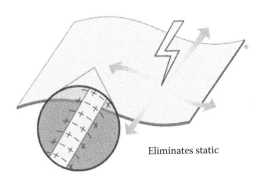

Eliminates static

FIGURE 4.15
Resists Static®.

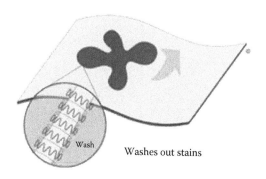

FIGURE 4.16
Release Stains®.

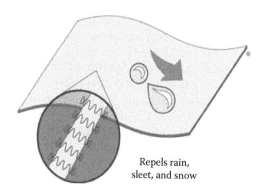

FIGURE 4.17
Resist spills.

Repels and Releases Stains® (Nano-tex™) is applied to cotton and polyester/cotton blends. As the name implies, Repels and Releases Stains has a built-in dual treatment. Once liquid comes in contact with the fabric, it beads up and rolls off. If liquids get around this barrier and into the fabric, the patented release technology frees it from the fibers during a typical home laundering cycle (Figures 4.16 and 4.17).

Technologists are working on nanocoatings that could possibly have self-healing property. Textile surfaces that can remove surface scratches and scuff marks, repel insects, and decolorize red wine spills are under development. Nanotechnology is being used to develop "sensorized" garments. Prototype garments with the ability to monitor functions such as body temperature and vital signs have already been developed. Sensorized garments could potentially be used in a wide variety of applications including hospital gowns and military uniforms. Research is focused on developing bioreactive plastic coatings that protect the wearer against biological and chemical attacks. The coating, embedded with antibodies and enzymes, decontaminates the textile surface as soon as pathogens or toxins are present. Scientists are also investigating ways to equip the coating with an alarm system that would alert the wearer to the invisible attack.

4.2.5 Coating and Lamination Technologies

Coated and laminated textiles are used in a very diverse range of clothing and technical applications. Use of coated fabrics in protective clothing and in industrial textiles is growing because the coatings can impart enhanced functionality with added value. Coated fabrics are engineered composite materials, produced by a combination of a textile fabric and a polymer coating applied to the fabric surface. The fabric component governs the tear and tensile strength, elongation, and dimensional stability, while the polymer mainly controls the chemical properties, abrasion resistance, and resistance to penetration by liquids and gases. PUs, silicones, and elastomers have been widely applied by various coating application technologies, for example, direct coating, transfer coating, hot melt coating, and powder coating.

4.2.5.1 Applications of Coating and Laminating in Industrial Textiles

The application of breathable fabrics is covering a diverse range of products such as special military protective clothing, clean room garments, surgical garments, rainwear, skiwear, sport footwear linings, hospital drapes, mattress and seat covers, specialized tarpaulins, packaging, wound dressings, and filters. Gore-Tex® materials are typically based on thermo-mechanically expanded PTFE and other fluoropolymer products with nine billion tiny micropores in every square inch; these are 20,000 times smaller than a drop of liquid water, but 700 times larger than a water molecule. It repels liquid water to pass through, but allows the passage of water vapor molecules. HyVent® is a waterproof/breathable PU-based membrane. Entrant® is a microporous PU coating produced by Toray Industries. PVC plastisol-coated HT woven polyester or nylon is used as tarpaulins. PVC is a versatile polymer, and coatings can be specially formulated for a wide variety of products, including luggage, handbags, toys, tents, aquatic sports items, flags and banners, and picnic ground sheets. Outdoor banners are still largely made of vinyl-laminated polyester secured by grommets, and indoor banners are often digitally printed on polyester, cotton canvas, flexible-face material, nylon, and mesh. PVC-coated fabric is still used for seat upholstery in some public transport vehicles, ferries, and other marine uses and in public buildings. Specialist fire protection and thermal insulation pipe coverings are made up of silicone-coated glass fabric.

Transporting bulk materials by conveyor belts dates back to approximately 1795; most of these early installations handled grain over relatively short distances. The coating applied is generally either a rubber specially formulated for the purpose or PVC that has the advantage of good fire resistance. PTFE coating is used in applications where high temperature and nonstick properties are required, such as in the food industry. Hot air balloons are produced from 300 g/m² PU-coated high-tenacity polyamide fabric, sometimes with a further coating in the form of a layer of PU resin as an extra barrier against UV light and air leakage. The envelope material in the Cargo Lifter is made from several layers of Vectran® fiber (by Celanese) and is coated with PU. Mylar film is laminated to the outside to reduce helium leakage, and Tedlar film is also used for protection from UV rays.

The rubber coating on printer blanket has to be able to withstand all the various solvents used in printing without swelling, which would cause distortions in print definition and quality. The other main requirements of printer blanket are dimensional stability and rubber resilience. Woven cotton fabric was used as the base material, because of its dimensional stability. A typical printer's blanket for newsprint is made up of three layers of woven fabric and many layers of specially formulated rubber compound. The three-layer blanket is produced by laminating together three separate layers of the single coated fabric. Aramid fibers are used in some of the latest printer blanket to attain the properties: lightweight, flexibility, temperature resistance, and dimensional stability.

Labels used for conveying information in textile articles can be produced by direct coating of soft water-based acrylic or PU resins or solvent-based PU resins. Acrylic and PU resins are currently used in filtration applications. Crushed foam coatings can modify the porosity of filter fabrics, depending on the add-on of resin and the severity of crushing. The roof is continuously exposed to various climatic conditions, and roofing materials must have excellent resistance to UV and light radiation, ozone and heat, moisture, and rotting. Coated fabric roofing material is generally produced from coated woven polyester or nonwoven glass fiber fabrics, and the coating materials are EPDM, neoprene and Hypalon, chlorinated polyethylene, and PVC. The main requirements of awnings are weather and sunlight durability, resistance to insects and microbes, and adequate physical strength for the purpose. PVC, polypropylene, and acrylic lacquers are applied on nylon, polyester, and polypropylene fabrics to improve UV and sunlight resistance of awning fabrics. Tent fabrics need to be strong enough to withstand high wind, resistant to rotting and microbes, and, most important, resistant to sunlight and UV degradation. Lightweight coated tents are generally made from PU coated fabric, the larger ones from PVC or neoprene. Glass fiber fabric coated with PTFE® was used in the construction of Millennium Dome at Greenwich, London. Dr. Kawaguchi, a professor at the University of Tokyo, designed a new concept, "a ceiling that will never lead to a serious accident, even if it collapses." The gypsum board used before as ceiling material was about 15 kg/m² with its subframe deteriorating easily, causing it to collapse even without an earthquake. He developed a safer fabric membrane ceiling using PVC-coated glass fiber fabric that is lightweight, flexible, and noncombustible.

Most of the automotive fabric are coated or laminated or in the form of composite materials, that is, seat covers, door panels, headliner modules, hoses and belts, tires, etc. The main durability requirement of car seat covering material, that is, resistance to abrasion, light, and UV radiation, must be met. Car seat fabric is almost invariably in the form of a trilaminate: face fabric/PU foam/scrim backing. This trilaminate car seat fabric is produced to impart a soft touch to the fabric surface and to prevent creases or "bagging" developing during the life of the car. Woven car seat fabric is sometimes back coated with acrylic, PU, or styrene butadiene rubber (SBR) to increase the abrasion resistance of the whole trilaminate. Flame retardancy of the car seat fabric is achieved by back coating with FR chemicals based on antimony/bromine synergy. Car seat fabric is usually laminated to PU foam by flame lamination, which is a quick and economical way of producing the triple laminate in a single process. Fabric to be used on door casings is sometimes coated with PVC latex to enable it to be HF welded to other materials used in door construction. Most door casings are produced by more traditional methods of molding and lamination, usually with fabric to foam laminates. Both PU and polyolefin foams are used for a soft touch and thermal and sound insulation. The modern car headliner is a multiple laminate of up to seven or more components all joined together. All layers are joined together by the action of the hot melt adhesives in a flatbed laminator, taking care not to damage the aesthetics of the decorative material or to reduce the thickness of the center core. Air bags manufactured from nylon 6,6 are coated with neoprene rubber and silicone by knife-on air coating. Bonnet liners are generally made from a laminate material, the main function of which is to absorb and dampen engine noise. Bonnet liners are made up of needle-punched polyester and polypropylene coated with specially formulated SBR latex. Carpets play a significant role in acoustic and vibrational control in car. Both tufted and needle-punched carpets are usually coated with SBR or acrylic latex, to stabilize them and to lock in the fibers. Neoprene is coated on drive belts, because of its overall good oil, chemical, and heat resistance properties.

4.2.5.2 Future Trends

UV-curable hot melts are used as alternatives to solvent-based and water-based coating systems that required energy-intensive thermal curing. Reduction of volatile organic compounds was important, and plasma technology could be utilized to provide versatile nanocoating, as well as smart coatings using nanoparticles. Temperature-sensitive coatings or photochromic sensors could be produced as well as low-surface-energy biomedical coatings. Self-cleaning properties and surface-engineered gecko-type adhesive textiles also offered many opportunities in the global market.

Bibliography

Alexander, R. E., Baugh, K. R., and, Monsanto Company, Process for softening non-woven fabrics, US Patent - 4329763, May 18, 1982.

Anderson, K., Nanotechnology in textile industry, January 2009, pp. 3–4. http://www.techexchange.com/library/Nanotechnology%20in%20the%20Textile%20Industry.pdf (accessed May 29, 2013).

Bellini, P., F. Bonetti, and E. Franzetti, G. Rosace, and S. Vago, *Finishing*, ACIMIT Foundation, Italy, 2002.

Böhringer, B., G. Schindling, and U. Schön, UV protection by textile, *Melliand International*, 3, 165 (1997).

Coojack, L., S. Davis, and N. Kerr, Textiles and UV radiation, *Canadian Textile Journal*, 111(3), 14–15 (1994).

Crow, R. M., Static electricity—A literature review, Defense Research Establishment Ottawa, Canada 1991, pp. 1–5.

Evans, M., Mechanical finishing, *American Dyestuff Reporter*, May, 1983, p. 36.

Fung, W., *Coating and Laminated Textiles*, Woodhead Publications, CRC Press Ltd., Cambridge, U.K., 2002.

Gorberg, B. L., Metallization of textile materials by ion plasma sputtering—New methods and possibilities, lvtechnomash Ltd., Russia, http://ivtechnomash.ru/en/articles.php (accessed May 29, 2013).

Gore-Tex, http:// www.gore-tex.com/remote/Satellite/home (accessed May 29, 2013).

Gupta, K., V. S. Tripathi, H. Ram, and H. Raj, Sun protective coatings, *Colourage*, 49(6), 35–40 (2002).

Hall, A. J., *A Handbook of Textile Finishing*, The National Trade Press Ltd, London, 1957.

Heywood, D., *Textile Finishing*, Society of Dyers and Colourists, Bradford, U.K., 2003.

Hilfiger, R. et al., Improving sun protection factors of fabrics by applying UV-absorbers, *Textile Research Journal*, 66(2), 61 (1996).

Holme, I. Technical textiles empowered—Innovative finishes and coatings bring versatile textiles, *ATA Journal for Asia on Textile & Apparel*, October 2010, pp. 48–49.

Holme, I., Innovations in coating and lamination gain new momentum, *ATA Journal for Asia on Textile & Apparel*, April 2011, pp. 30–32.

Holmes, D. A., Waterproof breathable fabrics, in *Handbook of Technical Textiles*, A. R. Horrocks and S. C. Anuad (eds.), Woodhead Publications, Cambridge, U.K., 2000, pp. 286–287.

Kaichiu, H., Plasma treatment technology for the textile industry, *ATA Journal for Asia on Textile & Apparel*, August 2007.

Kelly, P. J. and R. D. Arnell, Magnetron sputtering: A review of recent developments and applications, *Vacuum*, 56, 159–172 (2000).

Nair, L., Finishing of Technical textiles: I, Express Textile—Dyes and chemicals, November 18, 2004.

Nair, L., Finishing of Technical textiles: II, Express Textile—Dyes and chemicals, November 25, 2004.

Nano-tex, http:// www.nano-tex.com/technologies (accessed May 29, 2013).

Near chimica, http://www.anarkimya.com/images/pdf/fluorocarbons.pdf (accessed May 29, 2013).

Palacin, F., Textile finish protects against UV radiation, *Melliand International*, 3(3), 169 (1997).

Reinert, G. and F. Fuso, UV-protecting properties of textile fabrics and their improvement, *Textile Chemist and Colourist*, 29, 36–43 (1997).

Rouette, H.-K., *Encyclopedia of Textile Finishing*, Springer, New York, 2000.

Schindler, W. D. and P. J. Hauser, *Chemical Finishing of Textiles*, CRC Press, Boca Raton, FL, 2004.

Scott Bagley, R., ITT Technologies, Inc., ITMA 2003 mechanical finishing, *Journal of Textile and Apparel Technology and Management*, 3(3) (2003).

Senthil Kumar, R., Mechanical finishing techniques for technical textiles, in *Advances in the Dyeing and Finishing of Technical Textiles*, edited by Gulrajani, M.L., Woodhead Publications Ltd., U.K., 2013.

Shishoo, R., *Plasma Technologies for Textiles*, Woodhead Publications, Cambridge, U.K., 2007.

Sparavigna, A., Plasma treatment advantages for textiles, Dipartimento di Fisica, Politecnico di Torino Corso Duca Abruzzi 24, Torino, Italy, http://arxiv.org/pdf/0801.3727 (accessed May 29, 2013).

Tagawa, K., A safer fabric membrane ceiling, *Specialty Fabrics Review*, October 2011, http://specialtyfabricsreview.com/articles/1011_wv_fabric_ceiling.html (accessed May 29, 2013).

Tomasino, C., *Chemistry & Technology of Fabric Preparation & Finishing*, Department of Textile Engineering, Chemistry & Science College of Textiles North Carolina State University, Raleigh, NC, 1992.

Vazquez, F., Silicone softeners for stain repellent and stain release fabric finishing, Dow Corning Corporation, Greensboro, N.C., http://www.dowcorning.com/content/publishedlit/26-1277-01.pdf (accessed May 29, 2013).

5

Filtration Textiles

5.1 Introduction

There is hardly a human activity, industrial, commercial, or domestic, that is not affected by filtration. The general significance of environmental protection justifies mention of the topic here—because filtration has a major role to play in many of the schemes trying to achieve this protection. The market forces exerted by the imposition of environmental legislation are an important driver for the filtration market. Textile filter fabrics play an essential part in a countless industrial processes, contributing to product purity, savings in energy/production costs, and a cleaner environment.

5.2 Filtration: Market Share

The filtration industry today is a diverse and technically sophisticated business with annual sales reported to be in excess of $100 billion. The filtration business has evolved over time to become a complex industry with very specific requirements for each area of use. Performance standards for the media used in virtually every application have become very stringent. The usage of filtration and similar separation equipment is quite evenly spread throughout the economy, with the two largest end-use sectors being those in which the largest number of individual filters is found. The domestic and commercial sector with its many water filters (and coffee filters, and suction cleaner filters) is one, and the transport system sector with its huge number of engine filters for intake of air, fuels, and coolants is the other.

5.3 Filtration: Definition

Basically, filtration means the capture and retention of small particles from a moving stream of either gas or liquid, with minimum resistance to flow. Filtration conditions vary widely and also the equipment for filtration and the type of filter used.

5.3.1 Filtration: Terms and Definition

1. *Pressure drop* (Δp): Pressure drop through a filter is defined by the following expression:

$$\Delta p = P1 - P2$$

   where
   P1 is the pressure before the filtration
   P2 is the pressure after the filtration

2. *Filter efficiency E*: Filter efficiency is defined as a ratio between the quantity of particles retained in the filter and the number of dispersed particles found in the suspension.

3. *Filter capacity Q*: Filter capacity is defined by the amount of particles deposited in it (expressed in g or kg) and that accumulated before a drop in pressure begins. The capacity of a filter must be specified for each particle size.

4. *Cleaning efficiency*: It is the ratio of dust retained by fabric after cleaning to total dust deposited expressed in percentage.

5. *Degree of filtration*: This parameter defines the ratio between a certain size of particles that enter the filter and the particles of the same size that leave the filter.

6. *Porosity*: It is the ratio of the volume of voids to the volume of fabric.

$$\text{Porosity} = \frac{(\text{Volume of fabric} - \text{Volume of fiber})}{\text{Volume of fabric}} \times 100$$

$$\text{Porosity} = 1 - \frac{\text{Fabric density}}{\text{Fiber density}} \times 100$$

5.4 Filtration: Principles of Particle Retention

Filtration of particles relies on any one or more of the following principles:

1. Impaction
2. Diffusion
3. Straining
4. Electrostatics
5. Sedimentation
6. Interception

Particles can be influenced by any one of these principles, or all of them simultaneously.

1. *Impaction*

    As large particles move along with an air stream, their inertias prevent them from making abrupt changes in direction. If an obstruction such as a series of water droplets or fibers (glass, foam rubber, cloth, etc.) is placed randomly across the air stream path, there is a certain probability that a given particle will collide with the obstruction. As particle size and the number of particles increase, so does the probability of collision (Figure 5.1).

    Thus the efficiency of removing particles from an air stream by impaction is a function of particle size, fiber size, and the number

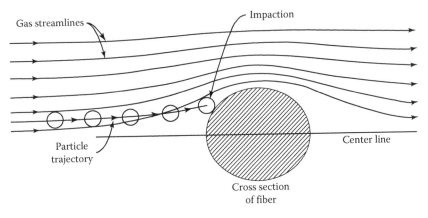

FIGURE 5.1
Particle retention by impaction.

of fibers. The greater the number of fibers in the filter fabric (thus the deeper the bed, the higher the pressure drop), the higher the filtration efficiency. In turn, as the dust particles collect, they themselves become part of the filter media, thereby increasing efficiency by adding to the number of possible collisions for other suspended particles. As the collected particles build up on the filter, system pressure drop increases, usually a good indication that the path through the filter, the equivalent filter depth, is increasing. This principle of filtration is most commonly found in fiber filters and certain wet collectors.

2. *Diffusion*

When particles become very small, their mass is so low that, should they collide with any air molecules, just the random motion of the air molecules will cause them to rebound randomly. This motion is commonly referred to as Brownian movement (Figure 5.2).

   If the velocity of the air stream is low, this diffusion movement will in turn cause random collisions with fiber or droplets in the way of the air flow. Hence, much like impaction, probabilities can be developed for collisions due to this diffusion. Key factors are fiber size, fiber quantity, and air stream velocity. As with impaction, as more and more particles collect, the probability of collision (efficiency) for other particles is enhanced, but with an associated increase in pressure drop.

3. *Straining*

If the width of a passage is smaller than that of the particle suspended in the air stream, then the particle will be stopped and held. However, as each particle plugs a hole, air resistance increases. Standard house screens are typical of this filter type. Small particles pass through it, but bugs cannot pass through. Very small particles are seldom collected using this method, which is primarily used only for specialized laboratory experiments (Figure 5.3).

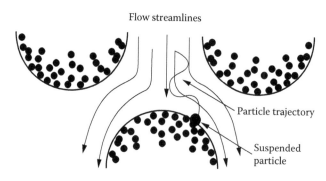

FIGURE 5.2
Particle retention by diffusion.

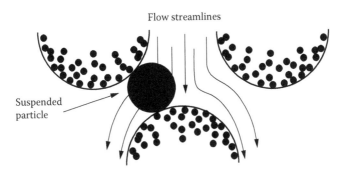

FIGURE 5.3
Particle retention by straining.

4. *Electrostatics*

If a charged particle passes through an electrostatic field, it is attracted to an oppositely charged body. Such charges can be generated and imparted to particles in an air stream in much the same way as static charges develop during the combing of one's hair or just walking across a rug. Electrons are stripped from large quantities of molecules with the net effect that particles of dirt not otherwise collected might be charged by friction as they pass through, then collected as they attach themselves to oppositely charged bodies. This effect can occur inside such filtering devices as fiber beds that operate primarily on the principles of impaction and diffusion but have their efficiencies enhanced by electrostatic effects (Figure 5.4).

Charges may be purposely induced onto air stream particles by applying energy to a special configuration of wires and plates stretched across the air stream. These devices, called electrostatic precipitators, form a special category of air filtration. Whether particle charges are induced by applying energy to a dirty air stream or occur naturally, they can be valuable tools in increasing air-cleaning effectiveness.

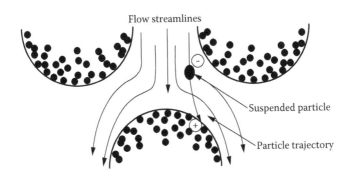

FIGURE 5.4
Particle retention by electrostatic attraction.

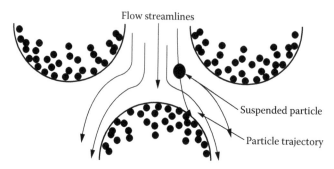

FIGURE 5.5
Particle retention by sedimentation.

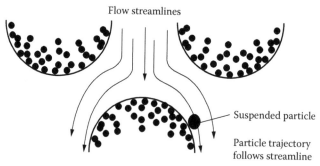

FIGURE 5.6
Particle retention by interception.

5. *Sedimentation*

   When the fluid flow is directed downward through a filter, gravitational sedimentation effects will cause particles to settle vertically through the flow streamlines, as the latter distort around the collector (Figure 5.5).

6. *Interception*

   If the suspended particle radius is greater than the distance between the flow streamline that contains the particle and the collecting media grain, then the suspended particle will contact the target, in the absence of any repulsive mechanisms (Figure 5.6).

5.4.1 Filtration: Collection Efficiency

The graph relating deep-bed filtration efficiency and the size of suspended particles can be explained in terms of the relative importance of diffusion, inertia, and straining. At low particle diameter, removal efficiency is mainly due to diffusion. This effect becomes less relevant at higher

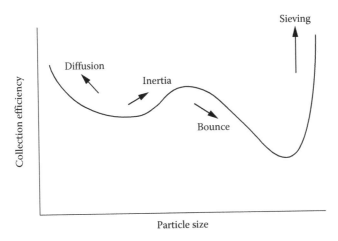

FIGURE 5.7
Capture efficiency as a function of particle size in a deep-bed filter.

diameters. However, as particle size increases, inertial impaction becomes more relevant and efficiency increases again with size. Eventually straining, or sieving, becomes the dominant mechanism (Figure 5.7).

5.5 Filtration Fundamentals

The selection of a suitable filter media is closely linked to efficient filtration operation. In many industrial filtration processes, most of the difficulties that arise are mainly related to interaction between the impinging particles and the pores in the filter medium. The ideal circumstance, where all separated particles are retained on the surface of a medium, is often not realized; particle plugged into the cloth pores leads to an increase in the resistance of the filter media to the flow of filtrate. This process can ensue to the level of total blockage of the system; such difficulties can be avoided if the pores in the medium are all smaller than the smallest particulate in the mixture processed.

The filtrate velocity V, through the clean filter medium, is proportional to the pressure differential ΔP imposed over the medium; the velocity is inversely proportional to the viscosity of the flowing liquid μ and the resistance of the medium. These relationships may be expressed mathematically as

$$V_0 = \frac{\Delta P}{\mu R_m}$$

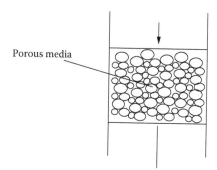

Porous media

FIGURE 5.8
Schematic diagram of porous media.

Filter cake resistances vary over a wide range, from free filtering sand-like particulates to high-resistance sewage sludges. Generally, the smaller the particle, the higher will be the cake resistance.

5.5.1 Fluid Flow through Porous Media

The fundamental relation between the pressure drop and the flow rate of liquid passing through a packed bed of solids, such as that shown in Figure 5.8, was first reported by Darcy in 1856. The liquid flows through the pores or voids within the bed. As it flows over the surface of the solid packing, frictional losses lead to a pressure drop. The amount of solids inside the bed is clearly important; the greater this is, the larger will be the surface over which liquid flows and, therefore, the higher the pressure drop will be as a result of friction. The volume available for fluid flow is called the porosity or voidage, and this is defined as follows.

In many solid–liquid separations, the use of solid concentration is often preferred to porosity. This is usually the volume fraction of solids present within the bed C; porosity is the void volume fraction so these two fractions sum to unity. Hence, solid volume fraction concentration is

$$C = 1 - \varepsilon$$

Darcy discovered that the pressure loss was directly proportional to the flow rate of the fluid.

5.5.2 Particle Settlement

The nature of the particle, in terms of its size and its relative density, is a critical factor in the creation and maintenance of dust-laden air or particle-contaminated water. This is because of the development of a constant terminal velocity by a particle falling freely through a fluid, whose value, as demonstrated by Stokes' law, is

- Directly proportional to the square of the particle diameter
- Directly proportional to the difference between particle and fluid densities
- Indirectly proportional to fluid viscosity

This relationship is strictly true only for spherical particles falling while isolated from one another by some distance. Nonspherical particles are accounted for by the use of an effective diameter (often determined by the reverse process of measuring a terminal velocity and calculating back to the diameter). Stokes' law holds well enough in the case of fluids with a low solid concentration such as that concerns a study of contaminant removal.

If the particle size is small enough so that the terminal velocity is very low, or if there are random movements in the fluid at velocities in excess of this terminal figure, then the particle effectively does not settle out, and a stable suspension results—which may then need decontaminating. The settling velocity in the air of a spherical particle, whose specific gravity is 1, is given by $1.8 \times 10^{-5} \times d^2$ m/min where d is the particle diameter, expressed in µm.

5.6 Filtration Types

1. *Wet filtration*

   Filter fabrics used for the separation of solid particles from liquids in the form of cake are known as wet filtration. In wet filtration, free flow of liquid through media is not restricted, but solid particles are easily stopped through the textile used.

2. *Dry filtration*

   In dry filtration, the dusts are removed by using bag filters. Large numbers of nonwoven or woven bags are used, and pure air is filtered out using a fan.

5.6.1 Classification of Filtration

Depending on the process of separation, filtration is classified as follows:

1. *Particle filtration*

   Particle filtration is the separation of particles having sizes above 10 µm. These can be filtered out easily without any usage of micro-porous membrane.

2. *Microfiltration*

Microfiltration process effectively retain contaminants from liquid or gaseous phase through a microporous membrane. The pore size ranges from 0.1 to 10 μm in a typical microporous membrane. The microfiltration process is basically different from reverse osmosis (RO) and nanofiltration (NF) because the latter-mentioned systems use pressure as a means of forcing water to go from low pressure to high pressure. Microfiltration can use a pressurized system, but it does not need to include pressure.

3. *Ultrafiltration*

In the ultrafiltration (UF) process, hydrostatic pressure forces a liquid against a semipermeable membrane. Ultrafiltration is a type of membrane filtration in which high molecular weight solutes and suspended solids are retained, while low molecular weight solutes and water pass through the membrane.

This filtration process is used in industry and research for purifying and concentrating macromolecular solutions, especially protein solutions. UF is not basically different from microfiltration or NF, except in terms of the size of the molecules it retains. Mostly, UF is applied in cross-flow mode, and separation in UF undergoes concentration polarization.

4. *Nanofiltration*

Nanofiltration (NF) is also a type of membrane filtration process recently used most often with low total dissolved solids in water such as surface water and fresh groundwater, with the purpose of softening (polyvalent cation removal) and removal of disinfection by-product precursors such as natural organic matter and synthetic organic matter. NF is also becoming more widely used in food processing applications such as dairy, for simultaneous concentration and partial (monovalent ion) demineralization. NF is a cross-flow filtration technology that ranges somewhere between UF and RO. The nominal pore size of the membrane is typically below 1 nm, thus NF. Nanofilter membranes are typically rated by molecular weight cutoff rather than nominal pore size. The trans-membrane pressure (pressure drop across the membrane) is required considerably lower than the one used for RO, reducing the operating cost significantly. However, NF membranes are still subject to scaling and fouling, and often modifiers such as antiscalants are required for use.

5. *Reverse osmosis*

RO is similar to the membrane filtration treatment process. However, there are key differences between RO and filtration. The predominant removal mechanism in membrane filtration is straining, or size exclusion, so the process can theoretically achieve perfect exclusion of particles regardless of operational parameters such as influent pressure and concentration. RO, however, involves a diffusive mechanism so that separation efficiency is dependent on influent solute concentration, pressure, and water flux rate. It works by using pressure to force a solution through a membrane, retaining the solute on one side and allowing the pure solvent to pass to the other side. This is the reverse of the normal osmosis process, which is the natural movement of solvent from an area of low solute concentration, through a membrane, to an area of high solute concentration when no external pressure is applied. Nonwoven wet-laid polyester substrates support RO membranes in spiral wrap modules in a $30 million worldwide market for nonwovens. The modules are found in systems predominantly located in arid regions where seawater is converted to potable water. Spun-bond fabrics are used as pleat supports and separators in virtually every microporous membrane cartridge sold, accounting for nonwoven sales of approximately $35 million/year.

5.6.2 Methods of Filtration

There are many different methods of filtration; all aim to attain the separation of substances. Separation is achieved by some form of interaction between the substance or objects to be removed and the filter. The substance that is to pass through the filter must be a fluid, that is, a liquid or gas. Methods vary depending on the location of the targeted material, that is, whether it is in the fluid phase or not.

1. *Solid gas separation*
   Example: Filters used in cigars, and filters used in AC systems.
2. *Solid liquid separation*
   Example: Filters used in sewage disposal plants, filters in chemical industries, and water purifiers.

5.7 Filter Media

A filter is a device used for separating the contaminant particle from liquid phase or gaseous phase—where separation is caused by mechanical means, without the involvement of a change in phase. Filtration is mostly a characteristic of the particle size, liquid or gas being separated. A successful filtration operation is highly dependent on the selection of an appropriate filter medium. The filtration mechanisms invoked in separations using such media will depend mainly on the mode of separation. Thus in "cake" filtration,

TABLE 5.1

Types of Filter Media and Particle Retention Characteristics

Main Type	Subdivisions	Smallest Particle Retained (μm)
Solid fabrications	Flat wedge-wire screens	100
	Wire-bound tubes	10
	Stacks of rings	5
Rigid porous media	Ceramics and stoneware	1
	Carbon	1
	Plastics	1
	Sintered metals	<1
Cartridges	Sheet fabrications	3
	Yarn wound	5
	Bonded beds	1
Metal sheets	Perforated	100
	Woven wire	5
Plastic sheets	Woven monofilaments	10
	Fibrillated film	5
	Porous sheets	0.1
Woven fabrics		5
Link fabrics		200
Nonwoven media	Filter sheets	0.5
	Felts and media felts	10
	Paper media—cellulose—glass	5
	Bonded media	2
		0.1
Loose media	Fibers	<1
	Powder	<0.1

ideally, impinging particles should be larger than the pores in the medium. The filtration efficiency of various filter medium is given in Table 5.1.

Efficient filtration process is largely accomplished with the aid of the appropriate filter medium, which is required to have different properties such as filtration characteristics, chemical resistance, mechanical strength, dimensional stability, dimensions, and wettability. The various factors considered in the selection of textile filter medium, which influences the filtration process, are as follows:

- Polymer type
- Temperature of the filtration process
- Nature of the particle
- Particle size and its distribution
- Concentration of the particle
- Loading tendency on filter fabric surface

The filter medium should possess greatest possible collection efficiency, low pressure collection efficiency, low pressure drop, small filtering area, low penetration of dust in the fabric, and optimum cost. In the case of fabric filters, the fabric performs very little of actual filtering; it provides substratum or matrix for the primary dust cake to form, which in turn collects the particulate and allows air to flow through the fabric. So the fabric should be able to permit the development of a loose and porous cake on its surface and also to release the cake during cleaning. As far as fabric is concerned, its abrasion resistance, chemical resistance, tensile strength, and permeability should be considered.

5.8 Filter Media Design/Selection Criteria

The primary factors that influence the design or selection of filter media may be listed as thermal and chemical conditions, filtration requirements, equipment consideration, and cost. The selection of the type and fineness of the fiber is largely governed by the following circumstances prevailing in filtration:

- Temperature
- Humidity
- Chemical conditions
- Composition and size distribution of dust particles

5.8.1 Thermal and Chemical Conditions

The polymer type of the fibers or filaments used in filter-fabric production is determined by the thermal and chemical conditions of liquid or gas being filtered. Conventionally, woven filter fabrics were manufactured from yarns spun from cotton fibers, which tend to swell on wetting, facilitating efficient filtration. Cotton fabrics are extensively used in gaseous phase filtration at temperatures below 80°C and where acid content is absent in gas. However, the life-expectancy level of cotton fabrics are restricted in the chemically aggressive filtration conditions. Wool fabrics are more resistant to acid than cotton and are used in the collection of the metallurgical fumes and for fine abrasive dust such as cement (Table 5.2).

By comparison, synthetic fibers are generally much more durable, but even so, it is still important to make the correct selection for the conditions that prevail in the filter. The excellent mechanical, physical, and chemical properties of synthetic fibers offer high-performance characteristics in the filtration process itself.

TABLE 5.2

Fibers and Their Properties

Fibers	Density (g/cc)	Maximum Operating Temp (°C)	Resistance to			
			Acids	Alkalies	Oxidizing Agents	Hydrolysis
Polypropylene	0.91	95	VG	VG	P	G
Polyethylene	0.95	85	VG	VG	P	G
Polyester (PBT)	1.28	100	G	P	F	P
Polyester (PET)	1.38	100	G	P	F	P
Polyamide 6.6	1.14	110	P	G	P	F
Polyamide 11	1.04	100	P	G	P	F
Polyamide 12	1.02	100	P	G	P	F
PVDC	1.70	85	VG	G	VG	G
PVDF	1.78	100	VG	VG	G	VG
PTFE	2.10	150+	VG	VG	VG	VG
PPS	1.37	150+	VG	VG	F	VG
PVC	1.37	80	VG	VG	F	VG
PEEK	1.30	150+	G	G	F	VG

VG, very good; G, good; F, fair; P, poor.

The use of filter cloth made of synthetic fibers has the following advantages:

- Greater filtrate purity and improved hygiene condition of filtration process
- Reduced fabric weight owing to the higher strength of constituent materials
- More efficient rinsing of the filter cloth in the filtering systems and washing in the washing machines
- Easier and more rapid drying
- Full resistance to rot during the out of operation of the filtering system
- Better resistance to effects of elevated temperature and moisture

If the operation temperature does not exceed 150°C, polyester is generally used. The most widely used fiber in dry filter media is polyester (~70%). If the resistance to hydrolysis of polyester is inadequate, then acrylic is used. Continuous exposure of polyamide fibers to strong acids leads to fiber degradation and, conversely, polyester fibers exposed to strong bases and prolonged hydrolytic conditions will be prone to degradation. Polyamide yarns exhibit good abrasion resistance but are sensitive to acid particulate. Aromatic polyamides (aramid) have gained some importance in the production of nonwoven filter materials with their high heat resistance.

The most widely used polymer in liquid filtration is polypropylene, belonging to the polyolefin family, which is relatively chemically inert; it too has a vulnerability with its susceptibility to attack from oxidizing agents. The presence of chlorine or heavy metal salt is the potential source of such attack. Polypropylene with its low thermal resistance is little used. Polytetrafluoroethylene (PTFE) is virtually resistant to chemicals and has a maximum service temperature of 280°C. Due to higher cost, PTFE fibers are mostly restricted in filtration applications, even though they have good resistance to most agents.

5.8.2 Filtration Requirements

In order to fulfill expectations of filtration requirements, the ideal filter medium will provide the following:

1. *Resistance to chemical/mechanical attrition*
   Polymer selection in relation to chemical conditions was discussed already. Mechanical conditions in the filtration process such as tensile forces acting on the filter fabric and the abrasive nature of the slurry to be filtered will be given due consideration while selecting suitable yarns and setting fabric's thread spacing. The abrasive forces in this context arise from the shape and nature of the particles in the slurry. Materials with sharp edges tend to abrade internally, fiber/filament breakage, and ultimately lead to pinhole formation in the filter fabric. The designing of filter fabric should be done in such a way to withstand the impact of such forces.

2. *Resistance to blinding*
   Closing of the filter medium pores resulting in reduced liquid or gas flow or an increased pressure drop across the medium is termed as "blinding." It is a familiar term, which relates to particulate matter getting captured, sometimes irretrievably, within the apertures of the fabric which leads to reduction in throughput. Blinding occuring in fabric may be temporary or permanent. If cloth may be rejuvenated by washing, either externally or in situ, then the blinding is temporary. If medium cannot be cleaned easily and the pores opened, it is called as "permanent blinding" or "plugging." One of the main reasons for the plugging is crystal growth from the process itself. An example of this type of plugging can be found in filtration processes dealing with gypsum, for example, on either horizontal belt or tipping pan filters in the production of phosphoric acid.

3. *Good cake discharge at the end of the filtration cycle*
   Adequate cake release is a fundamental prerequisite in efficient pressing operations, in maintaining a low downtime in the overall "batch" time. The ability of a medium to discharge its cake depends

very much upon the smoothness of the surface upon which the cake is residing, and hence upon the amount of fibrous material extending from the surface into the cake. Filter cake firmly adheres to the fabric surface and reduces the process efficiency, either by the reduction of filtration area or difficulty in mechanical removal of cakes which leads to manual cake removal, which is a time-consuming process. This typical problem has been addressed in recent years by manufacturers with the aid of brush cleaning and high-pressure cleaning systems. An understanding of the failure of the release mechanism follows consideration of the balance between the forces causing adhesion of the cake to the medium and the discharge forces. The adhesion of particles dispersed in liquids is mainly the result of electrostatic and Vander Waals interactions; chemical bonding also plays an important role. The effectiveness of discharge will depend on the following:

a. The strength of the bond between the cake and cloth is influenced by cake stickiness and yarn/weave characteristics. The bonding force depends on the mode of deposition of the first layer.

b. The internal strength of the cake. If the cohesion of the latter is less than the adhesion to the cloth, the cake will fail internally and leave solids on the cloth. The moisture content will vary across the depth of the cake.

c. The applied discharge force (e.g., gravity discharge from a vertical surface).

4. *Low cake moisture content*

Low cake moisture content is a significant factor in filtration processes where the cake has to be thermally dried. The cost of drying the cake with thermal energy makes it peremptory to squeeze the maximum amount of moisture from the slurry by mechanical means. While transporting filter cakes for landfill applications, the moisture content has to be controlled to meet environmental regulation requirements. Fabric and equipment have a vital part to play in this context.

5. *Filtrate throughput*

Throughput is the amount of fluid able to pass through a filter prior to plugging. The important objective of any filtration process will be the maximum throughput in minimum time and with low resistance, the operation often being critical to the balance of the total production cycle. Equipment parameters again play a vital role in this subject.

6. *Filtrate clarity*

The liquid which has passed through the filter is termed as "filtrate". There is always a balance between initial filtrate clarity and filtration rate. Absolute filtrate clarity may or may not be critical as the role of filter medium is to capture and retain the particles. The next destination of the filtrate and/or the ability of the process engineer to recirculate until satisfactory clarity is obtained need to be established and balanced against throughput requirements. Similarly, in certain screening operations, the fabric is designed to capture particles only of a specific size.

5.8.3 Equipment Considerations

In respect to equipment considerations, the ideal filter fabric will, in simplistic terms, provide a long trouble-free performance. Equipment considerations again focus on the amount of mechanical forces applied on filter fabric during cleaning cycle by the cleaning mechanisms. The following factors are considered while selecting the equipment for filtration process:

1. *Cloth shrinkage*
   Shrinkage of the filter fabric can create intense problems particularly with plate and frames and the larger recessed plates. Filter fabrics made up of polyamide fibers are prone to severe shrinkage problems particularly during repeated cloth laundering/drying cycles. For this type of fabric, storing under wet condition immediately after washing by avoiding drying process is preferred. Preshrinkage of fabrics is, therefore, widely practiced in order to retain dimensional stability in service.
   Preshrinkage can be affected in a number of ways:
   a. Treating filter medium with hot water in a relaxed state.
   b. Using conditioning oven, heat setting is performed to filter fabric which is held under tension in both warp and weft directions. This helps to maintain the filter fabric porosity, permeability, etc.
   c. Filter fabric is subjected to oven treatment with or without tension.

   The shrinkage process has to be carefully controlled, in view of the large structural changes of up to 15% shrinkage, which can ensue in relaxed conditions. Spun staple yarns are reported to shrink less than comparable fabric woven from filament yarns.

2. *Cloth stretching*
   Textile filters are prone to swelling while absorbing liquids. Dimensional changes occur in the filter fabric, that is, increase in fiber diameter and decrease in length leads to serious problems for closely fitted plates. This is particularly true for those yarns with poor absorption characteristics (nylon). The latter may absorb up to 4% by weight; this may be compared with 0.4% for terylene.
   The discharge of a relatively thin dust cake is carried out by peeling apart the cloth and solids with the help of automation and

also with high pressure squeezing during dewatering cycle. The complete cycle of filling, pressing, air blowing, discharge, and cloth washing may be short (5–6 min) in the modem auto-variable chamber unit. A filter cloth in the modern unit may be subjected to much stress compared to manually operated systems.

All these considerations call for cloths of great strength while retaining high-level filtration characteristics. For example, a traditional strong polyester fabric suitable for use on large plates (40 × 40 in.) in slow or manual systems would have, say, breaking load figures of 900 N/cm (warp) and 350 N/cm (weft). These figures may be compared with cloths of the same size in mechanized units: 1800 N/cm (warp) and 1600 N/cm (weft), respectively.

3. *Filter size*

The selection of filter size is to be done with regard to the acceptable pressure drop and the cycle time required between successive cleaning and element replacement. This is closely bound up with the type of element and filter medium employed. While performing filtration for the liquid containing heavy contamination, a filter fabric with high retention capacity may clog too quickly calling for a much larger size than normal, or alternatively a different type of element with better collecting properties, so that clogging is slowed down.

5.8.4 Cost

The design and development of filter cloth both in material selection and production involves considerable technology and although it may essentially contribute to the success of the filtration operation or the filter fabric quality, it is often accepted that the cost contribution of filter medium to the final product cost is extremely low. Filter fabrics are expected to provide the longest possible life before failure due to blinding or mechanical/ chemical damage.

5.9 Yarn Construction and Properties

5.9.1 Monofilament

Monofilaments are single filaments extruded from molten polymer through a spinneret and then subjected to drawing to orientate the molecules and thus provide the thread with the desired stress–strain characteristics. The monofilaments usually have round cross section although other profiles are possible (Table 5.3).

TABLE 5.3

Typical Filtration Characteristics of Different Fiber Forms

Fiber form	Maximum Retention	Maximum Production	Maximum Cake Moisture Reduction	Maximum Cake Discharge	Maximum Life	Maximum Resistance to Blinding
	Spun staple	Monofilament	Monofilament	Monofilament	Spun staple	Monofilament
	Multifilament	Multifilament	Multifilament	Multifilament	Multifilament	Multifilament
	Monofilament	Spun staple	Spun staple	Spun staple	Monofilament	Spun staple

Monofilament-based filter fabrics are characterized by their good blinding resistance, efficient cake release at the end of the filtration cycle, and higher throughput.

5.9.2 Multifilament

Multifilaments also have a similar route of production as monofilament, but the spinneret contains a large number of much smaller apertures. In this case, the diameter of the individual filament is in the range of 0.03 mm. Twisting is carried out to bind the filaments together after extrusion process. The twist impartation facilitates the filament assembly slightly stronger, rigid, and, if a high twist is imparted, can alleviate the tendency of blinding in yarn and fabric. Multifilament fabrics possess greater filtration efficiency, higher mechanical properties, and greater flexibility compared to monofilament fabrics are, in spite of having higher blinding tendency than the latter. The poor mechanical responses of monofilaments such as creasing, tearing, and stretching, instilled an interest in high-twist multifilament cloths woven in such a way as to ameliorate the related problems of cake release.

5.9.3 Staple Yarns

Short staple spinning technology is used to produce the staple spun yarns from short fibers. The filaments extruded from spinneret are cut to fiber length of 35–100 mm depending on the type of spinning system adopted. The characteristic advantages of woolen spun yarn compared to cotton spun yarns and multifilament yarns are less blinding tendency, greater throughput, and efficient particle retention. On the other hand, the blinding resistance of staple spun yarns and multifilament yarns are significantly inferior to monofilaments. High-twist yams are advantageous if compressed air is used for cake discharge (Table 5.4).

5.9.4 Fibrillated Tape Yarns

The narrow width polypropylene films are converted into coarse filaments by splitting the film to produce this yarn. Hence, it is also called as "split-film yarn". The usage of these yarns in filtration is limited. They are mainly used

TABLE 5.4

Effect of Twist of Yarn on Particle Flow

Twist	% Flow through Yarns
1.5–3.0	95–98
15	70
35	2

as secondary layers in the form of open-weave structures which provides support and drainage for the primary filter fabric.

5.10 Fabric Construction and Properties

5.10.1 Woven Fabric Filters

Fabrics make up the largest component of filter media materials. They are made from fibers or filaments of natural or synthetic materials and are characterized by being relatively soft or floppy, lacking the rigidity of dry paper, such that they would normally need some kind of support before they can be used as a filter medium (Figure 5.9).

The characteristics of the woven filter fabric depends on the type of the yarn used and fabric parameters in weaving. Yarns are available in several forms: monofilament, multifilament, staple fiber, and mixtures of the same. The fibers such as polypropylene, polyester, polyamide, etc. are used to produce yarns intended for filter fabrics. Fabrics of various weave patterns can be produced in the loom by varying the warp and weft interlacing pattern. The plain-weave monofilament cloth has been produced by warp and weft yarns of the same diameter, woven together in a simple one-under, one-over pattern (Figure 5.10).

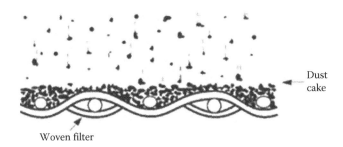

FIGURE 5.9
Woven fabric filter.

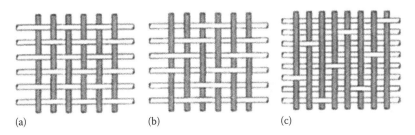

FIGURE 5.10
Types of weaves in woven fabric filter: (a) plain, (b) 2/2 twill, and (c) 8-end satin.

These cloths are available in a wide range of pore sizes from 5000 to about 30 μm, the lower limit being determined by the size of fiber available for the weaving process. Plain-weave monofilament fabrics are characterized by open pores that create little flow resistance, and many applications are found in areas where high flows are required, for example, in oil, paint, and water filtration and screening. Such cloths are readily cleaned by back-flushing. The fabric surface can be altered to enhance its performance by finishing; it involves calendering and heat treatment in order to reduce pore size and to flatten the surface, but preserve the construction by controlling dimensional stability during filtration and cleaning cycle.

Unfortunately, monofilament fabrics are light in weight and would be easily damaged if used directly in pressure filters. Thus the benefits of high throughput and ease of cleaning must be weighed against the fragile nature of the medium. Latest trends to find optimum solution are to produce composite weaves from fine and coarse monofilaments in the production of a surface layer with good release properties and nonblinding characteristics. The underside layer of coarser fibers provides support, assists drainage, and promotes attachment or caulking of the cloth onto the filter platform. In order to produce apertures finer than mentioned earlier, changes in the weave are available to modify the size (and shape) of the cloth pores. Fabric of this type, for example, satin weave, possesses very smooth surfaces that are optimal for cake release (Table 5.5).

Thus the more open weave fabrics will be superior in nonblinding characteristics, but may have poor particle retention. The latter will improve in the order monofilament < multifilament < staple fiber.

The relative amount of flow through and around the yarns in such cloths will depend on the degree of twist imparted to the yarn and the size of the apertures between yarns. The aperture size will, in turn, depend on the weave pattern: plain, twill, satin, etc. Swelling of fabrics can change the nature of flow in that closure of the cloth pores can force an increasing amount of flow through the yarns (Table 5.6).

TABLE 5.5

Effect of Weave Type on Properties of Filter Fabric

Property	Weave Type		
	Plain	Twill	Satin
Rigidity	Best	Satisfactory	Poor
Bulk	Poor	Best	Best
Initial flow rate	Poor	Good	Best
Retention efficiency	Best	Satisfactory	Poor
Cake release	Good	Poor	Best
Resistance to blinding	Poor	Good	Best

TABLE 5.6

Effect of Fabric Weave on Filtration Characteristics

	Maximum Retention	Maximum Production	Maximum Cake Moisture Reduction	Maximum Cake Discharge	Maximum Life	Maximum Resistance to Blinding
Weave	Plain	Satin	Satin	Satin	Twill	Satin
Pattern	Twill	Twill	Twill	Twill	Plain	Twill
	Satin	Plain	Plain	Plain	Satin	Plain

5.10.1.1 Properties of Woven Filter Fabric

Three properties by which a filter fabric medium may be judged are

1. Permeability of the unused filter medium (or, inversely the resistance)
2. Particle-stopping power of the filter medium while particle collides with it
3. Permeability (or resistance) of the used filter medium

The filtration process involves two principal resistances: (a) the resistance of the filter cake "α" and (b) the resistance of the medium "R_m." High levels of α, for example, greater than 1×10^{12} m/kg (characteristic of sludge-like material) changes in R_m, have little influence on overall productivity—at least in the range $1 \times 10^8 < R_m < 1 \times 10^1$. Thus a partially blinded medium may still function quite satisfactorily in a system controlled by α.

5.10.1.1.1 Filter Permeability

The permeability is the reciprocal of the resistance to flow offered by the filter; thus, high permeability represents a low resistance and vice versa. Permeability is usually expressed in terms of a permeability coefficient, which is directly proportional to the product of flow rate, fluid viscosity, and filter medium thickness, and inversely proportional to the product of filter area and fluid density, which gives the permeability coefficient the dimension of a length.

For a given flow rate, an increase in filter area will reduce the pressure drop across the filter, because the amount of fluid flowing per unit of filtration area is decreased (pressure drop is inversely proportional to filter area).

The operating temperature of the fluid will affect the pressure drop across the filter because the fluid viscosity will change. A less viscous fluid will experience less resistance to flow through the medium, and so a lower pressure drop will be needed to drive it. As a result, pressure drop is inversely proportional to temperature, with a decrease in temperature causing a rise in pressure drop.

TABLE 5.7

Typical Values of Air Permeability of Fabrics

Fabric Type	Air Permeability (m³/m³ s)
Nylon multifilament	0.030–1.52
Polypropylene monofilament	0.015–1.52
Polypropylene multifilament	0.005–0.508
Nonwoven cloth	0.002–1.27

The effect of prolonged filtration time is to produce a cumulative buildup of collected solids on or in the filter medium, thus reducing permeability (and increasing flow resistance) in direct proportion to the amount of solid collected (Table 5.7).

Thus a high permeability is taken as an indication of high porosity and, in turn, low particle retentivity.

5.10.1.1.1.1 Multifilament Cloth Permeability   In multifilament cloths, fluid flow may occur through or around the permeable yarns. If we define B_0, as the permeability of the porous yarns, and B_1, as the permeability of the cloth if the yarn were solid, that is, monofilament, it may be shown that

$$\Omega = \frac{B}{B_1} = \left(1 + 1.34\left(\frac{B}{B_0}\right)\right)^2 \quad \text{for} \quad \frac{B_0}{d_y^2} < 0.0017$$

where
  B is the overall permeability of the cloth
  d_y is the yarn diameter

The Ω index has been shown to vary in the range $1 < \Omega < 20$ within the order of accuracy of the experimental measurements necessary for the determination of B and B_0

$$\Omega = \frac{\text{Permeability of cloth}}{\text{Permeability of cloth composed of monofilament yarn}}$$

5.10.1.1.1.2 Monofilament Cloth Permeability   In the monofilament area, much more success in correlating permeability with cloth structure has followed the suggestions of Pedersen (1969), who adopted orifice-type formulae to correlate pressure-drop–flow information for various weave patterns.

A discharge coefficient was defined as

$$C_D = \left(\frac{v^2}{2\Delta P} \frac{\left(1-a^2\right)}{a^2} \right)^{0.5}$$

where
a (the effective fraction open area of the pore) = A_0(ec) (pc) in which
(ec) is the warp yarns per centimeter
(pc) is the weft yams per centimeter
A_0 is the effective area of orifice
v is the flow rate
ΔP is the pressure drop

The discharge coefficient was anticipated to be a function of the Reynolds number within the fabric.

5.10.2 Nonwoven Filtration Media

The filtration industry is burgeoning worldwide at the rate of 2%–6% per year above the gross domestic product. Nonwoven fabrics have seen tremendous growth with penetration into a number of filtration industry end-use market segments. Nonwovens offered a cost-effective alternative and often a distinct technical advantage by the basic attributes of the nonwoven construction. The factors such as fiber diameter, orientation, packing density, and web weight will determine the filter media properties. Nonwoven fabric and membrane filtration media together dominate the filtration media market, with more than 90% combined market share in terms of roll goods filtration media volume in comparison to all other material forms. Typically, nonwoven fabrics add backup support and/or mechanical strength to comparatively weak membrane media, allowing membranes to function at peak performance.

Nonwoven fabric filtration media have dominated in applications such as coolant filtration, bag house filtration media, vacuum cleaner bags, and many heating, ventilating, and air conditioning (HVAC) applications. In these and other applications, nonwovens are highly price-competitive. Air applications consume approximately 65%–70% of the nonwoven filtration media, with liquid uses consuming the remaining 30%–35%. Overall, 75% of synthetic nonwoven media go into commercial markets, such as manufacturing facilities, offices, theaters, hospitals, cruise ships, casinos, and other such markets, with about 25% found in residential and general consumer air filters.

The nonwoven fabric media vary by materials of construction, processing method, and performance characteristics. Nonwoven fabric production process consist of two steps: (i) web formation and (ii) web bonding. Nonwoven fabrics can be classified into two distinct types based on web formation method: (i) dry laid process and (ii) wet laid process. Dry laid processes include carded, air-laid, spun-bond, and melt-blown media. Wet-laid process is similar to paper-making process, where web formation is carried out by using water as medium. Web bonding can be done to impart strength to the nonwoven fabrics by thermal or mechanical or chemical means. Each process produces a media with unique properties that have advantages in different applications.

5.10.2.1 Dry-Laid Media

Dry-laid media which are isotropic in structure are produced with air-laid web laying technology in which fibers are suspended in air, then collected as a batt on a fine mesh screen. Anisotropic dry-laid media can be produced with carding technology. Dry-laid processes generally produce media with nominal ratings that are low cost and have high dirt-holding capacities. Melt-blown media are one of the most versatile nonwovens for liquid filtration. Melt-blown media are generally composed of a continuous network of self-bonded polypropylene, polyester, or nylon microfibers produced with a controlled fiber uniformity and density. The resulting medium has a uniform porosity, does not shed fibers, and contains no binders, adhesives, or surfactants. Melt-blown media have nominal ratings from 1 to 50 μm. Nonwoven melt-blown and spun-bond fabric along with nonwoven glass filtration media are the principal air filtration media for HVAC.

5.10.2.2 Wet-Laid Media

Wet-laid nonwoven technology involves uniformly dispersing shortcut fibers in water, transporting the slurry onto a continuous moving fine mesh screen called the wire, and then forming a mat as a result of the removal of water. The web then undergoes further water elimination through drying. A major objective of wet-laid nonwoven manufacturing in filtration is to produce structures with known pore size and filtration characteristics.

High-efficiency particulate air (HEPA) wet-laid glass nonwoven filtration media represent 90–100 million square meter nonwoven market. Air filters are found in end-use markets from general dust filtration to high-efficiency filtration in many different configurations. These filters are rated by a minimum efficiency reporting value (MERV) standard, which rates filters from 1 to 20 in terms of their degree of efficiency. At the high end, MERV 17- to 20-rated HEPA filters are typically used in situations that require absolute cleanliness for the manufacture of microchips, liquid crystal display screens, pharmaceutical production, and microsurgery in hospital operating rooms. HEPA filters are primarily constructed from wet-laid glass nonwoven filtration media. Wet-laid cellulosic and spun-bond polyester media that range from 200 to 300 g/m² are used in pleated dust collection cartridges. Pleated cellulose- or polyester-based filters offer significantly greater surface area than needle-felt filter bags for a given space as an alternative filter configuration in bag house applications.

5.10.2.3 Characteristics of Nonwoven Filter Media

The production of filter media with a wide range of thickness and pore size is highly possible with the nonwoven technologies. Needle-punched filter media is characterized by higher thickness and more number of pores per unit area. Melt-blown filter media is widely used in liquid filtration in which the presence of microfibers retain the micro-level contaminants. The highly porous nonwoven structure facilitates high retention capacity for depth filters. Nonwoven fabrics usually contain bonded fibers, bonding carried out by mechanical or thermal or chemical means. Thus, the wet strength and overall resistance to fiber shedding can be enhanced by bonding the fibers present in the web together, to produce a rigid network. Spun-bonded fabrics are inherently bonded during heat setting. These involve the extrusion of molten polymer into cylindrical filaments that are dispersed by hot gas flow into a tortuous, random array. The fiber mixture may include a small proportion of low melting point material. A wide variation in fiber diameter exists.

Examples are

1. 10 µm polyester (1.3 dtex where the latter unit is the weight in grams at 10,000 m of fiber)
2. 40 µm polypropylene (13.3 dtex)
3. 30 µm cellulose (may be fibrillated to produce fine fiber attachments or fibrils)
4. 0.03–8 µm glass (100% glass media used in laboratory liquid separation and in gas filtration)

Both the permeability and filtration characteristics of nonwovens are dependent on the felt porosity and fiber diameter. A medium that has been heavily calendered on both sides will possess the lowest porosity. Surface treatments and/or use of laminations of different porosities are aimed at improving cake filtration performance and cake release. Generally speaking, the filtration efficiency at a particular particle size is inversely proportional to the fiber diameter, other factors being the same.

5.10.2.4 Factors Influencing the Air Permeability of Nonwoven Filter

5.10.2.4.1 Effect of Fabric Weight

Weight of nonwoven filter media, which is usually measured in gm/m² plays a significant role on filtration performance. The increase in fabric weight due to more number of fibers per unit area reduces the air permeability of the filter media. Due to the increase in filter fabric's density, the air flow resistance increases. The usage of high gm/m² fabric increases the pressure drop which in turn improves the filtration efficiency. The tenacity at break increased with the fabric weight in both bias and cross direction. The reason attributed is due to the increase in the number of fibers in the web, leading to increase in the number of vertical loops and density and entanglement, thus causing less freedom of fiber movement and greater frictional resistance. But the breaking elongation decreases gradually in both bias and cross direction with the increase in fabric weight. The abrasion resistance of the fabric increases with the increase in fabric weight due to the increase in compactness and density of fabric. The bursting strength also increases with the increase in fabric weight due to the increased number of fibers that play an important role in resisting the bursting pressure.

5.10.2.4.2 Effect of Fabric Density and Thickness

If the thickness or density of the filter fabric increases, then there is a nonlinear reduction in air permeability. The density has a more significant influence on air permeability than either thickness or fiber size. Porosity and permeability have no general correlation because the permeability of a material is influenced by the capillary pressure curves. Air resistance increases with fabric thickness and fabric weight per unit area, but decreases with fiber fineness.

5.10.2.4.3 Effect of Synthetic and Cellulosic Fiber Blend

The cellulose/synthetic blends and the 100% synthetic media have similar particle retention, and the synthetic composite medium is slightly lower, but the cellulose medium has poor particle retention. The coarse particle retention does show a trend of improving as the synthetic content in the media increased. The dirt-holding capacity performed as expected based on the standard paper air flow or resistance testing. The dirt-holding capacity improves with the addition of 15% synthetic fiber to a cellulose-based medium.

5.11 Finishing Treatments

5.11.1 Heat Setting

Heat setting is a dry process used to stabilize and impart textural properties to filter fabrics. When fabric filters are heat set, the cloth maintains its shape and size in subsequent finishing operations and is stabilized in the form in which it is held during heat setting (e.g., smooth, creased, uneven).

5.11.2 Singeing

Fabrics produced from short staple fibers naturally possess a fibrous surface, which, in some cases, can impede cake discharge through mechanical adhesion of fiber to cake. Singeing is predominantly carried out on textile fabrics to achieve a smooth and fiber-free surface. Singeing enhances the smooth cake release.

5.11.3 Calendering

Calendering is a process where fabric is compressed by passing it between two or more rolls under controlled conditions of time, temperature, and pressure in order to alter its handle, surface texture, and appearance. It is done to improve the fabric's surface smoothness (for better cake discharge) and also to regulate its permeability and hence improve its collection efficiency.

5.11.4 Raising or Napping

Raising process is designed actually to create a fibrous surface, normally on the outlet side of the filter sleeve, to enhance the fabric's dust collection capability. Raised fabrics may comprise 100% staple fiber yarns or a combination of multifilament and staple fiber yarns, the latter being woven in satin style in which the face side is predominantly multifilament and the reverse side predominantly staple.

5.11.5 Antistatic Finish

Electrostatic charges within a filter dust cake can develop by friction during the processing and movement of gases and fine dust particles. These static charges can be sufficient to generate sparking giving an ignition source to initiate an explosion of combustible gases and dust particles. Antistatic filter media have electrically conductive substrates designed to safely dissipate these charges. The unique microporous structure of the PTFE membrane comprises millions of randomly connected fibrils giving an effective pore size many times smaller than that can be seen by the naked eye. The result is a surface able to capture very fine particulate, while allowing air and static charges to freely permeate the media. Antistatic medium is used to dissipate electrostatic charges where explosive hazards exist or where charged dust particles resist release from nonconductive filter media.

5.12 Nanofiltration

Nanofibers are defined as fibers with diameters in the range of 100 nm. Nanofibers can be produced by interfacial polymerization and electrospinning. This process has received a great deal of attention in the past decade because of its ability to consistently generate polymer fibers that range from 5 to 500 nm in diameter. If the fiber diameter is reduced from micrometers to nanometers, very large surface area to volume ratios are obtained and flexibility in surface functionalities and better mechanical performance may be achieved. The nanofibers are mostly produced through electrospinning process that has been used to spin fibers for the past eight decades. Electrospinning is a process of generating polymer fibers from an electrostatically driven jet of polymer solution. In this process, a charged polymer melt is extruded through a small nozzle. The charged solution is drawn toward a grounded collecting plate. As the jet of charged melt travels, the solvent evaporates, leaving a nonwoven nanofiber mat on a substratum. The nanofiber nonwoven webs produced by this process typically have nanoscale pore sizes, high and controllable porosity, high specific surface areas, and exceptional flexibility with respect to materials used and surface modification of fibers. Nanofibers are characterized as having a high surface area-to-volume ratio and a small pore size in fabric form. The high surface area-to-volume ratio and small pore size allow viruses and spore-forming bacterium such as anthrax to be trapped. Filtration devices and wound dressings are just some of the applications in which nanofibers could be utilized.

Studies have shown that a low fiber diameter allows a filter element with similar operational characteristics but with much higher filtration performance. Any process that is dependent on surface area, such as active filtration, will benefit from the incorporation of nanofibers into the process. The first successful commercial use of electrospun fiber was claimed by Donaldson Inc. for filtration elements. Donaldson Co. Inc. has been developing electrospun nanofiber filtration elements for dust collection, gas turbine air filtration, and air filters for heavy-duty engines. Active filtration implies that the entrapment method is based on chemical attraction rather than simple physical entanglement. The advantages of this method are a lower resistance to flow across the filter element and the possibility of selectivity so that particular elements can be removed during filtration.

The lightweight synthetic nanofibers are deposited over the cellulosic or polyester wet-laid nonwoven base substrate to produce a composite nanofilter, which find use in many industrial filtration applications especially in pleated dust collection and engine air-intake filters. The nanofibers are as fine as 200–300 nm in diameter, with the amount of nanofiber add-on being quite thin in cross section and typically weighing less than 1–2 g/m^2. The nanofibers are laid down over what will become the upstream side of the substrate using an electrospinning process and, in one case, an ultrafine melt-blown process. These nanofibers create a labyrinth of fibers with pores finer than particles in the incoming air stream. Particulate deposits reside on the surface of the fine nanofiber web, allowing the user to clean the filter by shaking off loose particles from the surface or by using an automated clean-air back-pulse system.

Bibliography

Anderson, K., Nanotechnology in textile industry, *(TC)^2 Newsletter*, Techexchange. com/library (accessed May 29, 2013).

Ballew, H. W., *Basics of Filtration and Separation*, Nuclepore Corporation, Pleasanton, CA, 1978.

Darcy, H. P. G., *Les Fontaines Publiques de la Ville de Dijon*, Victor Dalmont, Paris, 1856.

Deitze J. M., Kosik W., McKnight S. H., Beak Tan N. C., Desimone J. M., and Crette S., Electrospinning of polymer nano-fibers with specific surface chemistry, *Polymer*, 43, 1025–1029 (2002).

Dickenson, T. C., *Filters and Filtration Handbook*, 4th edn., Elsevier Science, New York, 1997.

Ehlers, S., The selection of filter fabrics re-examined, *Industrial Engineering Chemistry (International)*, 53(7), 552–556 (1961).

Grafe, T. H. and Graham, K. M., Nanofiber webs from electrospinning, Donaldson Company Inc., Minneapolis, MN, *Nonwovens in Filtration—Fifth International Conference*, Stuttgart, Germany, March 2003.

Gregor, E. C., Nonwoven fabric filtration, *The Textile World*, 159, 32 (2009).

Hardman, E., Some aspects of the design of filter fabrics for use in solid/liquid separation processes, *Filtration and Separation*, 31(60), 813–818 (1994).

Hutten, I. M., *Handbook of Nonwoven Filter Media*, Elsevier, Oxford, U.K., 2007.

Krcma, R., *Manual of Nonwovens*, Textile Trade Press, Manchester, U.K., 1971.

Masumo Inc., http://www.masumoincwatertreatment.com/reverse_osmosis_plants.html (accessed May 29, 2013).

Mayer, E. and H. S. Lim, New nonwoven microfiltration membrane material, *Fluid-Particle Separation Journal*, 2(1), 17–21 (1989).

Patil, U. J. and P. P. Kolte, Filtration in textile: A review, *Indian Textile Journal*, May, 69–72 (2011).

Pederson, G. C., *AIChE 64th Annual Meeting*, New Orleans, 1969.

Pedicini A. and Farris R. J., Thermally induced color change in electrospun fiber mats, *Journal of Polymer Science Part B: Polymer Physics*, 42, 752–757 (2004).

Purchas, D. B., Art, science and filter media, *Filtration and Separation*, 17(4), 372–376 (1980).

Purchas, D. B. and K. Sutherland, *Handbook of Filter Media*, Elsevier, New York, 2002.

Rushton, A., Effect of filter cloth structure on flow resistance, blinding and plant performance, *The Chemical Engineer*, 237, 88 (1970).

Rushton, A. and P. V. R. Griffiths, Filter media, in *Filtration Principles and Practices*, Part 1, C. Orr (ed.), Marcel Dekker, New York, 1977, pp. 169–252.

Rushton, A., A. S. Ward, and R. G. Holdich, *Solid–Liquid Filtration and Separation Technology*, Wiley-VCH Publication, New York, 1996, reproduced with permission.

Sandstedt, H. N., Nonwovens in filtration applications, *Filtration and Separation*, 17(4), 358–361 (1980).

Shields, C., Submicron filtration media, *Nonwoven Perspective, INJ*, Fall, 33–35 (2005).

Sutherland, K., *Filters and Filtration Hand Book*, 5th edn., B-H Publications, Poole, U.K., 2008.

Tiller, F. M., *Theory and Practice of Solid-Liquid Separation*, University of Houston, Houston, TX, 1978.

Wakeman, R. J., *Filtration Dictionary and Glossary*, The Filtration Society, London, U.K., 1985.

6

Textiles in Hoses

6.1 Introduction

Textile fabrics are used as reinforcement in many applications. Hose is one of such products where the fabric is reinforced in its structure to attain desirable properties. The reinforcement of any hose structure may comprise many materials or combinations of materials, dependent on the end use of the item. The reinforcement may be braided, woven, or in wound form, and it may be in single or multiple plies. The reinforcing members include both natural and man-made textiles. Obviously, the choice of the most advantageous material to be used will be dictated first by the end use of the products and second by economics.

6.2 Hose: Definition

A hose is a flexible link on pipe capable of use with gases, liquids, solids, or admixtures of such under positive or negative pressures.

6.2.1 Factors Governing Hose Selection

6.2.1.1 Pressure

The amount of pressure to which a hose will be exposed is one of the important factors governing hose selection. The pressure exerted within a hose depends on the volume a pump can push forward and the diameter of the hose. The more fluid that is pushed into the hose, the more it pushes out on the wall. This internal pressure is measured in pounds per square inch (psi). Typically, pressures fall into any of three general groups:

1. <250 psi (low-pressure applications)
2. 250–3000 psi (medium-pressure applications)
3. 3000–6000 psi (high-pressure applications)

To withstand higher pressures, hose manufacturers use various reinforcement materials to hold the core together. Reinforcement materials can be made of stainless steel wire or textile materials combined into single or double braids, depending on expected hose pressures. Greater reinforcements increase the pressure ratings. In addition to the pressure rating of a hose, vacuum rating is also considered for some applications. Vacuum rating of a hose refers to suction hose applications in which the pressure outside the hose is greater than pressure inside the hose. It is important to know the degree of vacuum that can be created before a hose begins to collapse. Surge pressure is a sudden increase in pressure, usually for a short time, which will shorten hose life. Surge pressures should be lower than the maximum operating pressure in order to prevent hose deformation.

6.2.1.2 Temperature

The temperature of fluid and ambient temperature both static and transient should not exceed the limitation of the hose. The inner cover and reinforcement material should be selected in such a way to overcome the heat due to higher temperature.

6.2.1.3 Fluid Compatibility

The compatibility of hose tube and cover with the fluid or gas transmitted is essential in deciding the durability and life of the hose. The hose that is incompatible with the liquid or gas it carries will deteriorate the hose lining. Rubber hose is used to convey petroleum products both in the crude and refined stages. Aromatic materials in contact with rubber tend to soften it and reduce its physical properties. This can lead to weak areas in hoses that will be unable to withstand pressure surges and will fail.

6.2.1.4 Size

Hoses have inside (ID) and outside diameters (OD). The ID is the more important of the two. An application system's delivery rate is a function of its pump's rated capacity and its hose ID. The size of the components must be adequate to keep pressure losses to a minimum and avoid damage to the hose due to heat generation or excessive turbulence. The OD is used to determine the size of the clamp that will secure the hose to the fitting.

6.2.1.5 Environment

Environmental conditions such as ultraviolet light, ozone, salt water, chemicals, and air pollutants can cause degradation and premature failure and, therefore, must be considered while the selection of hose. Once a hose dries out, it becomes hard, brittle, and loses its ability to expand under pressure.

In extreme cases, dried-out hoses will even crack. Extreme cold can actually freeze the rubber compound in hoses, causing them to crack when bent.

6.2.1.6 Mechanical Loads

External forces can significantly reduce hose life. Mechanical loads that must be considered include excessive flexing, twist, kinking, tensile or side loads, bend radius, and vibration. Flexibility and minimum bend radius are important factors in hose design and selection if it is known that the hose will be subjected to sharp curvatures in normal use. Bend radius measures how much a hose can bend without kinking and compromising its integrity. The bend radius does not necessarily reflect the force required to bend the hose to this radius, which is a major factor in flexibility. For instance, a hose with a 3 in. bend radius can bend around an object that is 6 in. in diameter (or more) without kinking. The properties that keep hoses from kinking include wall thickness, reinforcement material, and construction material. The larger the bend radius, the less bending it can tolerate before it kinks. The hose should be able to conform to the smallest anticipated bend radius without overstress. Textile-reinforced hoses have a tendency to kink as the bend radius is reduced. Generally, a helix of wire is used when a hose must withstand severe bends without flattening or kinking.

6.2.1.7 Abrasion

While a hose is designed with a reasonable level of abrasion resistance, care must be taken to protect the hose from excessive abrasion that can result in erosion, snagging, and cutting of the hose cover. Exposure of the reinforcement will significantly accelerate hose failure.

6.2.1.8 Electrical Conductivity

Static electricity is generated by the flow of material (even some liquids) through a hose. As the material flows, molecules collide and generate friction, which creates minute amounts of electrical charge (excess electrons). The amount of charge increases with material volume and linear velocity, coarseness of the material, and length of the hose. If not properly grounded, the accumulated charge (potential energy) will seek its own ground. The charge will be attracted to external materials in proximity (such as a steel storage container); the electrons may jump to the external material, igniting volatile materials in the hose. Electrically conductive reinforcements and conductive rubber components are used in hose to prevent static electricity build-up and discharge as a spark. Nonconductive hose constructions are those that resist the flow of electrical current. In some specific applications, especially around high-voltage electrical lines, it is imperative for safety that the hose be nonconductive.

6.2.1.9 Hose Length

Changes in ambient temperatures, internal temperatures, and vibrations require hoses to be flexible. Hoses shrink and expand slightly based on ambient or internal temperatures. Loads also may shift, so hoses should have slack in them so they will not be stretched, kinked, twisted, or even pulled off.

6.3 Hose: Construction

A hose has three parts (Figure 6.1):

1. *Cover:* It is the outermost layer of the hose. The prime function of the cover is to protect the reinforcement from damage and the environment in which the hose will be used. Covers are designed for specific applications and can be made to be resistant to oils, acids, abrasion, flexing, sunlight, ozone, etc.

2. *Body or carcass:* It is the reinforcement supporting structure of the hose. Reinforcement can be textile, plastic, or metal, alone or in combination, built into the body of the hose to withstand internal pressures, external forces, or a combination of both.

3. *Tube or lining:* It is the innermost element of the hose and is in contact with the material being carried. The tube may be placed over reinforcing elements. For suitable service, the tube must be resistant to the materials it is intended to convey. The characteristics of the rubber or plastic compound from which the tube is made and the thickness of the tube are based on the service for which the hose is designed.

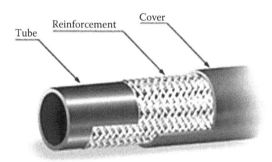

FIGURE 6.1
Construction of hose.

6.3.1 Reinforcement and Its Functions

The purpose of reinforcement is

- Primarily to withstand pressure
- To prevent under vacuum conditions for providing medium against kinking
- To resist against external damage
- To conduct electrostatic charges to the earth
- To increase heat resistance
- To enable couplings to be anchored securely

6.3.2 Fibers Used in Hose Reinforcement

Cotton, asbestos, glass, polyester, nylon, rayon, high-tensile steel wires, and various stainless steels are used as hose reinforcements. Reinforcement materials are chosen depending upon service requirements and economical aspects.

The industrial uses of the hose are innumerable, and cotton is being increasingly supplemented by man-made fiber to give special qualities such as higher bursting strength, higher flex resistance, abrasion resistance, and rot resistance, ease of handling that may be connected with low moisture absorption, greater flexibility, and weight reduction due to using high-tenacity fiber. The reinforcement guarantees the hose integrity during its lifetime. The higher the working pressure is, the more critical the reinforcement becomes. A way to show this is to modify the well-known bursting pressure equation for single braids in such a way that all relevant parameters for the reinforcing material in this equation are represented by one symbol "C" (Table 6.1).

Modified bursting pressure equation,

$$P_b{}^* D_v = 3.7 \ C$$

where
C is $T / \{(1 + E/100)^2 * \sqrt{(Titer/\rho)}\}$, no unit
D_v is the reinforcement diameter (mm)
E is the elongation (%)
P_b is the bursting pressure (N/mm^2)
T is the breaking force (N)
Titer is the weight per length dtex (g/1000 m)
ρ is the specific weight (kg/cu.m)

Rayon and polyamide show the lowest C value. The high elongation of polyamide offsets the higher strength. Polyester is roughly 50% stronger in

TABLE 6.1

"C" Value for Textile Reinforcing Materials

S. No	Reinforcing Material	"C" Range	"C" Mean
1	Rayon	46–88	67
2	Nylon 66	50–85	68
3	Polyester	65–120	93
4	PVA	101–147	124
5	Steel wire	230–280	255
6	Aramid	270–388	329

hose than rayon or polyamide. PVA is on its turn some 25% stronger than polyester, whereas aramid is the strongest reinforcing material. For the same thickness of the reinforcing layer and 100% coverage, aramid offers four to five times more strength than rayon or polyamide.

In case if high adhesion levels are required, on the same level as the other reinforcing materials, polyester and aramid need normally double bath dipping. The moisture sensitiveness is important for the residual strength or corrosion. The high temperature resistance is given by aramid and steel wire. Polyester monofilament fiber reinforcement provides slightly higher pressure resistance and better resistance to hose kinking at a small bend radius. Glass fiber reinforcement offers higher temperature resistance, up to 570°F (300°C) intermittent exposure, but is not recommended for dynamic or high-frequency pulsating pressure applications.

Aramid fiber reinforcement is suitable for dynamic applications up to 570°F (300°C) intermittent exposure. Aramid fiber is the best all-round reinforcement with regard to resilience and high burst pressure but is more expensive than glass or polyester. The Nomex family of reinforcement fabrics was developed for relatively high-temperature applications. Hydrolysis, alkali, and oxidative resistance are all excellent. The Nomex family is considered a premium class of fabrics for rubber reinforcement. Nomex has also been used as a reinforcement fabric in hose construction (Table 6.2).

6.3.3 Yarn Structure in Hose Reinforcement

Yarns are used as reinforcement in hose for reinforcing tube material to provide strength to impart the desired resistance to internal pressure or to provide resistance to deformation, or both. The basic property requirements in yarn used for hose reinforcement are adequate strength, acceptable heat resistance, dynamic fatigue resistance, and satisfactory processability for the various methods of reinforcing hose. Other special properties such as stiffness, adhesion, and conductivity may be developed depending upon the specific hose application. Yarn is available in two basic forms: staple (sometimes referred to as spun yarn) and filament.

TABLE 6.2

Properties of Hose Reinforcement Fibers

Fiber	Properties
Meta-Aramid	Exceptional heat resistance with low shrinkage
Para-Aramid	Exceptional strength with low elongation. High heat resistance
Cotton	Natural vegetable fiber used in hose. Gains strength with increased moisture content. Requires protection against chemical and fungal activities
Glass	Very high strength compared to other fibers. Low elongation; mainly used in high-temperature applications
Nylon	High strength and elongation with good resistance to abrasion, fatigue, and impact. Low moisture absorption and excellent moisture stability. High resistance to fungal activity
Polyester	High strength, good resistance to abrasion, fatigue, and impact. Low moisture absorption and excellent moisture stability. High resistance to fungal activity
PVA	High strength, low shrinkage, and good chemical resistance
Rayon	Similar to cotton in chemical and fungal resistance. Moisture absorption higher than cotton. Dry strength is substantially greater than cotton. Strength is reduced with increased moisture content but retains a wet strength level above cotton

6.3.4 Reinforcement Fabric Structures

Textile fabrics used as reinforcement in hose construction provide the strength to achieve the desired resistance to internal pressure or to provide resistance to collapse, or both. Fabric properties are influenced by fiber used, yarn and fabric construction particulars, and type of weave adopted during production process. The most common weave is known as "plain weave." This is done on a relatively simple loom. Other weaves used, though to a lesser degree, are twill, basket weave, and leno. Leno weave is preferred for fabric that undergo distortion in the hose as in certain types of curved hoses. Leno fabrics also render a means for better adhesion with rubber compared to other weave patterns. The fabric is either frictioned or coated with a thin layer of rubber. Before rubberizing, some fabrics are treated with liquid adhesive (Figures 6.2 through 6.5).

6.3.5 Hose Components Manufacture

6.3.5.1 Tube

6.3.5.1.1 Extruded Tubes

For the tube extrusion process, an uncured rubber or thermoplastic compound ribbon or pellets are fed into the extruder, through the screw or auger with proper temperature controls, and finally forced through a pair of metal dies, where the cylindrical tube is formed. In the noncontinuous process, the tube is then cooled, lubricated to minimize tackiness, and stored in coils on pans, reels, or rigid mandrel poles. Extrusion temperatures are

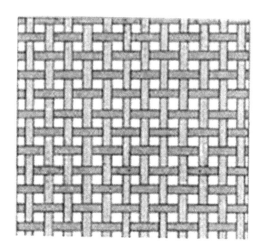

FIGURE 6.2
Plain weave.

FIGURE 6.3
Twill weave.

typically between 200°F and 275°F for rubber compounds and 300°F–600°F for thermoplastics. Normally, extrusion is the preferred method for the tubing process on hoses with IDs up to 1-1/2 in. when built on a flexible mandrel, to 4 in. for rigid mandrel. Beyond these dimensions, wrapped is usually employed.

6.3.5.1.2 Wrapped Tubes

For the larger diameter rigid mandrel rubber hose constructions, the wrapped tube process is utilized. Here, the rubber compound is calendered to a specific thickness and width, then spirally wrapped on the rigid mandrel with sufficient overlap to form the tube. With the wrapped

FIGURE 6.4
Basket weave.

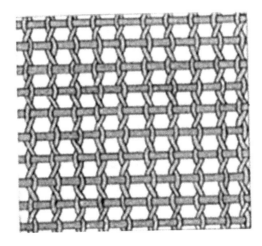

FIGURE 6.5
Leno weave.

process, the challenge is to provide good bonding at the tube overlap area to prevent tube delamination.

6.3.5.2 Reinforcement: Types

The strength component of the hose, designed to handle the entire pressure load with appropriate safety factors, is the reinforcement. In most cases, it is located between the tube and the cover. Occasionally, there are hoses, applications not requiring a cover, in which case the reinforcement also acts as the outer protective layer.

When multiple plies of reinforcement are required to meet working pressure performance levels, typically they are applied one over the other normally separated with a rubber layer (friction or jacket) to fill voids, to prevent adjacent reinforcement abrasion, and to maintain adequate hose component adhesion levels.

Hose reinforcements are either textile, both synthetic polymeric and natural, or wire. Methods of applying these reinforcements are braid, spiral, knit, wrap, and woven. Combinations, such as spiral/knit, are available. Selection of reinforcing equipment is dependent on pressure rating, size, fitting requirements, flexibility, and crush resistance levels.

6.3.5.2.1 Braid Reinforcement

Braiding is probably the most common and traditional method of reinforcing hose. Braid is a complex structure or pattern formed by intertwining three or more strands of flexible material such as textile yarns or wires. Braiding machine may be horizontal or vertical, based on the direction the tube progresses through the machine during braiding process. The major types of braiding machines are "maypole" or tubular type and rotary type. (Figure 6.6).

Maypole-type, as the name implies, braid is formed from multiple carriers, each carrying a reinforcement package traveling in a serpentine maypole fashion generally with a two over–two under pattern. The common carrier varieties available are 20, 24, 36, 48, and 64. They are utilized in vertical or horizontal, single or multiple deck arrangements.

In a rotary braiding machine, the carriers holding yarn package are fixed on two counter-rotating decks and do not rotate in and out in a serpentine path like the maypole type. The deflection of the yarn strands from the outside deck under and over two carriers on the inside deck, repeating the motion continuously during rotation, forms the braided pattern. The rotary braiding machine is faster than the maypole braider. Because of the simpler travel of the carriers, output speeds can be as much as 200% faster than an equivalent maypole type.

6.3.5.2.2 Spiral Reinforcement

This is similar to braiding and is used mainly for the larger bore high-pressure hoses, from 50 up to 200 mm bore diameter. The spiral process is

FIGURE 6.6
Braided reinforcement.

FIGURE 6.7
Spiral reinforcement.

accomplished by wrapping a specified number of yarns through the rotating yarn guide; as the hose passes through at a controlled rate, ideally with sufficient individual yarns or wires being laid together so that the angle of wrap, giving complete cover of the yarns, equals the neutral angle (Figure 6.7).

Spiraling is carried out horizontally with two opposing decks revolving in opposite directions each holding clusters of yarn spindles. Each strand of yarn is fed through an array of tensioning devices to the center point of the decks where they are applied to the tube in a parallel array. The feed is determined by taking the circumference of the hose divided by the desired spacing between the yarns. The linear coverage is determined by the distance the hose moves through the spiral head in one revolution divided by the feed. The spirals should be maintained in multiples of two in order to produce a balanced hose construction that will have minimum distortion under pressure. The yarn angle is a determinant factor in the burst strength of the hose. Spiral is produced by using a continuous line process at a rate of 60–200 feet per minute. Textile spiral is well suited for nonmandrel or flexible mandrel constructions, with low to medium pressure ratings.

6.3.5.2.3 Knit Reinforcement

A flexible hose with knitted reinforcement, consists of at least one inner tubular layer made of polymer material and at least one layer of knitted reinforcement. The mesh wales of the first series of yarns are superimposed on the mesh wales of the second series to define a single-layer reinforcement knitting. The yarn is fed from cone packages (usually four or eight) through a series of guides and latch-type needles onto the hose. The total feed determines the number of courses of knit on the diameter of the hose per revolution. The needles determine the lines of knit linearly with the hose. The number of needles around the circumference of the hose determines the number of loops. Knit is produced by using a continuous line process at a rate of 30–60 feet per minute. Plain stitch reinforcement is produced by a single track cam with 4, 6, or 8 lobes. Lock Stitch is produced by using a double track cam with 8 or 12 lobes. Lock Stitch can run at a higher speed and has less expansion than the plain stitch. Although the knitted hose is easily shapeable for coolant hose applications, it is a very inefficient reinforcing method restricted to low-pressure applications. Nevertheless, the hose with knitted reinforcement is quite sensitive to variations of the internal pressure and reacts to the same by axially rotating, creating a number of difficulties.

6.3.5.2.4 *Wrap Reinforcement*

The multiple plies of reinforcement are wrapped spirally to a rigid mandrel hose tube with the direction of lay reversed with each succeeding ply. The tire cord is the most common fabric reinforcement used, which has strength only in the cord direction. To compensate for its unidirectional strength, plies are usually applied in multiples of two. The rubberized fabric is usually wrapped on the hose tube, thereby resulting in hoses in the lower working pressure range. The addition of a wire or thermoplastic helix or helices to the wrapped construction will prevent the hose from collapse or kinking.

6.3.5.2.5 *Woven Hose*

The reinforcement for woven hose is a seamless tubular textile jacket woven on a loom. This produces a strong, lightweight hose that is flexible for flat storage. Because the longitudinal warp yarns are parallel to the axis, woven hose tends to kink more easily than other hose constructions. Fire hose consists of a tube and seamless circular woven jacket or jackets, either separate or interwoven.

6.4 Hose: Manufacturing

The principal methods used to manufacture hose will be classified as (1) nonmandrel, (2) flexible mandrel, and (3) rigid mandrel, which describe how the various components of the hose are supported during processing into a finished product. Hose is manufactured in the unvulcanized state by forming a cylindrical tube over which a reinforcement and cylindrical cover are applied.

6.4.1 Nonmandrel Style

Nonmandrel hose is manufactured by passing long lengths of extruded tube material through a machine which adds the reinforcement in braided or spiraled layers. This manufacturing method is preferred for lower working pressure (less than 500 psi), smaller diameter (2 in. and under), textile-reinforced hoses not requiring stringent dimensional tolerances. Frequently, low-pressure air is used inside the tube for minimal support, keeping the tube from flattening during the reinforcing process. Most smooth bore thermoplastic hoses are extruded nonmandrel. The higher rigidity of most thermoplastics eliminates the need for mandrel support. Typical hose products in this category would include garden, washing machine inlet, and multipurpose air and water styles.

6.4.2 Flexible Mandrel Style

Flexible mandrels are used for the hoses that require moderate tube-processing support and more accurate dimensional tolerances. The flexible mandrel method combines the long-length advantage of nonmandrel hose with the close inside diameter tolerances and high-pressure ratings of rigid mandrel hose. These mandrels are made of rubber or flexible plastic, sometimes with wire core to minimize distortion. This style process may be used for mid-range working pressures (up to 5000 psi). Of the three flexible mandrel styles, solid rubber offers minimal support, while rubber with wire core and thermoplastic versions provides good dimensional control. In all cases, the flexible mandrel is removed from the hose with either hydrostatic pressure or mechanical push/pull after processing. Examples of this style product are power steering, hydraulic, wire-braided, and air-conditioning hoses.

6.4.3 Rigid Mandrel Style

In larger hose sizes, where flexible mandrels become quite cumbersome to handle, working pressures are high, or stringent dimensional control is required, the rigid mandrel process is the preferred technique. Hose produced by this method is supported on a rigid metal mandrel made up of aluminum or steel. While a rigid mandrel limits the hose length, it ensures good control of the inside diameter. It also offers sufficient support to the tube that either wire or textile reinforcement may be applied at high tensions, which is necessary in high-pressure constructions. The hose tube may be extruded on the mandrel, pneumatically pulled onto the mandrel, or wrapped in sheets onto the mandrel.

6.5 Application of Textile-Reinforced Hoses in Different Sector

Hose types in the market include:

S. No	Hose Type	Percentage in Market
1	Hydraulic hose	25
2	Chemical hose	20
3	Water hose	20
4	Automotive hose	10
5	Gas	9
6	Fire and irrigation	5
7	Petrol	4
8	Food	1
9	Others	6

Of the total hoses, 75% have less than 50 mm bore and the remaining 25% have large bore (more than 50 mm). Of the hydraulic hoses, 90% are used on all sorts and types of mechanical equipment, 5% are for the aeronautical and military end use, and the remaining is for mining equipments.

6.6 Pressure and Bursting Pressure in Hose

The pressure exerted within a hose depends on the volume a pump can push forward and the diameter of the hose. The more fluid that is pushed into the hose, the more it pushes out on the wall. This internal pressure is measured in psi. Greater reinforcements increase the pressure ratings. All hoses have two pressure ratings: a working pressure rating and a bursting pressure rating. The working pressure rating represents common pressures that a hose experiences each time the equipment is operated. Bursting pressure ratings are normally four times higher than working pressure ratings.

Bibliography

ENKA Product information catalogue, the Netherlands, 2003, http://www.polyamide-hp.com/www/download/PHP_Yarn_Properties_2003_09.pdf (accessed May 29, 2013).

Good Year Industrial Hose Catalog, http://www.goodyearep.com/ (accessed May 29, 2013).

Hose Handbook, The Rubber Manufacturers Association, Inc., 2003, pp. 11–24.

Purdue Extension, The selection and inspection of hoses, PPP 89.

Teijin Group, http://www.teijinaramid.com/applications/reinforced-thermoplastic-pipes (accessed May 29, 2013).

Warwickmills, http://www.warwickmills.com/FabricReinforced-Rubber.aspx (accessed May 29, 2013).

Wootton, D. B., *The Application of Textiles in Rubber*, Rapra Technologies Ltd., U.K, 2001, pp. 187–193.

7

Textiles in Transmission and Conveyor Belts

7.1 Introduction

The invention of transmission belt and conveyor belt revolutionized the manufacturing process for many industries around the world. Transmission belts are used as a source of motion, to transmit power efficiently, or to track relative movement. Conveyor belts enable factories to move parts, equipment, materials, and products from one place to another, cutting labor costs and saving time. Conveyor belts appear in many forms in everyday life. The textile structures are used as reinforcement material in both belts mentioned earlier.

7.2 Transmission Belts

A belt drive is a method of transferring rotary motion between two shafts. A belt drive includes one pulley on each shaft and one or more continuous belts over the two pulleys. The motion of the driving pulley is, generally, transferred to the driven pulley via the friction between the belt and the pulley. Transmission belts are classified into V-belt, flat belt, and timing belt.

7.2.1 V-Belts

V-belts are the most widely used belts. V-belt drives achieve drive efficiencies of about 95%. The selection of the type of V-belt depends on the power capacity of the drive and the small pulley's shaft speed (rev/s), acceptable limits of the speed ratio, pitch length of the belt(s), and diameters of the two pulleys, etc. The industrial belts account for 55%–60% of the total market. Through years, vast improvements have been made in the materials used in V-belt construction and in cross-sectional shape as well. Originally, V-belts were manufactured using prime quality cotton cord as tensile members along with natural rubber compounds. Steel cable was introduced as a

V-belt-reinforcing member during World War II. Later, high-tenacity rayons replaced cotton as tensile members because of their much greater strength capacity. Due to the deficiencies of cotton and rayon tensile members, today, polyester, fiberglass, and aramid fibers are the predominant tensile members on all high-capacity V-belts.

7.2.1.1 V-Belt Construction

V-belts are generally manufactured from a core of high-tensile cord in a synthetic rubber matrix enclosed in a fabric-reinforced rubber lining. Woven fabric or cord that is reinforced as ply in the drive belt is primarily made of polyester, nylon, and cotton (Figure 7.1).

The characteristic of the V-belt system is that the belt is wedged in the groove of the pulley but does not bottom in the pulley groove. As the tension is carried by the cords in the top of the belt, when under tension and supported only at the edges, there is a tendency for the center of the belt to be distorted downward.

7.2.2 Flat Belts

Flat power transmission belting in continuous lengths continues to be made in textile-reinforced rubber constructions in steady quantities. There are

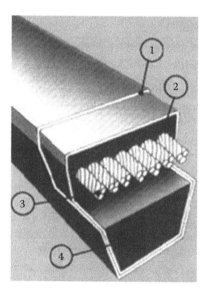

FIGURE 7.1
Cross section of V-belt : (1) cover—protective fabric cover impregnated with rubber, (2) tension section—synthetic rubber compounded to stretch as belt bends, (3) cords—synthetic fiber cords carry the power and minimize stretch, and (4) compression section—synthetic rubber compound to support cords evenly and compress as belt bends.

Fabric
plies

FIGURE 7.2
Flat belt—cross section.

two basic types: plied and solid-woven. These are basically very similar to conveyor belting and are essentially produced in the same manner. The solid-woven constructions have the advantage that they are resistant to delamination and edge fraying and have the best characteristics for holding metal belt fasteners (Figure 7.2).

7.2.3 Timing Belts

Timing belts are parts of synchronous drives that represent an important category of drives. Characteristically, these drives employ the positive engagement of two sets of meshing teeth. Hence, they do not slip, and there is no relative motion between the two elements in mesh. Timing belts are basically flat belts with a series of evenly spaced teeth on the inside circumference, thereby combining the advantages of the flat belt with the positive grip features of chains and gears (Figure 7.3).

The load-carrying elements of the belts are the tension members built into the belts. These tension members can be made of any one of the following:

1. Spirally wound steel wire
2. Wound glass fibers
3. Polyester cords
4. Kevlar

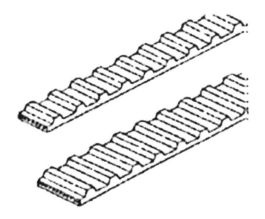

FIGURE 7.3
Toothed belt.

The tension members are embedded in neoprene or polyurethane. The neoprene teeth are protected by a nylon fabric facing, which makes them wear resistant. The contributions of the construction members of these belts are as follows:

1. *Tensile member*—provides high strength, excellent flex life, and high resistance to elongation.
2. *Neoprene backing*—strong neoprene bonded to the tensile member for protection against grime, oil, and moisture. It also protects from frictional wear if idlers are used on the back of the belt.
3. *Neoprene teeth*—shear-resistant neoprene compound is molded integrally with the neoprene backing. They are precisely formed and accurately spaced to assure smooth meshing with the pulley grooves.
4. *Nylon facing*—tough nylon fabric with a low coefficient of friction covers the wearing surfaces of the belt. It protects the tooth surfaces and provides a durable wearing surface for long service.

7.2.3.1 Characteristics of Reinforcing Fibers

a. Polyester
 - Tensile strength—160,000 lb/in.^{2}
 - Elongation at break—14.0%
 - Modulus (approx.)—2,000,000 lb/in.^{2}

One of the main advantages of polyester cord over higher tensile cords is the lower modulus of polyester, enabling the belt to rotate smoothly over small-diameter pulleys. Also the elastic properties of the material enable it to absorb shock and dampen vibration. In more and more equipment, stepping motors are being used. Polyester belts have proven far superior to fiberglass- or Kevlar-reinforced belts in these applications. High-speed applications with small pulleys are best served by polyester belts under low load.

b. Kevlar
 - Tensile strength—400,000 lb/in.^{2}
 - Elongation at break—2.5%
 - Modulus—18,000,000 lb/in.^{2}

High tensile strength and low elongation make this material very suitable for timing belt applications. Kevlar has excellent shock resistance and high load capacity.

c. Fiberglass
- Tensile strength—350,000 lb/in.^{2}
- Elongation at break—2.5%–3.5%
- Modulus—10,000,000 lb/in.^{2}

The most important advantages are

1. High strength
2. Low elongation or stretch
3. Excellent dimensional stability
4. Excellent chemical resistance
5. Absence of creep, 100% elongation recovery

The disadvantages are

1. High modulus (difficult to bend).
2. Brittleness of glass. Improper handling or installation can cause permanent damage.
3. Poor shock resistance. No shock-absorbing quality when used in timing belts (Table 7.1).

7.3 Conveyor Belts

Belt conveyors have been used for more than 150 years. The basic design of this common manufacturing apparatus consists of a rubber belt stretched between two cylindrical rollers at either end. Conveyor belts enable factories to move parts, equipment, materials, and products from one place to another, cutting labor costs and saving time. Belting was therefore "tailored" to suit different applications such as larger particle sizes that would introduce greater impact loads into the belt; higher material temperature resistance to convey warm products in process plants; and oil-resistant belting that is suitable for transporting oil-contaminated products including foodstuff.

7.3.1 Conveyor Belt: Construction

Conveyor belts have two basic components: the carcass, or strength member, and the rubber, which protects the carcass. As the heart of the conveyer system is the conveyer belt, so too is the carcass the heart of the latter, which has to meet the stresses already mentioned (Figure 7.4).

TABLE 7.1

Comparison of Different Belt Reinforcement Materials

Belt Requirements	Nylon	Polyester Cont. Fil. Yarn	Polyester Spun Yarn	Kevlar–Polyester Mix	Kevlar Cont. Fil. Yarn	Kevlar Spun Yarn	Glass	Stainless Steel	Polyester Film Reinforcement
Operate over small pulley	E	G	E	F	P	F	P	P	G
High pulley speed	E	E	E	F	P	F	P	P	G
High intermittent shock loading	F	G	G	E	E	E	P	G	F
Vibration absorption	E	G	E	G	F	F	P	P	F
High torque low speed	P	P	P	F	G	F	E	E	F
Low belt stretch	P	P	P	P	G	F	E	E	G
Dimensional stability	P	P	P	F	G	G	E	E	G
High temperature 200°F	P	P	P	P	E	E	E	E	F
Low temperature	F	G	G	G	G	E	E	E	G
Good belt tracking	E	G	E	G	F	G	F	P	E
Rapid start/stop operation	F	G	E	G	P	G	P	E	G
Close center-distance tolerance	P	P	P	P	G	F	E	E	G
Elasticity required in belt	E	G	E	G	P	P	P	P	P

E, Excellent; G, Good; P, Poor.
Source: Courtesy of Chemiflex, Inc.

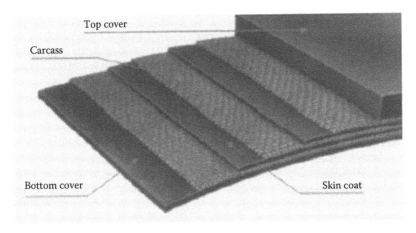

FIGURE 7.4
Cross section of conveyor belt.

There are two basic categories into which all conveyor belting falls: fabric belting and steel cord belting. Externally, both types of belt appear to be identical; however, the difference is in the internal structure of the belt. The internal structure or carcass of a belt dictates the tensile strength of the belt. In fabric belts, the carcass comprises "plies" or mats of reinforced fabric separated by cushioning layers. Steel cord belts on the other hand have a series of steel cables embedded into the belt, separated by rubber. When tension is applied to the belt, the carcass absorbs the force. The greater the required tensile force to move the transported material, the greater is the required strength of the belts' carcass. In both fabric and steel cord belts, the carcass is covered by rubberized covers to protect the carcass or cables.

7.3.2 Conveyor Belt: Types

1. *Fabric plied belting*
   Fabric plied belt consists of a single- or multilayered series of synthetic fabric layers interlaced between rubber-based shock-absorbent layers. The "top" and "bottom" sides of the belt consist of hard wearing, abrasion-resistant and cut-resistant rubber covers. These covers protect the belt from damage, especially at the loading points of the conveyor (Table 7.2).

2. *Steel cord belting*
   Steel cable belting consists of steel and rubber only. Steel cable belts consist of steel cables manufactured from high-tensile steel wire. These steel cables are surrounded by a layer of high-grade rubber to facilitate adhesion to the outer covers and to improve lateral tear resistance.

TABLE 7.2

Properties of Fibers Used as Reinforcement in Conveyor Belts

Property	Aramid	Nylon 66	Polyester	Steel
High belt strength	Good	Moderate	Moderate	Good
Dimensional stability	Good	Poor	Moderate	Good
Impact resistance	Moderate	Good	Good	Poor
Flexibility	Good	Good	Good	Poor
Low belt weight	Good	Moderate	Moderate	Poor
Low belt thickness	Good	Moderate	Moderate	Poor
Flame resistance	Good	Moderate	Moderate	Good
Hygienic test	Good	Good	Poor	Good
Rust resistance	Good	Good	Good	Poor

7.3.3 Conveyor Belt: Property Requirements

a. *Breaking strength*

An important design parameter is the breaking strength, that is, the force required to break the yarn. This force is expressed in Newton (1 kg force = 9.81 N). Since the total breaking strength is influenced by the yarn count, calculations are based on the specific strength or tenacity, which is expressed in mN/tex. The relative strength is expressed both in tenacity (mN/tex) and in tensile strength (N/mm²). Belt breaking strength is the parameter that decides the selection of the conveyor belt. The belt breaking strength can be calculated as

$$B_s = \frac{C_r \times P_p}{C_v \times V}$$

where
   B_s is in Newton
   C_r is the friction factor
   C_v is the breaking strength loss factor
   P_p is the power at drive pulley in Newton
   V is the belt speed in m/s

The breaking strength (in longitudinal direction) of a belt is largely determined by the type and amount of reinforcement material used in longitudinal direction. The safety factor of a conveyor belt system is 6–10. It varies with the type and size of the material to be transported, the method of loading, belt splicing, and the condition of the conveyor installation. The maximum working load is about 10%–15% of the breaking load of the conveyor belt.

b. *Elongation*

Elongation at break (i.e., the percentage elongation at the moment when the yarn breaks) is also a frequently used property. The load at a certain elongation is known as the modulus.

c. *Resistance to atmospheric conditions*

Conveyor belt has to be resistant to atmospheric conditions such as heat, light, mildew, etc. Cover material plays the essential role in meeting the various environmental conditions. Nylon and polyester, in contrast to cotton and rayon, are fully resistant to rotting and the influence of moisture; steel will corrode after a period of time. A distinction should be made between brass-coated or galvanized steel. Brass-coated steel is affected a lot sooner by moisture. In cut-edge belts, the reinforcement also shows sufficient resistance to atmospheric conditions.

d. *Growth*

The criterion of low growth of a belt arises from the required high-dimensional stability in longitudinal direction of the belt, which gives it a service life without frequent delays for readjusting the correct pretension to avoid slippage between belt and driven pulley. The belt growth is determined by the reinforcement of the belt in longitudinal direction by factors such as type of reinforcement, fabric construction, and heat treatments. In underground mining, low growth is essential because only limited space is available for take-up device.

e. *Impact resistance*

High impact strength is required to absorb the impact forces on the belt in the loaded area. This impact resistance is proportional to the breaking energy of the belt. Breaking energy is determined by stress–strain behavior of the reinforcing material. Construction of the top cover also plays an important role.

f. *Troughability*

In order to increase the loading of a conveyor belt, idler angles up to 45° are applied. Good troughability in transverse direction is essential for the conveyor belt. Troughability of the conveyor belt is influenced by the following factors of the reinforcement fabric used:

- Weft yarn
- Fabric construction
- Number of plies
- Distance between the plies
- Matrix material

g. *Bending resistance and buckling*

The requirement of a low bending resistance in longitudinal and transverse directions and no buckling are directly related to the durability of the belt. To ensure a long service, no compression ribs caused by the compression of the inner fabric should occur in rounding the pulleys. Buckling occurs only if more than one reinforcement layer is used. Buckling can be overcome by the correct choice of

- Warp yarn (modulus)
- Fabric construction (warp crimp)
- Belt construction (distance between the number of layers)

h. *Effective mechanical splicing*

Mechanical fasteners are used for light belts and in those cases where the length of the conveyor belt has to be changed frequently. Mechanical fasteners can be used only if the belt has been reinforced with fabrics. The mechanical splicing properties are determined by fabric type, weft yarn, and weft density. The efficiency of joints made with mechanical fasteners relating to the nominal belt strength shows figures up to 90%.

i. *Adhesion*

In a conveyor belt system, the force required to move the belt is transmitted from the driven pulley via a matrix interlayer to the reinforcing cords, cables, or fabrics. Furthermore, when the belt is in operation, it is continuously subjected to flexing in both longitudinal (bending over pulleys) and transverse (troughing) directions. In order to overcome the flexing, adhesion between the reinforcement and the matrix material should be adequate to prevent stripping of covers or ply separation (Table 7.3).

TABLE 7.3

Factors Influencing Properties of Conveyor Belt

Properties	Influencing Material in Conveyor Belt
Strength	Fabric warp and weft
Growth	Fabric warp or cord or cable
Impact resistance	Fabric warp and weft, matrix
Troughability	Fabric weft, matrix
Bending resistance and buckling	Fabric warp, matrix
Fastener holding	Fabric weft, matrix
Resistance to atmospheric conditions	Matrix
Adhesion	Fabric warp and weft, matrix
Safety requirements	Matrix

7.3.4 Carcass Constructions

Conveyor belt fabric is made of warp yarns that run lengthwise and weft yarns or filling, which run transversely. Four types of weave patterns are commonly used: plain weave, straight-warp weave, solid-woven weave, and woven-cord weave.

1. *Plain weave:* Most belt carcasses are formed with a plain weave, that is, the warp and weft yarns alternating across each other. Since the warp yarns are the tension- or load-carrying members, the fabrics are designed with the dominant strength in this direction. This can be accomplished by using a greater number of ends per unit width of warp; by using larger, stronger yarns in the warp; or, in some instances, by using a combination of both of these techniques (Table 7.4).

2. *Straight-warp weave:* In this weave, the tension-bearing warp yarns are essentially straight with little or no crimp. Fill yarns are laid above and below the warps, and the warps and the fills are held together with binder warp yarns.

3. *Solid-woven weave:* This weave consists of multiple layers of warp and fill yarns held tightly together with binder warp yarn. This is a carcass where three plies are interwoven to form a single one. The center-ply made from high-tensile polyamide or polyester carries the tension load and is protected by top and bottom plies made from wear-resistant cotton, thus serving in a way as protective covers. It is usually impregnated and covered with PVC with relatively thin top and bottom covers, generally less than 1 mm. Abrasion resistance is provided by the combination of PVC and the top and bottom yarns of the fabric. Being fire resistant, it generally serves for underground coal mining only (Tables 7.5 and 7.6).

TABLE 7.4

Reinforcement Fabric Weave Particulars

Type of Weave	Derivatives	Fabric Weight
Single weaves	Plain 1/1	Under 1000 g/m^2
	Plain 2/1	Medium 1000–1500 g/m^2
	Plain 3/1	Heavy, over 1500 g/m^2
Double weave	—	Over 1000 g/m^2
Multi warp systems	—	Over 1000 g/m^2 (straight-warp fabrics)
	—	Over 1500 g/m^2 (solid-woven fabrics)

TABLE 7.5

Properties of Different Reinforcement Fabrics

Properties	Cord Fabric	Straight-Warp Fabric	Solid-Woven Fabric	Cable
Strength efficiency	Moderate	Moderate	Poor	Good
Transverse strength	Poor	Good	Good	Poor
Splice performance	Poor	Poor	Moderate	Good
Ease of dipping	Good	Good	Moderate	Moderate

7.3.4.1 Factors Influencing Carcass Selection

To select the optimum plied belt carcass, five properties must be considered:

1. The belt width.
2. The service conditions under which the belt will operate.
3. The maximum operating tension (T_{max})—both steady-state condition and peak.
4. The minimum number of plies required to support the load.
5. The maximum number of plies beyond which transverse flexibility is reduced and the troughing efficiency is affected. This varies with the belt width, trough angle, and the idler roll arrangement.

7.3.5 Belt Covers

Belt covers may be thought of as protection for the belt carcass. The covers must withstand the wear, cutting, and gouging of the material being conveyed. The covers also may have to resist heat, oil, or chemical deterioration. In a few cases, these conditions may be so moderate that no protection and no belt covers are required. Rubber or rubber-like compounds are used for the top and bottom covers of conveyor belting and for banding together various components of the belt carcass. These compounds are provided by mixing rubbers or elastomers with various chemicals in order to obtain reinforcement and to develop the physical properties necessary for service conditions.

7.3.6 PVC Impregnation

After weaving, the roll of carcass is vacuum impregnated with PVC plastisol containing a careful blend of polymer, plasticizers, stabilizers, fire retardants, and special additives, with special attention being given to viscosity control in order to ensure full impregnation of the woven structure. While the textile elements fix many of the belt properties such as tensile strength and elongation in service, the properties of the plastisol are equally

TABLE 7.6

Carcass Materials: Particulars

Carcass Type	Carcass Materials		Strength Range Kilonewtons per Meter Width	Features and Applications
	Warp (Longitudinal)	Weft (Transverse)		
Plain weave	Polyester	Nylon	315–2000 kN/m (150–400 kN/m/ply)	Low elongation Very good impact resistance Good fastener holding An excellent general-purpose fabric
Crow's foot weave	Polyester	Nylon	630–2500 kN/m (315–500 kN/m/ply)	Low elongation Good impact resistance Very good fastener holding Excellent rip resistance For high-abuse installations
Double weave	Polyester	Nylon	900 and 1350 kN/m (450 kN/m/ply)	Low elongation Excellent impact resistance Excellent fastener holding For high-abuse installations
Plain weave	Polyester	Polyester	Up to 900 kN/m (120 and 150 kN/m/ply)	Used in special applications where acid resistance is needed
Plain weave	Nylon	Nylon	Up to 2000 kN/m (150–450 kN/m/ply)	High elongation, mostly replaced by polyester–nylon Used in special applications where low modulus needed or in high pH environment
Solid woven	Nylon/cotton or polyester/cotton	Nylon/cotton	600–1800 kN/m	Main use in underground coal mining Good fastener holding and impact resistance Used for bucket elevators
Steel cord	Steel cord	None (special reinforcement available)	500–7000 kN/m	Very low elongation and high strength Used for long haul and high-tension applications
Aramid nylon (Kevlar)	Polyaramide	Nylon	630–2000 kN/m	Low elongation, high strength, low weight Used on high-tension applications and on equipment conveyors

Source: Dunlop, F., *Conveyor Belting—Technical Manual*, 2011.

important, and its formulation will influence not only the fire performance properties but also operational factors such as troughability and the ability to hold fasteners.

7.3.7 Conveyer Belts: Applications

a. *Transbelt*

Transbelt moves loose and lump materials under typical working conditions. The carcass is made of polyamide or polyester fabrics.

b. *Steel belt*

Steel belt transports loose and lump materials on heavy-duty conveyors over long distances and under arduous conditions. The carcass is a high-strength steel cord placed in one plane.

c. *Shock belt*

Shock belt transports unsorted lump materials under heavy-duty conditions with an improved resistance to impact fracture. The carcass consists of one or two plies. The warp is made of a combination of polyester and polyamide. The shock belt series with a breaker includes conveyor belts with a textile or steel cord carcass, protected by one or two textile breakers.

d. *Fire belt*

Fire belt is specially designed to handle loose and lumpy materials in explosion-hazardous locations. The fire belt should meet all requirements of fire protection standards and has improved fire resistance and antistatic properties. The carcass is formed from rubber supplies textile plies of polyamide and steel cords placed in one plane.

e. *Thermal belt*

Thermal belt transports hot lump and bulk materials of up to 175°C. The carcass is formed from polyamide or polyester plies and protected by rubber covers designed to withstand thermal stress generated by hot materials.

f. *Frostbelt*

Frostbelt transports materials at very low operating temperatures or handles of frozen materials up to −60°C. Two types of frostbelts are produced: with a rubber textile carcass or with steel cord reinforcement.

e. *Ecotubelt*

Ecotubelt transports ecologically dangerous materials that cause severe pollution of the environment. The carcass consists of two or three polyamide or polyester plies. The ecotubelt covers convey abrasive materials at temperatures from −25°C to 60°C.

f. *Oil belt*

Oil belt transports oily and greasy materials containing nonpolar organic solvents and fuel. The carcass is formed from polyamide or polyester plies.

g. *Chemical belt*

Chemical belt transports materials containing inorganic acids and bases. The carcass is formed from polyamide or polyester plies.

7.3.8 Specialty Conveyor Belts

a. *High-performance aramid belts*

Conveyor belts reinforced with aramid fibers are as light as other synthetic fibers like polyester or polyamide, but as strong as steel. It has low elongation, no creep, and excellent resistance against heat and chemicals. If increased impact resistance is required, aramid conveyor belts can be produced with additional polyamide breaker plies. The cord fabric consists of straight aramid cords in longitudinal direction. The straight-warp fabric contains additional transverse polyamide cords that protect the aramid cords from both sides. As there is a single fabric ply only, the carcass is light and flexible with optimum strength utilization. Aramid conveyor belts are fatigue resistant throughout their lifetime. The reinforcement does not corrode or rot and is resistant against chemical influences.

b. *Low-noise silent belts*

Low-noise silent belts are used in areas where noise levels must be kept to a minimum. These conveyor belts are equipped with a highly wear-resistant special fabric that makes an extremely low noise level possible due to its exceptional construction and special yarns used. In passenger areas, noise level of ~55 dB (A) must be often maintained. Noise abatement and labor regulations also stipulate noise levels <60 dB (A) for other work areas.

c. *Flame-retardant/self-extinguishing belts*

Flame-retardant belts are designed to keep fire from spreading. All over the world and to an ever-increasing degree, conveyors are being used to carry dangerous goods. Furthermore, conveyors are often installed near potential hazards. Flame-retardant belts provide an added margin of safety, both in industry (e.g., chemical manufacture) and in public facilities.

d. *High friction incline belts*

This type of belt guarantees smooth operations and a more efficient flow of goods and no slippage of products. The conveying angle is dependent upon a variety of factors such as the type of goods being conveyed, the top face coating or surface texture of the belting used, and external influences like dust, moisture, etc. These belts have top face coatings and patterns that are suitable for incline conveying.

e. *Highly conductive conveyor belts*

With their highly conductive surfaces, these belts carry sensitive electronic components safely and ensure smooth and efficient operations. Conveyor belts made of synthetics are prone to accumulate electrostatic charges on their surfaces and in their carcass. The amount of static buildup depends upon various factors such as

- Relative humidity of ambient air
- Belt speed
- Type of material of which belt is made
- Amount of relative motion between belt and underlying surface(s)

With their highly conductive surface coatings and carcass fabrics, these belts are designed to dissipate static charges before they can accumulate. The degree of conductivity of a material is an intrinsic property of the material's electrical resistance.

Bibliography

Chemiflex, http://www.chemiflex.com/technical-bulletins (accessed May 29, 2013).

CKIT: The Bulk Materials Handling Knowledge Base, *Troughed Conveyor Belting–Beginner's Guide*, http://www.ckit.co.za/secure/conveyor/troughed/belting/belting_basics.html (accessed May 29, 2013).

Davies, G., *Aspects of Conveyor Belting*, SAIMH, South Africa, http://www.saimh.co.za/beltcon/beltcon1/paper14.html (accessed May 29, 2013).

Dunlop, F., *Conveyor Belting—Technical Manual*, 2011, http://www.dunlopconveyorbelting.com/textile-belting/ (accessed May 29, 2013).

Matador Conveyor Belts, Slovakia, http://www.miningworld-centralasia.com/pages/documents/Matador.pdf (accessed May 29, 2013).

Scribd, http://www.scribd.com/doc/16607087/Timing-belts (accessed May 29, 2013).

SDP/SI, http://www.sdp-si.com/D265/PDF/D265T010.pdf (accessed May 29, 2013).

Senthil Kumar, R., Conveyor belts, Scribd.com, http://www.scribd.com/doc/31189457/Conveyer-Belt (accessed May 29, 2013).

Sommer, J. G., Engineered Rubber Products, Hanser Publications, Cincinnati, OH, 2000.

Woottoon, D. B., *The Application of Textiles in Rubber*, Rapra Technology Ltd., Shropshire, U.K., 2001.

8

Textiles in Ropes

8.1 Introduction

Ropes are one of the oldest human artifacts. It has been in use for a long time and has played a useful role in the progress of civilization. One of the earliest examples of artificial cordage is a piece of a fishing net made 10,000 years ago in Mesolithic times. Most of the early ropes were relatively short and hand twisted or braided. The expansion of shipping and the increase in ship size drove the necessity for longer ropes. It has a unique property of resisting large axial load in comparison to bending and torsional loads. A variety of natural fibers has been used as a basic element in rope construction for centuries. Modern climbing ropes came about with the production of high-grade nylon in the 1950s and 1960s. Prior to nylon, most ropes were made from natural fiber, like manila, hemp, or sometimes silk. Today, with the advent of synthetic fibers, the field of application has widened with associated improvement in performance. Nylon allowed construction of lighter weight ropes having great impact absorption and the ability to hold upward of 5000 lb. With the progress in manufacturing technology, newer constructions are being made with improved performance. As a result, a variety of rope now finds its way in household, industrial, civil construction, and defense sectors for performing a variety of task. Textile fiber rope has advantages over its metallic counterpart for a number of reasons, and it is increasingly being preferred in many applications over metallic ropes.

8.2 Ropes: Definition and Types*

Rope: Rope is defined as a product obtained when three or more strands are twisted or braided or paralleled together to provide a composite cordage article larger than 4 mm in diameter.

---

* Courtesy of Denton, M. J. and Daniels, P. N., *Textile Terms and Definitions*, The Textile Institute, Manchester, England, 2002.

Braided rope or sennit rope: A cylindrically produced rope made by inter-twining, maypole fashion, several to many strands according to a definite pattern with adjacent strands normally containing yarns of the opposite twist.

Cable-laid rope: A rope formed by three or more ropes twisted to form a helix around the same central axis. The ropes that become the secondary strands are "S" lay and the finished cable is "Z" lay, or vice versa.

Combined rope: A rope in which the strand centers are made of steel and in which the outer portions of each strand are made from fibrous material.

Double-braided rope: A rope in which a number of strands are plaited to form a core and around which are plaited further strands to form a sheath. The core lies coaxially within the sheath.

Eight-strand plaited rope: A rope normally composed of four pairs of strands plaited in a double four-strand round sennit.

Hard-laid rope: A rope in which the length of lay of the strands and/or the rope is shorter than usual, resulting in a stiffer and less flexible rope.

Hawser-laid rope: A rope in which the single and first-ply twist are in the same direction and the second-ply twist is in the opposite direction, and S/S/Z or Z/Z/S construction.

Laid rope: A rope in which three or more strands are twisted to form helixes around the same central axis.

Shroud-laid rope: A four-strand rope with or without a core with the strands twisted to form a helix around the central axis.

Soft-laid rope: A rope in which the length of lay of the strands and/or the rope is longer than usual resulting in a more flexible rope that is easily deformed.

Spring-lay rope: A rope made with six strands over a main core, each strand of which has alternating wire and fiber components laid over a fiber core.

8.3 Fibers Used in Rope Construction

Every sort of flexible strand has been used to make ropes in some place at some time. The choice of fiber depends on (1) the performance requirements for a particular end use such as strength, durability, flexibility, and handle, and (2) availability and economics. Both natural and synthetic fibers are used in rope. Cotton was used for cheaper, soft ropes, where strength and durability were less important. Synthetic fibers can be used either in staple or continuous filament form. Some typical values of fineness, extension at break, and tenacity of most commonly used fibers are given in Table 8.1.

The primary materials used for making ropes are fibers. The positive and negative attributes of natural and synthetic fibers used in ropes are given in Table 8.2.

TABLE 8.1

Properties of Fiber Used in Rope

Fiber	Fineness dtex	Extension (%)	Tenacity (cN/dtex)
Cotton	1.5	8	3.6
Flax	1.6	5.9	
Jute	1.5	3.0	
Abaca	139–273	2.8	5.8
Henequen	362–383	4.7	2.7
Coir	180–540	42	1.2
Polyethylene	>5	20	3
Polypropylene	>5	20	3
Polyester	>6	15	8
Nylon	>6	20	8.5
Kevlar (Aramid)	<2	<5	20

TABLE 8.2

Fibers: Positive and Negative Attributes

Fiber	Positive Attribute	Negative Attribute
Polyamide	High extension, elastic, flexible, high energy-absorption capacity, good abrasion resistance in dry state, little diameter ratio restriction while working on sleeves or pulleys	Attribute poor abrasion resistance in wet condition, swelling in water results in strength loss by 10%–20%, very high extensibility under large load in the case of twisted structure, possibilities of kink formation after a sudden retraction
Polyester	Strong, elongation less than nylon, wet abrasion resistance, and fatigue behavior better than nylon	
Polypropylene	Cheaper than nylon and polyester, density less than water, soft in handling	30% weaker than nylon and polyester, poor fatigue resistance, degrades in sunlight, poor creep behavior, necessitates larger and heavier rope for similar application
Kevlar	High strength-to-weight ratio, high modulus, extension less than nylon and polyester but more than steel wire, remains unaffected by conventional corrosion in ocean, low creep, nonconductivity	Extension less than nylon and polyester, poor compressive properties, poor abrasion resistance, low damage tolerance
Natural fibers	Biodegradable, cheap	Not durable, weaker than their synthetic counterparts, absorbs water, poor mildew resistance

The ropes made out of these fibers are especially suitable for certain end uses/application as stated as follows.

1. *Nylon rope:* Suitable for climbing since it gives adequate protection from fall due to high energy-absorption capacity and less peak load to be experienced by the body. Good for accommodating high-amplitude motion in mooring.

2. *Polyester rope:* Suitable for mooring application for ship and buoy, hauling, lifting, cable recovery, supporting antenna, etc.

3. *Polypropylene rope:* Suitable for those applications where the rope is demanded to remain in floating condition in water.

4. *Kevlar rope:* Its high modulus facilitates anchoring floating oil platform in sea in place. It can be used as a replacement of wire rope in suspension bridge.

8.4 Rope Construction

Ropes are structures made of textile fibers. A rope is made of several strands twisted, plaited, or braided together to form a coherent assembly. The particular arrangement of the strands that makes up the structure of the rope is called the construction. The strands are made of yarns that consist of fibers, filaments, or tapes. Nylon was first used in the traditional three-strand construction. Synthetic fibers are continuous filaments of effectively infinite length. In principle, a collection of parallel filaments could act as a tension member. If fibers lie at an angle to the rope axis, there is inevitably a loss of strength and stiffness, compared to the fiber values. Essentially, the characteristics of rope are a function of the fiber, or raw material, from which it is made, and the way those fibers are formed together, or its construction. Broadly speaking, there are two common methods of construction—twisting and braiding, and each results in rope with very different characteristics.

8.4.1 Twisted Ropes

The twisting of bundles of individual yarns together to form three strands that are then again twisted themselves together to form the twisted ropes. As the successive strands are twisted together, the direction of the twisting is alternated such that the torque resulting from twisting in one direction is balanced against the torque resulting from twisting in the other direction, thereby counteracting the tendency of the three strands to unwind. These ropes are recognized by their spiral shape. Some larger ropes may be made

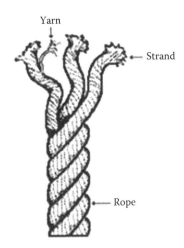

FIGURE 8.1
Twisted rope construction.

up of more than three strands. Good construction design, and balanced twisting, will spread load evenly over all three strands (Figure 8.1).

Twisted ropes are typically less expensive than braided ropes because the manufacturing process is faster. They are easily spliced. Despite the balancing of torque achieved by alternating the direction of twist, they do nevertheless retain some torque and do have the tendency to hockle and rotate under load.

8.4.2 Braided Ropes

Braided ropes come in various forms and patterns but always consist of bundles of fibers that are formed into strands, which are then intertwine or twined together by passing each strand over and under other strands. All braided ropes, including 8-strand and 12-strand single braids, double braids, and core-dependent double braids, are constructed from an equal number of "S-strands" and "Z-strands". This creates a balanced, or torque-neutral construction that will not naturally twist while under load. Braiding produces round rope structure compared to twisted ropes which produces spiral shape. This makes them well suited for use with hardware such as pulleys, winches, and rope grabs (Figure 8.2).

These ropes are inherently relatively torque free and nonrotating. Braided ropes are more expensive than twisted ropes due to slower process of braiding. There are a number of variables, which the manufacturer can utilize to alter characteristics such as strength, elongation, flexibility, and durability. The following will describe the main characteristics of the common types of braided ropes.

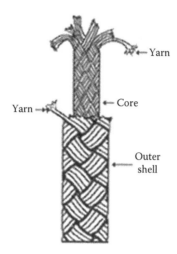

FIGURE 8.2
Braided rope construction.

FIGURE 8.3
Solid braid construction.

8.4.2.1 Solid Braid Ropes

Solid braid ropes are also called as "sash cord" because this pattern was used in sash windows. It is formed by braiding 12 or 18 strands in a reasonably complicated pattern with all the strands rotating in the same direction on the braider. The individual stitches are oriented in the same direction as the rope (Figure 8.3).

The center may contain a filler core. These ropes perform exceptionally well in pulleys and sheaves due to the control of round shape. They tend to have high elongation and are generally less strong than other forms of construction and are difficult to splice.

8.4.2.2 Diamond Braid

Diamond braid or single braid or hollow braid is produced on the braiding machine by rotating one half of the strands in one direction while the other half is rotating in another direction. Braid is formed as the strands cross alternately over and under each other (Figure 8.4).

Braiding pattern created by this method is simple and efficient, although the rope tends to flatten quite a bit. A filler is incorporated into the core of

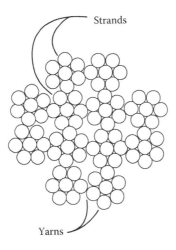

FIGURE 8.4
Diamond braid construction.

the rope to maintain the rope rounder and firmer or to produce a rope of a desired size. This addition of filler may affect other characteristics of the rope and is more common on smaller ropes that have less critical applications.

8.4.2.3 Double Braid

Double braid is made of two hollow braided ropes, one inside the other. The core is made up of large single yarns in a slack braid. The cover is also made up of large single yarns but in a tight braid that compresses and holds the core. The inner rope and outer rope are designed to share the load fairly evenly. Double-braided rope is usually made of a combination of man-made materials. Nylon, polyester, and Dacron are often used due to their resilience and resistance to wear and UV radiation. These ropes are generally very flexible and strong and easy to eye splice these ropes (Figure 8.5).

Double braid ropes are widely used in ships, sailing vessels, horse leads, oil fields and fire rescue operations. "Milking" is a condition closely associated with double braid rope, which is the slippage of the core and sheath of a rope in opposite directions. When double braid ropes are run over pulleys, where the outer rope may slide along on the inner rope and bunch up leads to "milking". This condition will cause dramatic loss of strength due to the

FIGURE 8.5
Double braid.

FIGURE 8.6
Kernmantle rope.

inner rope bearing the entire load, because the sheath is bunched up and not under the same tension as the inner rope.

8.4.2.4 Kernmantle Ropes

Kernmantle ropes are made by braiding a cover (mantle) over a core (kern). The core may be made of filaments of fiber lying essentially parallel inside the rope or it may be twisted into little bundles much like miniature twisted ropes. In some cases, it is made of small braided ropes. Kernmantle ropes are always designed, however, so that the inner core is taking most of the load (often all of the load) and the outer cover serves primarily to protect the inner core. If "milking" occurs on these ropes, therefore, it does not affect strength very much because the rope is designed such that the inner core is the load-bearing member (Figure 8.6).

These ropes are very strong and durable, and can be made to have very low elongation. By having the load-bearing part of the rope inside the protective outer cover, it is well protected from abrasive action, dirt, and ultraviolet rays, which are harmful to all ropes. All other forms of ropes have the load-bearing fibers exposed, thereby resulting in more rapid deterioration.

8.5 Properties of Rope

8.5.1 Rope Size

Rope size is best expressed in terms of its linear density (often called "rope weight"), that is, mass per unit length.

Rope linear density in kg/m = 10^{-6} × Number of textile yarns in rope

× Linear density of textile yarns in Tex

× Contraction factor due to twisting or braiding

The contraction factor is a measure of the reduction in length as the yarns follow helical paths in the rope.

The effective linear density for ropes submerged in water is given as

$$\text{Submerged linear density} = \frac{(\text{Fiber density} - \text{Water density})}{\text{Fiber density}}$$

where water density = 1.00 for freshwater and 1.04 for seawater (g/cm³). The relations to other dimensions are

$$\text{Rope density}\left(\text{kg/m}^3\right) = \frac{\text{Rope linear density}\left(\text{kg/m}\right)}{\text{Rope area}\left(\text{m}^2\right)}$$

Rope area is determined from nominal diameter or circumference as listed in specifications or by measurement done by wrapping a narrow tape around the irregular outer surface designating the value as the circumference of the rope and calculating the area or diameter as if it were a circle.

$$\text{Packing factor} = \frac{\text{Rope density}\left(\text{kg/m}^3\right)}{\text{Fiber density}\left(\text{kg/m}^3\right)}$$

where the packing factor is the fraction of the rope's nominal cross-sectional area occupied by fiber.

8.5.2 Rope Strength

Tensile strength is the most desirable property in a rope since it has a strong association with durability. The strength of a rope mainly depends upon raw material, construction, and test environment. Fiber rope engineering usually normalizes strength in terms of weight, so that strength and weight should be considered together. Rope mass is determined by weighing a sample of rope whose length has been measured at a reference load.

The mass of a length of rope is determined from

$$\text{Total rope mass(kg)} = \text{Rope linear density(kg/m)} \times \text{Rope length(m)}$$

For most ropes, the load is calculated as

$$\text{Reference load}\left(\text{kg}\right) = \frac{D^2}{8}$$

where D is the rope nominal diameter (mm).

The following are the important relationships with respect to strength:

$$\text{Breaking stress}\left(\text{in MPa}\right) = \frac{\text{Breaking load}}{\text{Fiber area}}$$

$$\text{Fiber area}\left(\text{m}^2\right) = \frac{\text{Rope linear density}\left(\text{kg/m}\right)}{\text{Fiber density}\left(\text{kg/m}^3\right)}$$

$$\text{Specific strength}\,(\text{rope tenacity})\,\text{in N/tex} = \frac{\text{Breaking load}\,(N)}{\text{Linear density}\,(\text{tex})}$$

$$\text{Rope break load in MN} = \text{Linear density}\,(\text{kg/m}) \times \text{Tenacity in N/tex}$$

$$\times \text{Strength conversion efficiency}\,(\text{decimal})$$

Rope strength decreases with the amount of twist induced into the rope. The effect of twist varies with the fiber type, diameter, and construction of the rope. The three-strand laid rope shows lowest strength due to the presence of large twist used in making the rope. Braided rope shows higher strength compared to three stranded ropes due to the presence of less twist in the structure. Parallel arrangement of strands or filaments in core produces the strongest rope, since the load is evenly shared among the strands of the rope. As the twist adversely affects strength, a tightly twisted hard-laid rope will be weaker than the corresponding normal-laid rope.

Rope of any type exposed to elevated temperature or heat has a pronounced effect on the strength of rope because it either causes fiber embrittlement or softening of the polymer. The ropes made up of natural fibers and polyolefin-based fibers are prone to progressive strength loss at elevated temperatures. Nylon and polyester fiber ropes do not lose strength until 140°C. An abrupt failure occurs at higher temperatures. Ropes for high-temperature applications are often made up of aramid, fiber glass, and ceramics. PyroRope™ is an E-glass braided or knitted rope with a thick coating of iron oxide red silicone rubber will withstand up to 260°C under continuous exposure and up to 1650°C when intermittently exposed.

Wetting has a significant effect on the structure and properties of rope depending upon the fiber or filament used. The fibers that are hygroscopic and swell, tend to shrink in length of the rope, cause the angle of twist to increase and hence the hardness. Strength may increase in some cases, especially for those fibers that show an increase in strength due to wetting.

Strength translation efficiency (STE) is a measure of the rope's ability to convert yarn strength into rope strength. The strength of a yarn is generally less than the algebraic summation of the strength of constituent fibers across the yarn cross section. Strength translation from fiber to rope is never 100%, which is due to obliquity effect in the case of twisted structures and variability in breaking extension of constituent fibers. The larger diameter ropes have lower STE values. The STE value for polypropylene is highest followed by polyethylene, polyamide, and polyester. Polypropylene having the lowest modulus can easily yield and adjust itself within the structure when stress acts on it. As a result, it can easily orient itself toward the direction of application of load and, therefore, can share the load more uniformly. A stiff fiber will not be able to adjust itself so easily within the structure and will remain in distorted configuration. Strength-to-weight ratios decrease as rope size increases. This is especially true of high-modulus ropes in a braided construction.

8.5.3 Rope Elongation

Elongation reflects the ability of a rope to extend under the application of load. All types of ropes elongate, when loaded. Low elongation ropes are generally called static ropes. It is important for static ropes to have minimal elongation and maximum strength. The ropes meant for mountain climbing applications are generally called dynamic ropes, where a certain degree of elongation is important, as it is related to the resultant impact force absorbed by the climber's body and the system that breaks the fall. Rope elongation is predominantly influenced by constituent fiber elongation and structural parameters such as twist and size. Manila rope is much more extensible than a steel wire rope, and nylon is still more extensible. A soft-laid rope with low twist is less extensible than a tightly twisted hard-laid rope. In wet state, rope breaking extension is more than that in dry state. Natural fiber rope shrinks in wet state, causing higher elongation at break. Small-size ropes, due to compactness and less inclination of strands with respect to rope axis, lead to less extensibility compared to larger size ropes. Rope elongation also depends upon pretension applied for eradicating constructional looseness. Rope elongation also gets affected by service condition and weathering.

Rope extension consists of several components:

1. *Elastic extension:* This is the recoverable component of the rope's extension and is immediately realized upon release of the load.

2. *Viscoelastic extension:* The contraction of a rope does not follow the same path as the rope's extension. This results in an element of extension that is not immediately recoverable but will recover if relaxed for sufficient time. If the load on the rope is cycled, a hysteresis loop is formed that will exacerbate this element of stretch.

3. *Permanent extension:* This is nonrecoverable. When the rope is initially loaded, all the plaits, strands, and yarns become "bedded in." This results in a small permanent extension. Most of these constructional effects occur within the first few loadings and have little effect on the rope after this time. In addition to this, there are some permanent molecular changes that occur to the material that result in creep.

8.5.4 Energy Absorption

Ropes that stretch absorb energy. The area described under the stress–strain curve pertains to the force acting through distance (or the work required to break it). The energy-absorption capacity of a rope primarily depends on its linear density and the fiber type. Dyneema® SK60 has the highest value of the specific strength, impact strength, and modulus among commercialized organic strong fibers. A 10 mm diameter rope of Dyneema® SK60 can bear up to a 20 ton (theoretical value)-weight load. Nystron® rope made of nylon/polyester provides improved shock absorption capabilities and reduces the chance of failure with a dropped load. The energy-absorbing capacity of a length of rope can be calculated as follows:

$$\text{Energy (ton-m)} = \text{Area under curve} \times \text{Breaking strength (ton)}$$

$$\times \text{Length (m)} \times 10^{-4}$$

where area under the curve is measured graphically or from curve matching.

8.5.5 Rope Recovery from Stretch

The rope starts recovery from stretch immediately after removing the load applied, part of the recovery being almost instantaneous, whereas the remaining proceeds progressively with time, with a part remaining unrecovered permanently. The amount of recovery depends upon the magnitude of applied load, size of rope, structure, and nature of rope material. The recovery is more if the load applied is less, and the smaller-size rope recovers more readily than the larger-size rope. A rope made of elastic fiber like nylon or polyester is expected to recover more than natural fiber rope. The recovery rope made up of ultra high elongation, high tenacity nylon with PU coating can stretch 30%–40% from its initial length, resulting in less "jerk" during a recovery. The recovery rope is loosely braided therefore it will recover quicker after use. Repeated loading improves the recovery behavior of rope since with repeated loading, the structure gets compacted and starts showing complete recovery provided the load applied does not exceed the load used to prestretch the rope.

8.5.6 Creep

Creep is the length or rate at which the fibers in the rope stretches irreversibly over time. Due to this potential failure, creep should be one of the main concerns for applications that require long-term static loading. Creep of fibers may be either recoverable (primary creep) or nonrecoverable (secondary creep). The propensity of rope to creep depends upon the ultimate elongation of the fiber, type of rope structure, size, and the magnitude of applied load. Creep is accompanied by a reduction in the elongation at the breaking strength. However, as the rope creeps under tension, it will eventually stretch to a point of complete failure called creep rupture. Polyester, aramid, and liquid crystal polymer (LCP) fiber ropes have very low levels of creep. Polypropylene is also subject to creep and should not be held at high tension (excess of 20% of breaking strength) for long periods of time. Nylon can fail in less than a day at the 50% load level. Nylon 6 fiber exhibits lower creep rates than nylon 6,6. All synthetic fibers exhibit some degree of creep and not all irreversible elongation of a new rope is due to creep.

8.5.7 Fatigue and Flexing Endurance

Fatigue is the progressive and localized structural damage that occurs when a material is subjected to cyclic loading. Fatigue occurs when the individual fibers lose their ability to bend, and many times is caused by some form of constraint that prevents the rope from moving freely and smoothly, and working in conjunction with each other. Fatigue results in loss of strength and possibly an increase in axial stiffness. Ropes that are cycled under load are subject to tensile fatigue. Fatigue may be accelerated by abrasion, nicking, and other types of damage such as kinking. When fatigue breaks occur and there is no sign of wear, it is usually caused by bending stresses. The performance of the polyester is seen to be very superior to the steel wire. The relationship between sheave diameter and rope diameter is critical in determining a rope's fatigue resistance or relative service life.

When a rope moves over a pulley, it gets repeatedly flexed. Flexing is a common source of fatigue and is often wrongly ignored if tensions are relatively low. Flexing endurance becomes an important property in such application. In the case of twisted rope, flexing endurance improves with twist and an SSZ construction has better flexing properties than the ZSZ construction. Flexing endurance can be improved by impregnating the rope strand with lubricant as it reduces interstrand and yarn friction.

8.5.8 Abrasion Resistance and Friction

Abrasion, being the dominant factor in decreasing strength and service life, consists of external and internal abrasion. The ability to resist damage from abrasion due to rubbing on exterior surfaces is an important property. Nylon behaves well in dry condition whereas high-modulus polyethylene (HMPE) is very abrasion resistant. LCP has very good abrasion resistance. Ropes made up of manila fiber gives fair performance, but it is biodegradable. An important aspect of rope design that plays a critical role in the rope's abrasion resistance is the braid's cycle length. The length of a single braid cycle in a rope, which is controlled by the braid angle, can vary greatly. The loosely constructed braid had a significant decrease in their ability to resist abrasion damage. Abrasion resistance of a rope can also be enhanced by coating technology that protects the rope from both internal and external abrasion. The coefficient of friction between ropes and other surfaces is an important consideration in many applications. Frictional heat can cause fibers with low melting points (HMPE and polypropylene) to melt, which can lubricate the friction surface, reduce the coefficient to near zero, and create a dangerous situation. Fiber-to-fiber friction is also essential to make a splice work.

TABLE 8.3

Knot Retention Rating of Ropes

Fiber	Construction	Knot Retention Rating
Manila	Braided	Excellent
Manila	Eight-strand plaited ropes that are tightly twisted and plaited	Good
Polypropylene	Staple ropes	Good
Nylon	Three-strand braids	Poor
Polyester	Continuous filament braids	Fair
Polyester	Staple (fuzz) on the surface	Good
Polypropylene	Monofilament three-strand braids	Poor

8.5.9 Environmental Protection

Ultraviolet inhibitors are used in most synthetic fibers designed for use in rope. This is especially important for nylon, polyester, aramid, and polypropylene. Nylon and polyester should be limited to 90°C and polypropylene to 60°C. Higher temperatures are acceptable for aramid. HMPE should not exceed 50°C because of creep considerations. Natural fiber ropes degrade relatively quickly in damp conditions and are attacked by microorganisms. Synthetic fiber ropes are resistant to attack by most chemicals, unless they are highly corrosive and/or at elevated temperatures.

8.5.10 Shrinkage, Spliceability, and Knot Retention

Nylon ropes can be "shrunk" by processing the rope through steam. Polyester fiber can be shrunk by heating. The ability to be spliced with relative ease can be an important property of any rope. Knot retention can be an important property for certain applications. The knot retention rating of some of the ropes are given in Table 8.3.

8.6 Production of Rope

Rope making involves twisting textile yarns into rope yarns, rope yarns into strands, and strands into ropes. For braided rope, the yarn is braided rather than being twisted into strands. Plaited rope is made by braiding twisted strands. Other rope construction includes combinations of these three techniques such as a three-strand twisted core with a braided cover. The concept of forming fibers or filaments into yarn and yarn into strands or braids is fundamental to the rope-making process. The basic steps involved in making a rope can be given as follows:

1. Yarn manufacture
2. Strand manufacture
3. Rope manufacture

8.6.1 Production Routes of Modern Ropes

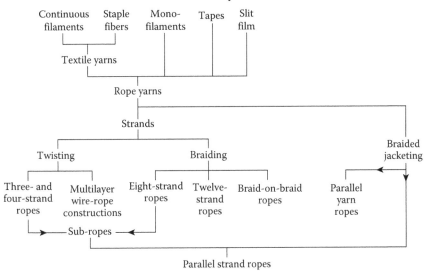

Rope may be made either from natural fibers, which have been processed to allow them to be easily formed into yarn, or from synthetic materials, which have been spun into fibers or extruded into long filaments. Most of the yarns used by the rope industry are multifilament yarns, which are composed of synthetic fibers in the form of continuous filaments. Fibers and filaments are first formed into yarn. The yarn is then twisted, braided, or plaited according to the type of rope being made. The diameter of the rope is determined by the diameter of the yarn, the number of yarns per strand, and the number of strands or braids in the finished rope.

8.6.1.1 Yarn Manufacture

If the rope is manufactured using natural fibers, the fibers are first opened and cleaned to remove the trash particles present in the raw fiber bundle. The cleaned fiber tuft is converted into sliver in carding process. The carded slivers are doubled and drafted further to enhance the uniformity of the output sliver. Combing operation is performed to remove short fibers present in fibrous material. This produces a loose, continuous ribbon of fibers called a sliver. The fibers in the sliver have been aligned along the long axis of the ribbon. Synthetic fibers of staple fiber form follow a similar process, but tend to align more easily. If the rope is produced from long filaments of synthetic

FIGURE 8.7
Yarn twist type.

material, several filaments are grouped together in a process called doubling or throwing. This produces a sliver of multiple plies of filaments. The sliver is run through the rollers of a drawing machine to compress it before it is twisted into yarn (Figure 8.7).

Yarn that has a right-hand twist is said to be "Z" twist, and yarn that has a left-handed twist is said to be "S" twist. The finished yarn is wound on spools called bobbins. At this point, the yarn may be dyed in various colors to produce a strand, or an entire rope, of a particular color.

8.6.1.2 Strand Manufacture

The bobbins of yarn are set on a frame known as a creel. For three-strand, right-hand twist rope, Z-twist yarns would be used to make each strand (Figure 8.8).

The ends of the yarns are fed through a hole in a register plate that keeps the yarns in proper relation to each other. The ends of the yarns are then fed into a compression tube. As the yarn is pulled through the compression tube, the tube twists it in the S-twist direction, opposite of the yarn twist, to produce a tight strand.

8.6.1.2.1 Twisted Rope Manufacture

The strands are either transferred to strand bobbins or fed directly into the closing machine. For common three-strand rope, three S-twist strands would be used. The closing machine holds the strands firmly with a tube-like clamp called a laying top. The end of each strand is then passed through a rotating die that twists the strands in the Z-twist direction, locking them together. This process is called closing the rope. The finished rope is wound onto a reel. When the end of the strands has been reached, the finished coil of rope is removed from the reel and tied together with bands of smaller rope.

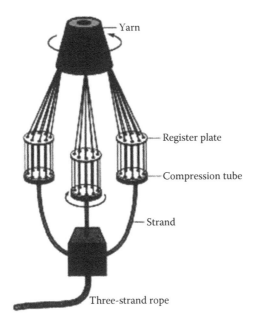

Yarn

Register plate

Compression tube

Strand

Three-strand rope

FIGURE 8.8
Strand manufacture.

The ends are either taped or, if the rope is a synthetic material, melted with heat to prevent them from unraveling.

8.6.1.2.2 Braided Rope Manufacture

Braided ropes are commonly made from synthetic materials. The bobbins of yarn are set up on several moving pendants on a braiding machine. Each pendant travels in an oscillating pattern, weaving the yarn into a tight braid. A set of rollers pulls the braid through a guide to lock, or set, the braid and keep tension on the rope (Figure 8.9).

In some machines, the braiding process is accomplished by feeding the yarns through separate counter-rotating register plates. One yarn is woven in one direction followed by another in the opposite direction, and so on, to form an interlocked braid. If a double-braided rope is being formed, the first braid becomes the core, and the second braid is immediately woven on top of it to form the outer covering, called the coat. As the rope emerges from the rollers, it is taken up on a reel. The finished coil is then removed and banded, and the ends are taped or melted.

8.6.1.2.3 Plaited Rope Manufacture

Eight-plaited rope consists of four S-twist strands and four Z-twist strands. The strands are paired together with one S-twist and one Z-twist in each pair. These pairs are then held together and braided with the other pairs.

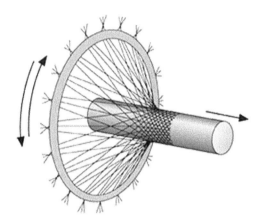

FIGURE 8.9
Braided rope manufacture.

The manufacturing process first follows the twisted rope process to make the strands and then the braided rope process to form the final rope.

Bibliography

ASCE, *Glossary of Marine Fiber Rope Terms*, American Society of Civil Engineers, New York, 1993.

ASME, B30.9 Slings, *Synthetic Fibre Rope Slings*, American Society of Mechanical Engineers, New York, 1996, Chapter 4.

ASTM, *D–123 Standard Terminology Relating to Textiles*, American Society for Testing Materials, West Conshohocken, PA, 1993.

Backer, S., Yarns, cords and ropes, in *Proceedings of the Mechanices of Flexible Fibre Assemblies*, edited by J. W. S. Hearle, J. J. Thwaites, and J. Amirbayal, NATO Advanced Study Institute, 1979, 535–542.

Chattopadhyay, R., Textile rope-a review, *Indian Journal of Fibre & Textile Research*, 22, 360–368 (1997).

CND Rope and Industrial Supply Ltd., Canada, http://www.cndrope.com/services-rope-construction.html (accessed May 29, 2013).

Cordage Institute, CI B–1.4: Fiber rope technical information manual, Cordage Institute, Wayne, PA, 1993.

Cordage Institute, CI 1403, Terminology for fiber rope, Cordage Institute, Wayne, PA, 1996.

Cordage Institute, CI 2003: *Fiber Properties (Physical, Mechanical and Environmental) for Cable, Cordage, Rope and Twine*, Cordage Institute, Wayne, PA, 2003.

Denton, M. J. and Daniels, P. N., *Textile Terms and Definitions*, The Textile Institute, Manchester, England, 2002.

Hearle, J. W. S., Ropes, *Textile Horizon*, May 12–15, June–July 28–31 (1996).

Himmelfarb, D., *The Technology of Cordage Fibres and Ropes*, Leonard Hill, London, U.K., 1957.

McCorkle, E., R. Chou, D. Stenvers, P. Smeets, M. Vlasblom, and E. Grootendorst. Abrasion and residual strength of fber tuglines. *ITS 2004: The 18th International Tug and Salvage Convention*, Paper tw2, Day 2, May 2004.

McKenna, H. A., J. W. S. Hearle, and N. O'Hear, *Hand Book of Fibre Rope Technology*, Woodhead Publication, Cambridge, U.K., 2004.

Sterling Rope—Guide to Rope Engineering, Design, and Use, Sterling Rope Company, Scarborough, ME, Volume 1.

Volpenhein, K., Abrasion and twist effects on high-performance synthetic, *Ropes for Towing Applications*, Samson Rope Technologies, USA, http://www.samsonrope. com/documents/technical papers (accessed May 29, 2013).

9

Textiles in Civil Engineering

9.1 Introduction

Civil engineering applications have a widespread demand with growing and emerging economies building capacity and infrastructure rapidly. The civil engineering industry plays a predominant role in the advancement of human society as they are involved in the planning, designing, building, and maintenance of infrastructure. Historically, major technological advancements in civil engineering have been possible only through the applications of innovative construction materials. Textiles, polymers, and composites are increasingly being utilized in the construction sector. Textiles play a significant role in building infrastructures, offering desirable properties such as lightweight, strength, and resilience as well as resistance to many factors such as creep, degradation from chemicals, sunlight, and pollutants. The use of textiles in building design and construction creates a lot of interest and growth potential that can be driven by a desire for greener, cleaner, lighter, higher performing, and sustainable structures. A niche in the textile industry provides high-strength, high-modulus textile fabrics to the construction industry as a potential replacement for more traditional building materials such as wood, concrete, masonry, and steel. An important contribution of the textile industry to the civil engineering sector is what is, according to their use, referred to as construction textiles or geotextiles. Geotextiles help in reducing energy consumption and improving performances in the construction sector. Textile structures are also used in architectural sector to improve the aesthetics of buildings.

9.2 Geotextiles

Geotextiles are technical fabrics used in civil engineering construction projects such as road pavements, dams, embankments, drains, and silt fencing for the purpose of soil reinforcement and stabilization, sedimentation and erosion control, drainage and support, and many other applications. Geotextiles were one of the first industrial textile products used. Geotextiles have proven to be among the most versatile and cost-effective ground modification materials. The usage of geotextiles increased rapidly, encompassing nearly all areas of civil, geo-technical, environmental, coastal, and hydraulic engineering. Conventionally, geotextiles were produced from natural fibers or vegetation mixed with soil to enhance road quality, especially for roads constructed on unstable soil. With the evolution of synthetic fibers, geotextiles offers many desirable and peculiar properties required for modern civil structures. Geotextiles today are highly developed products that must comply with numerous standards.

9.2.1 Classification of Geotextiles Based on Manufacture

According to ASTM D-4439, geotextile is defined as "A permeable geosyn-thetic comprised solely of textiles. Geotextiles are used with foundation, soil, rock, earth, or any other geotechnical engineering-related material as an inte-gral part of human-made project, structure, or system." Geotextiles can be manufactured by weaving or knitting or nonwoven technologies. Geotextiles are classified as follows:

1. *Woven geotextiles* are produced with the interlacement of two sets of yarns at right angles in the weaving process. Woven geotextiles have high strengths and modulus in the warp and weft directions and low elongations at rupture. Woven geotextiles are produced with a simple pore structure and narrow openings between fibers. Plain weave is most commonly applied in geotextiles, sometimes made by twill weave or leno weave. Woven geotextiles can be composed of monofilament or multifilament. Woven geotextiles constructed with multifilaments have superior strength and modulus compared to all other fabric constructions. Monofilament can be used in the form of slit film or ribbon filament for the production of woven geotextile.

2. *Knitted geotextiles* are produced with the interlooping of one or more yarns in the knitting process. These geotextiles are highly extensible and have relatively low strength compared to woven geotextiles, which limits its usage.

3. *Nonwoven geotextiles* are thicker than woven and are made either from continuous filaments or from staple fibers. The fibers are gener-ally oriented directionally or randomly in the web sheet and bonded with thermal or mechanical or chemical means. In the spun-bonding process, filaments are extruded and laid directly on a moving belt to form the mat, which is then bonded by any one of the following bonding techniques:

   a. *Needle punching:* Mechanical interlocking of fibers by penetrating many barbed needles through one or several layers of a fiber mat normal to the plane of the geotextile.

b. *Thermal bonding:* Incorporation of binder component in the form of fiber or powder or film or web that has a lower melting point in the fiber web and application of heat melts the lower melting point fiber that acts as binding agent.

c. *Chemical bonding:* Chemical binder is introduced into the fiber web, coating the fibers and bonding the contacts between fibers.

4. *Stitch-bonded geotextiles* are produced by interlocking fibers or yarns or both, bonded by stitching or sewing. Even strong, heavyweight geotextiles can be produced rapidly. Tubular geotextiles are manufactured in a tubular or cylindrical fashion without longitudinal seam.

5. *Geogrids* are materials that have an open grid-like appearance. The principal application for geogrids is the reinforcement of soil (Figure 9.1).

6. *Geonets* are open grid-like materials formed by two sets of coarse, parallel, extruded polymeric strands intersecting at a constant acute angle. The network forms a sheet with in-plane porosity that is used to carry relatively large fluid or gas flows (Figure 9.2).

7. *Geomembranes* are continuous flexible sheets manufactured from one or more synthetic materials. They are relatively impermeable and are used as liners for fluid or gas containment and as vapor barriers (Figure 9.3).

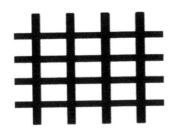

FIGURE 9.1
Geogrid.

FIGURE 9.2
Geonets.

FIGURE 9.3
Geomembranes.

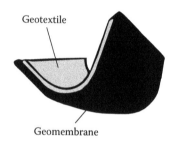

FIGURE 9.4
Geocomposites.

8. *Geocomposites* are made from a combination of two or more geosynthetic types. Examples include geotextile-geonet; geotextile-geogrid; geonet-geomembrane; or a geosynthetic clay liner (GCL) (Figure 9.4).

9. *GCLs* are geocomposites that are prefabricated with a bentonite clay layer typically incorporated between a top and bottom geotextile layer or geotextile bentonite bonded to a geomembrane or single layer of geotextile. Geotextile-encased GCLs are often stitched or needle-punched through the bentonite core to increase internal shear resistance. When hydrated, they are effective as a barrier for liquid or gas and are commonly used in landfill liner applications often in conjunction with a geomembrane (Figure 9.5).

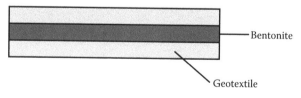

FIGURE 9.5
Geosynthetic clay liners.

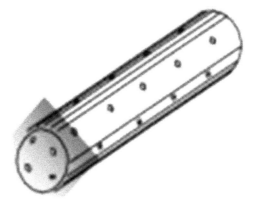

FIGURE 9.6
Geopipes.

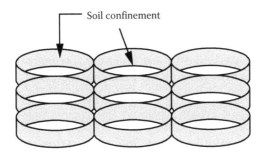

FIGURE 9.7
Geocells.

10. *Geopipes* are perforated or solid-wall polymeric pipes used for drainage of liquids or gas (including leachate or gas collection in landfill applications). In some cases, the perforated pipe is wrapped with a geotextile filter (Figure 9.6).

11. *Geocells* are relatively thick 3-D networks constructed from strips of polymeric sheet. The strips are joined together to form interconnected cells that are filled with soil and sometimes concrete. In some cases, 0.5–1 m wide strips of polyolefin geogrids have been linked together with vertical polymeric rods used to form deep geocell layers called geomattresses (Figure 9.7).

12. *Geofoam* blocks or slabs are created by expansion of polystyrene foam to form a low-density network of closed, gas-filled cells. Geofoam is used for thermal insulation, as a lightweight fill or as a compressible vertical layer to reduce earth pressures against rigid walls (Figure 9.8).

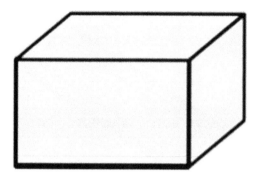

FIGURE 9.8
Geofoams.

9.2.2 Functions of Geotextiles

Geotextiles have numerous application areas in civil engineering including pavements, filtration and drainage, reinforced embankments, railroads, erosion and sediment control, moisture barrier, silt fencing, and earth-retaining walls. They always perform one or more of the earlier functions when used in contact with soil, rock, and/or any other civil structures. The basic functions of geotextiles are as given in the following sections.

9.2.2.1 Separation

Separation is the process of preventing undesirable mix-up of two dissimilar materials. The geotextile acts as a separating layer between fine aggregates and coarse aggregates or soils that have different particle size distributions to avoid undesirable mix-up. Separators also help to prevent fine-grained subgrade soils from being pumped into permeable granular road bases thereby keeping the structural integrity and functioning of both materials intact (Figures 9.9 and 9.10).

In the construction of roads over soft soil, a geotextile can be placed over the soft subgrade and then gravel or crushed stone placed on the geotextile. The geotextile separator reduces the rut depth that occurs due to soil migration because of vehicle load. Design of separator is largely influenced by the grain size of the soils involved. The ideal geotextile to be used for roads, railways, foundations, and embankments would be a continuous polymeric sheet of a high-strength geotextile or a geocomposite consists of geogrid and geotextile. California bearing ratio (CBR) test is performed on the subgrade soil, so that the selection of the primary function from reinforcement and separation can be done empirically. Soft subgrade soil has a low CBR value and vice versa. The property requirement of geotextile for the separation function is given in Table 9.1.

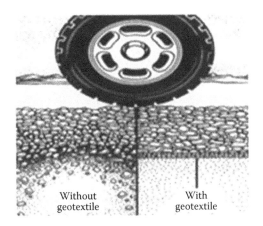

FIGURE 9.9
Geotextile in separation function (without and with geotextile).

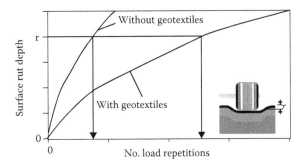

FIGURE 9.10
Rut depth (without and with geotextile).

TABLE 9.1

Property Requirements of Geotextile for Separation Function

Period	Property Requirement of the Geotextile for Separation		
	On Mechanical Conditions	On Hydraulic Conditions	Long-Term Performance
During installation	Impact resistance, elongation at break	Apparent opening size, thickness	UV resistance
During construction	Puncture resistance, elongation at break	Apparent opening size, thickness	Chemical stability, UV resistance
After completion of construction	Puncture resistance, tear propagation resistance, elongation at break	Apparent opening size, thickness	Resistance to decay, chemical stability

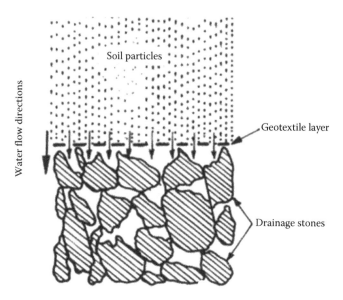

FIGURE 9.11
Geotextile in filtration.

9.2.2.2 Filtration

Geotextile is placed in contact with and down gradient of soil to be drained. The plane of the geotextile is positioned normal to the expected direction of water flow. The geotextiles function as a sand filter by allowing water with limited migration of soil particles across its plane over a projected service period (Figure 9.11).

The capacity for flow of water normal to the plane of the geotextile is termed as permittivity. Geotextile filters have to attain criteria as shown as follows that assure that the base soil will be retained with unimpeded water flow.

9.2.2.2.1 Filter Opening Size FOS ≤ n D$_{ss}$

The FOS is the filter opening size of geotextile, which is associated with pore and constriction sizes in the geotextile, n is a number that depends on the criterion used, and D$_s$ is a representative dimension of the base soil grains (usually D$_{85}$, which is the diameter for which 85% in weight of the soil particles are smaller than that diameter). The geotextile openings should be sized to prevent soil particle movement.

The filter has also to be considerably more permeable than the base soil throughout the project lifetime.

$$K_G \geq N k_s$$

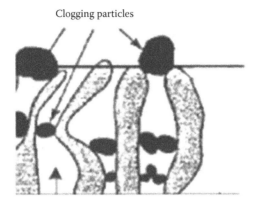

FIGURE 9.12
Clogging.

where

K_G is the geotextile coefficient of permeability
N is a number that depends on the project characteristics (typically varying between 10 and 100)
k_s is the permeability coefficient of the base soil

The voids of a geotextile filter are progressively filled by solid matter to
the point that the passage of water compromised is termed as "clogging"
(Figure 9.12). Both the granular filter and the geotextile filter must allow
water to pass without significant buildup of hydrostatic pressure.

9.2.2.3 Drainage

The geotextile acts as a drain to carry fluid flows through less permeable
soils. The application of geotextiles in drainage applications has improved
the economical usage of blanket and trench drains under and adjacent to the
pavement structure, respectively (Figures 9.13 and 9.14).

For example, geotextiles are used to dissipate pore water pressures at
the base of roadway embankments. For higher flows, geocomposite drains
have been developed. Geotextiles have been used as slope interceptor
drains, pavement edge drains, and abutment and retaining wall drains. The
relatively thick nonwoven geotextiles are the products most commonly used.
The essential property of the geotextile drain is transmissivity, which is the
capacity for in-plane flow.

9.2.2.4 Moisture and Liquid Barrier

The protection of civil structures from the effects of seeping water is a common
need. The geotextiles acts as a relatively impermeable barrier to prevent the
penetration of liquids or moisture over a projected service period (Figure 9.15).

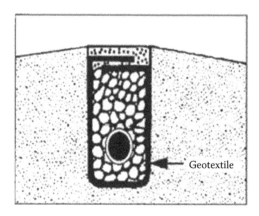

FIGURE 9.13
Geotextile in drainage.

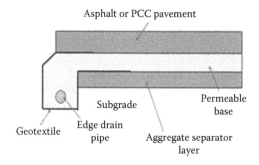

FIGURE 9.14
Geotextiles in edge drain.

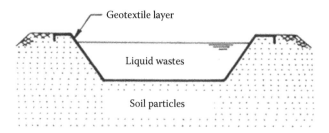

FIGURE 9.15
Geotextile as liquid barrier.

For example, geomembrane, thin film geotextile composites, GCLs, and field-coated geotextiles are used as fluid barriers to impede flow of liquid or gas. Geotextiles of both woven and nonwoven construction can serve as moisture barriers when impregnated with bituminous, rubber–bitumen, or polymeric mixtures that reduce both the cross-plane and in-plane flow capacity of the geotextiles to a minimum. These materials can also be used in asphalt pavement overlays, encapsulation of swelling soils, and waste containment to perform the same function.

9.2.2.5 Erosion Control

Erosion is the process by which soil and rock are removed from the earth's surface by exogenetic processes such as wind or water flow, and then transported and deposited in other locations. Erosion collapses the soil structure mainly due to the insufficient protective cover on steep slopes or in drainage channels that have been designed to rely on vegetation for long-term erosion control. The maximum use of vegetation in erosion and sediment control is often referred to as green engineering (Figures 9.16 and 9.17).

The geotextile anchored in steep slope protects soil surfaces from the tractive forces of moving water or wind and rainfall erosion. Limited-life geotextiles are used for temporary protection against erosion on newly seeded slopes. The geotextile is anchored to the seeded slope holding the soil and seed in place until the seeds germinate and vegetative cover is established.

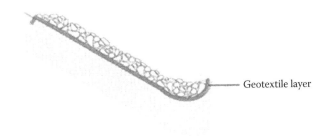

FIGURE 9.16
Geotextile in erosion control.

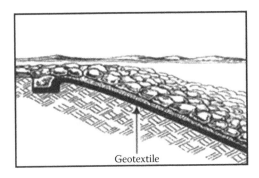

FIGURE 9.17
Geotextile in erosion control.

For example, temporary geosynthetic blankets and permanent lightweight geosynthetic mats are placed over the otherwise exposed soil surface on slopes. Geotextile silt fences are used to remove suspended particles from sediment-laden runoff water.

9.2.2.6 Reinforcement

Permanent roads carry larger traffic volumes and typically have asphalt cement concrete surfacing over a base layer of aggregate. The subgrade consists of soft clays, silts, and organic soils degrading the road structure whenever the subgrade gets wet. The geotextiles act as a reinforcement element within a soil mass or in combination with the soil to produce a composite that has improved strength and deformation properties over the unreinforced soil. The geotextile interacts with soil through frictional or adhesion forces to resist tensile or shear forces (Figure 9.18).

Geotextile meant for reinforcement function must have sufficient strength to overcome the tensile forces generated, and the strength must be developed at sufficiently small strains (i.e., high modulus) to prevent excessive movement of the reinforced structure. Woven geotextile is recommended to reinforce embankments and retaining structures, because it can provide high strength at small strains. The design of geotextile for the reinforcement function will be done by considering the factors such as in situ soil strength, design of load, and thickness of the aggregate. In situ soil strength is determined using on-field CBR test or cone penetrometer. The benefit of the geotextile reinforcement in the road pavement is assessed by determining "traffic benefit ratio" (TBR), which relates the ratio of reinforced load cycles to failure (excessive rutting) to the number of cycles that cause failure of an unreinforced road section. TBRs have a range of 1.5–70, depending on the type of geotextile, its location in the road, and the testing scenario. The geotextile material enhance the overlay life by relieving stress and/or providing

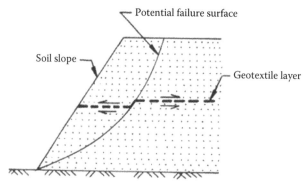

FIGURE 9.18
Geotextile in reinforcement function.

TABLE 9.2

Property Requirements of Geotextiles for Reinforcement Function

	Mechanical	Hydraulic	Long-Term Performance
Base failure	Shear strength of bonding system	Hydraulic boundary conditions	Chemical and decay resistance
Top failure	Tensile strength of geotextile, geotextile/ soil friction	Hydraulic boundary conditions	Chemical and decay resistance
Slope failure	Tensile strength of geotextile, geotextile/ soil friction	—	Creep of the geotextile/ soil system, chemical and decay resistance

reinforcement and by enabling moisture barrier in pavement. Table 9.2 gives the property requirement of geotextile for reinforcement function.

9.2.2.7 Survivability of Geotextiles

The ability of the geotextiles to resist installation conditions without significant damage, is termed as survivability conditions. Survivability affects the strength properties of the geotextile required. The various factors that affect survivability of the geotextiles in the field are given as follows:

- Protruding objects present on ground surface
- Subgrade strength (i.e., CBR)
- Thickness of first lift placed over geotextile
- Ground contact pressure of construction equipment
- Angularity of gravel placed over geotextile

9.2.3 Design of Geotextiles

The criteria for designing geotextiles for specific function have been determined by reviewing the site conditions and the purpose of geotextiles. In peculiar situations where soil conditions are in undesirable manner or high-risk applications, project specific design may be required. Soil survey should be conducted to evaluate the soil properties which helps to judge the site conditions. Soil in the vicinity of the proposed geotextile location to be tested for texture, grain size and its distribution, type of soil, depth, soil structure, liquid limit and plasticity index of soil, pH content, etc., for better designing of the geotextile. The physical and chemical properties of soil such as moist bulk density, permeability, shrink swell potential, organic matter, and erosion factor are also to be studied before the designing of geotextiles.

The filtration properties of a geotextile are the function of the soil type. For underground drainage applications, if the subgrade soil is relatively clean gravel or coarse sand, a geotextile is probably not required. Mechanical properties for the underground drainage geotextile depend on the survivability level required to resist installation stresses. Soil stabilization geotextile is placed directly on the soft subgrade material, even if some over-excavation of the subgrade is performed. Soil stabilization geotextile should not be used for roadway fills greater than 5 ft high or when extremely soft and wet silt, clay, or peat is anticipated at the subgrade level. The need for a permanent erosion control geotextile depends on the type and magnitude of water flow over the soil being considered for protection, the soil type in terms of its erodibility, and the type and amount of vegetative cover present. If groundwater cannot escape through the geotextile, an erosion control system failure termed *ballooning* (resulting from water pressure buildup behind the geotextile) or soil piping could occur.

The designing of geotextiles for any type of specific function has taken two parallel routes: "design by specification" and "design by function." In general, design by specification is applied for ordinary and noncritical applications, while design by function is applied for site-specific and generally critical applications.

Design by specification is a very familiar concept extensively applicable to public agencies and many private owners as well. In this method, several application categories are listed in association with various physical, mechanical, hydraulic, and/or endurance properties. The application areas are usually related to the intended primary function.

Design by function consists of assessing the primary in-service function to be performed by the geosynthetic and then calculating the required numerical value of a particular property for that function. By dividing this value into the candidate material's allowable property value, a factor of safety (FS) results:

$$\text{Factor of safety (FS)} = \frac{\text{Allowable (test) property}}{\text{Required (design) property}}$$

where

allowable property is the numerical value based on a laboratory test that models the actual situation or is adjusted accordingly by reduction factors

required property is the numerical value obtained from a design method that models the actual situation

FS is the factor of safety against unknown loads and/or uncertainties in the analytical or testing process, sometimes called a global factor of safety

If the FS is sufficiently greater than 1.0, the candidate geosynthetic is acceptable.

9.2.4 Properties of Geotextiles

The properties and test methods of geotextile material arose first from previously existing materials that resemble geotextiles, such as textile materials.

The property requirements and characteristics of geotextiles used for specific application depend on their purpose and the desired functions. For example, geotextile meant for reinforcement require good mechanical responses while filtration and drainage functions depend on hydraulic properties. The applications involve transport and storage of materials, construction in particularly harsh environments, and the necessity for a long service life, for which endurance and durability properties are important.

9.2.4.1 Physical Properties

The physical properties of geotextiles which are of primary concern are fiber composition, areal density, thickness, stiffness, and specific gravity. In geogrids and geonet, the properties such as structure type, junction type, aperture size, and shape are of prime importance. The structure of the geotextile often dictates the application area for which the material is appropriate. The physical properties of the geotextile are more temperature and humidity dependent.

a. *Specific gravity:* The specific gravity of a geotextile is determined on the basic polymeric fiber materials used to form the geotextile. The introduction of additives in the base polymer during spinning will alter the specific gravity of the resulting polymer. The specific gravity of polypropylene (PP) and polyethylene (PE) is less than 1.0, which is a drawback when working with geotextile in underwater applications.

b. *Areal density:* The areal density of a geotextile is measured in terms of mass per unit area. The mass per unit area is determined by cutting from a roll a minimum of 10 specimens, each at least 100 mm^2, and then weighing the specimens on an accurate balance. Typical values for geotextiles lie between 130 and 700 g/m^2, while for geogrids, the values range from 200 to 1000 g/m^2.

c. *Thickness:* The thickness of a geotextile is the distance between the upper and lower surfaces of the material. The compressive stress applied normal to the geotextile material through a circular pressure plate with a cross-sectional area of 2500 mm^2 is generally 2.0 kPa for geotextiles and 20 kPa for geogrids and geomembrane, for 5 s.

d. *Stiffness:* The stiffness refers to the ability of the geotextile to resist bending under its own weight, which is determined by cantilever principle. Flexural rigidity describes the material's capability of providing a suitable working platform during installation.

9.2.4.2 Mechanical Properties

Mechanical properties are important for the applications where a geotextile is required to perform as a load-bearing member under applied loads and where it is required to survive on-field installation related stresses. Mechanical properties of geotextile are often categorized as in-isolation properties and performance properties. In-isolation properties are determined on the geotextiles itself with the absence of surrounding soil. Performance properties are determined on geotextiles in the presence of site-specific soil.

a. *Tensile properties:* The determination of tensile properties, such as tensile strength and tensile modulus, of geotextile is important and is carried out by wide-width strip tensile testing. Geotextile specimen of at least 200 mm wide with a gage length of 100 mm is clamped within the compressive jaws of a tensile testing machine, which is capable of applying the load at a constant rate of strain. The load at rupture and the corresponding elongation are measured.

b. *Compression:* The compressibility of a geotextile is measured by a decrease in its thickness at increasing applied normal pressures. This test is important for nonwoven geotextiles especially needle-punched geotextiles, which are most compressible. The compressibility of other geotextiles is comparatively lower than needle-punched geotextile.

c. *Puncture strength:* Geotextile specimen is clamped without tension over an empty cylinder, and a solid steel rod is pushed through the fabric. A load indicator attached to the rod measures the force required to cause rupture. This test can also be used to assess the fabric's resistance to aggregate penetration, particularly in separation applications.

d. *Seam strength:* Geotextile may be joined by sewing, stapling, glueing, or melting. More secure seams can be produced in a manufacturing plant than in the field. The portable sewing machines used for field sewing of geotextiles can produce either the single-thread or two-thread chain stitches. Sewing thread for geotextiles is usually made from Kevlar, polyester, PP, or nylon.

e. *Durability:* Exposure to sunlight degrades the physical properties of polymers. Polymer materials become brittle in very cold temperatures. The durability of the geotextiles can be determined by accelerated weathering test. All polymers gain water with time if water is present. High-pH water can be harsh on polyesters while low-pH water can be harsh on polyamides.

f. *Tear strength:* The ability of a geotextile material to withstand stresses causing to continue or propagate a tear. Stresses acting on the geotextile during installation may propagate tear. Tearing strength of geotextiles can be determined by trapezoid tearing strength test. A typical range of trapezoid tearing strength of geotextiles is 90–1300 N.

g. *Soil–fabric friction:* Shear resistance between a geotextile and a soil is important in reinforcement applications. Shear strength can be determined by measuring the force required to cause sliding between the fabric and the soil for different normal stresses. This test is performed in direct shear test machine with the soil of normal confinement.

9.2.4.3 Hydraulic Properties

The hydraulic properties of geotextiles are those that relate directly to filtration and drainage functions of geotextiles. The filtration function of a geotextile requires the pore size sufficiently small to retain the erodible soil particles and permeability adequate to allow the free passage of seepage from the protected soil. Filtration properties are controlled by pore sizes, pore size distribution, and porosity of the fabric. The permeability of the geotextile is determined by measuring the rate of flow of water through the fabric in the direction normal to the plane of the fabric. Porosity, permittivity, and transmissivity are the most important hydraulic properties of geotextiles, geonets, and geocomposites, which are commonly used in filtration and drainage applications.

a. *Porosity:* Porosity is defined as the ratio of void volume to total volume of the fabric. The parameters such as apparent opening size (AOS) and percentage open area are related to porosity. For filtration applications, AOS should be higher to avoid clogging of soil in the geotextile structure. The parameter AOS is widely used in the United States and FOS (filtration opening size) is used in European countries.

b. *Apparent pore size distribution:* The pore size distribution of the fabric is determined by sieving dry spherical solid glass beads for a specified time at a specified frequency of vibration and then measuring the amount retained by the fabric sample. The test is carried out on a range of sizes of glass beads.

c. *Percent open area for woven geotextiles:* Using a planimeter, the magnified open spaces can be measured and expressed as a percentage of whole area. The test is primarily applicable to monofilament woven fabrics. The test provides information on pore size openings, which is important in assessing a geotextile's soil filtration capability.

d. *Permeability and permittivity:* Ability of a geotextile to transmit a fluid is called permeability. Measurement of the quantity of water that can pass through a geotextile (normal to the plane) in an isolated condition is termed as "permittivity."

Permittivity of geotextile

$$= \frac{\text{Cross-plane permeability of the geotextile}}{\text{Thickness of geotextile}}$$

Constant-head and a falling-head permeability test are performed to measure permittivity of the geotextile under zero normal stress confinement.

e. *Transmissivity:* The ability of fluid flow within the plane of a geotextile or a composite drain is termed as "in-plane permeability". Transmissivity is the product of in-plane permeability and geotextile thickness.

$$\text{Transmissivity} = \text{In-plane permeability} \times \text{Material thickness}$$

This test is performed in constant head test under varying normal stress confinement.

f. *Soil retention:* Retention of erodible soil particles is done in filter fabric. The ideal retention criteria for fabrics should specify appropriate pore structure to eliminate piping through the fabric, to provide adequate fabric seepage rate and to assure resistance to clogging.

9.2.4.4 Endurance, Degradation, and Survivability Properties

The endurance, degradation, and survivability properties (e.g., creep behavior, abrasion resistance, long-term flow capability, durability—construction survivability, and longevity) of geotextiles are related to their behavior during service conditions, including time.

a. *Creep:* Creep is the time-dependent increase in accumulative strain in a geotextile resulting from an applied constant load. Creep and stress relaxation phenomena occur simultaneously in a geotextile-reinforced soil structure. Recent developments allow for accelerated determination of creep characteristics through stepped isothermal methods. Creep and stress relaxation are most significant for PP and PE geotextiles and less significant for polyester and polyamide geotextiles. Creep rate of geotextiles depends on temperature.

b. *Abrasion:* Wearing away of the surface of the material by rubbing against another material. Too much abrasion leads to loss of strength. The resistance offered by the geotextile against soil surface is termed as abrasion resistance. This test can be done in Martindale abrasion tester. Abrasion resistance is expressed as percentage weight loss or strength loss as a result of abrasion.

c. *Long-term flow capability:* The ability of a geotextile to separate soils having varying grain size distributions while still allowing for flow makes them ideally suited for erosion control and filtration applications. The long-term flow test, in which soil is placed above the geotextile and a constant head of water, is applied above the soil. A geotextile that neither clogs nor pipes soil fines is considered compatible with the soil.

d. *Construction survivability:* The selection of geotextile either for temporary or permanent applications depends on survivability criteria. Geotextiles that meet or exceed these survivability requirements can be considered acceptable for most projects. The mechanical properties such as tearing strength, impact strength, puncture strength, fatigue strength, and bursting strength are connected to construction survivability.

- *Tearing strength* is the resistance offered by the geotextile to withstand stress causing to propagate a tear. This test can be done in trapezoidal tear tester.

- *Puncture strength* is the resistance offered by the geotextile to withstand localized stresses generated by penetrating or puncturing objects such as aggregates or roots. This test can be evaluated by CBR plunger test.

- *Impact strength* is the ability of a geotextile to withstand stresses generated by sudden impact during installation process in the construction site. This test can be evaluated by cone drop test method.

- *Bursting strength* is the resistance offered by the geotextile to withstand pressure applied perpendicular to its plane. This test can be done in multiaxial tensile tester.

- *Fatigue strength* is the resistance offered by geotextile to withstand repetitive loading before undergoing failure. It can be measured by change in mechanical and physical properties as a function of repeated loading.

9.2.4.5 Durability

The durability of a geotextile tells about the ability of geotextile to maintain requisite properties against environmental or other influences over the selected design life. The various factors affecting the durability of the geotextile are temperature, oxidation, hydrolysis, chemical degradation, and ultraviolet (UV) light. Geotextile are more prone to degradation at elevated temperatures. PP and polyester are the most susceptible polymers that deteriorated due to oxidation. Nylon tends to suffer when exposed to low-pH liquids. Geotextiles exposed continuously to sunlight may suffer from UV radiation, which results in property degradation of material. UV resistance is imparted to geotextiles by adding additives such as carbon black and other stabilizers to the polymer in which geotextiles are made.

9.2.5 Raw Materials of Geotextiles

Geotextiles are made from PP, polyester, PE, nylon, polyvinylidene chloride, and fiberglass. Thermoplastic polymers, which may be amorphous or semi-crystalline, are mostly used to manufacture geotextiles. Molecular weight of the polymer meant for geotextile manufacturing can significantly influence physical and mechanical properties, durability, and heat resistance of geotextile material. The physical and mechanical properties of the polymers are also influenced by the bonds within and between chains, the chain branching, and the degree of crystallinity. PP and polyester are the most widely used fibers in many geotextiles applications. The physical properties of these fibers can be modified by incorporating additives in the composition during melt spinning and by changing the spinning methods used to convert molten polymer into filaments. PP is lighter than water (specific gravity of 0.9), strong, and very durable, and has excellent chemical resistance. Another reason for preferring PP in geotextile applications is its low cost. With the addition of UV stabilizers, geotextiles have good resistance toward UV radiation. PP filaments and staple fibers are used in manufacturing woven yarns and nonwoven geotextiles. High-tenacity polyester fibers and yarns are also used in the manufacturing of geotextiles. Polyester is heavier than water, has excellent strength and creep properties, and is compatible with most common soil environments.

9.2.6 Applications of Geotextiles

The applications of geotextiles are numerous and continuously growing steadily. The various detrimental factors deteriorating the service life of roads and pavements including environmental factors, subgrade conditions, traffic loading, utility cuts, road widening, and aging. In an asphalt concrete pavement system, the geotextile provides a stress-relieving interlayer between the existing pavement and the overlay that reduces and retards reflective cracks under certain conditions and acts as a moisture barrier to prevent surface water from entering the pavement structure that increases the life of the pavement. The presence of the geotextile restricts lateral movement of both the aggregate and the subgrade, improving the strength and the stiffness of the road structure.

Geotextile-reinforced retaining walls consist of geotextile layers as reinforcing elements in the backfill to help resist lateral earth pressures. The construction of embankments over soft soils is a challenge for civil engineers. The placement of geotextile layer over the soft soil and construction of the embankment directly over it will consolidate the soft soil. Nonwoven paving fabrics have high elongation and low tensile strength and are used for stress relief. When saturated with asphalt, the flexible interlayer allows considerable movement around a crack but nullifies or at least lessens the effect the movements have on the overlay. The geotextile-reinforced foundation soils are being used to support footings of many structures including warehouses, oil drilling platforms, platforms of heavy industrial equipments, parking areas, and bridge abutments.

The control of water is critical to the performance of buildings, pavements, embankments, retaining walls, and other structures. Drains are used to relieve hydrostatic pressure against underground and retaining walls, slabs, and underground tanks and to prevent loss of soil strength and stability in slopes, embankments, and beneath pavements. Geotextiles retain the surrounding soil while readily accepting water from the soil and removing it from the area. Railway tracks serve as a stable guideway to trains with appropriate vertical and horizontal alignment. Geotextiles are used to stabilize the track.

Geotextiles are widely used in the construction of sports turf. Playing fields are synthetic grass surfaces constructed of light-resistant PP material with porous or nonporous carboxylated latex backing pile as high as 2.0–2.5 cm. Another synthetic turf sport surface made of nylon 6,6 pile fiber knitted into a backing of polyester yarn provides high strength and dimensional stability. Geotextiles are used for the improvement of muddy paths and trails those used by cattle or light traffic; nonwoven fabrics are used and are folded by overlapping to include the pipe or a mass of grit.

9.2.7 Limited-Life Geotextiles

Geotextiles made from synthetic fibers are nonbiodegradable, and it is suitable for long-term applications. Biodegradable natural geotextiles are deliberately manufactured for short-term applications. They are generally used to control soil erosion until vegetation can become properly established on the ground surface. The technical requirements of limited-life geotextiles could be satisfied by fabrics made up of natural vegetable fibers. Natural fibers in the form of paper strips, jute nets, wood shavings, or wool mulch are being used as geotextiles. It is possible to manufacture designed biodegradable jute geotextile, having specific tenacity, porosity, permeability, transmissibility according to need, and location specificity. Furthermore, after degradation of jute geotextiles, lignomass is formed, which increases the soil organic content, fertility, and texture, and also enhances vegetative growth with further consolidation and stability of soil. There are many situations where stability or functional requirement is very short after the construction. The usage of geotextiles paves an effective way for the civil engineer to solve countless geotechnical problems.

9.3 Textile-Reinforced Concrete

Concrete is a composite construction material made primarily with aggregate, cement, and water. In plain concrete, structural cracks develop even before loading particularly due to drying shrinkage or other causes of volume change. When loaded, the microcracks propagate and open up, the development of such microcracks is the main cause of the inelastic deformation in the concrete. Reinforced concrete is a composite material in which concrete's relatively low tensile strength and ductility are counteracted by the inclusion of reinforcement which provide strength and ductility. Reinforced concrete (RC) is successfully used as building material by the civil engineering industry in the past and present century. Steel bar reinforcement concrete is an almost ideal composite material, which is extremely powerful, durable, and cost-effective. Steel has been used for a long time in construction of roads and also in floorings, particularly where heavy wear and tear is expected. A minimum concrete cover of 20–70 mm per layer is necessary to protect steel bar reinforcements from corrosion during a building's lifetime. RC structures could be more lightweight, more elegant, and more efficient with the help of textile structures.

Textile in the form of fiber, roving, yarn, and fabric are used as reinforcement in concrete. Fiber reinforced concrete (FRC) is concrete containing fibrous material which increases its structural integrity. The fibers used as reinforcement are categorized as natural and synthetic fibers. Synthetic fibers are predominantly used as reinforcement in concrete due to its superior mechanical properties compared to natural fibers. Natural fibers are mostly preferred for low-strength applications. Textile reinforcement structures produced from carbon fibers represent an excellent alternative and complement existing reinforcement materials made from steel. Textile reinforcements are produced from continuous yarns or rovings that are processed in a planar structure by a textile technique to produce an optimal alignment and arrangement of fibers in structural members (Figure 9.19).

Carbon textile reinforcements do not need any concrete cover for the corrosion protection since the reinforcing materials used do not corrode under normal environmental conditions. As there is no concrete cover,

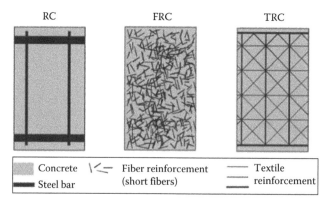

FIGURE 9.19
Textile-reinforced concrete.

the lightweight concrete structure can be manufactured easily. Carbon textile reinforcements have a much larger surface area than traditional steel bar reinforcements. Glass fiber is inexpensive and corrosion proof, but not as ductile as steel. Glass fiber concretes are mainly used in exterior building façade panels and as architectural precast concrete. Graphite reinforced plastic fibers, which are nearly as strong as steel, lighter weight, and corrosion proof, are used in some applications. Acrylic fibers are added to conventional concrete at low volumes to reduce the effects of plastic-shrinkage cracking. Nylon is effective in imparting impact resistance and flexural toughness, and sustaining and increasing the load carrying capacity of concrete following first crack. Aramid fibers are two-and-a-half times as strong as glass fibers and five times as strong as steel fibers, per unit mass. Due to the relatively high cost of aramid fibers, aramid fiber-reinforced concrete has been primarily used as asbestos cement replacement in certain high-strength applications. Thus, very high bond forces can be introduced into the concrete. The current generation of textile reinforcements has reached strengths well over 1500 N/mm^2. The use of additional textile-reinforced concrete (TRC) strengthening layers in RC structures has proved to have a positive influence on subsequent concrete cracking.

In recent years, the need to strengthen and retrofit existing RC structures has increased. A variety of methods can be used to increase the load-carrying capacity. The use of TRC can be an ingenious alternative, particularly since it is closely related to traditional, established strengthening procedures like externally bonded carbon fiber-reinforced polymers. Reinforcement with carbon fiber textile products has been used in hundreds of projects worldwide over the past few years. For applications in TRC, heavy tow carbon fiber improves cost-efficiency in several ways. Carbon fiber is an excellent material for seismic retrofitting because of its strength and lightweight. In terms of strength, carbon fiber increases the in-plane shear and out-of-plane flexural strength of unreinforced masonry structures. Other applications of textiles in building construction include self-healing concrete, localized crack repair, the reinforcement of critical walls, the wrapping of existing columns, or protection against earthquake or hurricanes.

9.4 Textiles in Architecture

Architectural textiles will increasingly serve structural roles in buildings, providing durable shelter and thermal protection for occupants; air-supported structures will be an important part of this adaptable future. Convertible and foldable roof construction is a growing trend in architecture. Textile structures are mostly used as a cost-effective solution for providing shade and shelter. The added bonus is that they are festive in nature and can be used for temporary or permanent applications. The cost of the membrane on fabric structures can vary significantly as it is based on the complexity of the design and how much or little steel there is in the project. However, fabric structures are usually chosen for a project to perform a function, such as shade, signage, or shelter. The chosen material is selected for specific reasons, such as light translucency and durability. Social benefits could also limit or narrow a material choice but rely more on the performance of the end product in action to determine its success. Membranes also come with different coatings that provide different forms of protection. The two most common materials chosen in the market are PTFE and PVC. Compared to coated glass fiber fabrics, PTFE yarn coated with fluoropolymer is more pliable and can be used for retractable structures, a growing trend in the industry.

Textile facades are usually designed with a metal clamping system attached to the building with a variety of devices to tension the membrane. They have social attributes that are hard to calculate, such as improved worker performance and more use of public shaded spaces, but they also have environmental attributes that can contribute to hard number savings in utility and energy costs. Sun and weather protection as well as light and temperature regulation are the main requirements for textile applications in sport facilities. ETFE fluoropolymer membranes allow 98% light transmission, water repellency, and insulating properties controlling interior temperature and humidity of large sport buildings.

Energy savings and sustainability are driving the need for smarter new and retrofitted stadium roofs. A new layered composite called Tensotherm™ by Birdair Inc., Amherst, N.Y., uses Nanogel® aerogel, a half-inch thick, feather-light insulation layer that traps air to prevent heat loss and solar heat gain, and allows daylight harvesting by transmitting and diffusing natural light. A layer of aerogel sandwiched between two layers of PTFE, rather than the standard one- to two-foot space required for fiberglass insulation, can boost thermal insulation value.

AeroLite fabric Ltd. create a three-layer fabric composite called AeroLite™ to address the need for greater thermal efficiency without sacrificing the thin profile necessary for free-form design and translucency in its fabric structures. A thin layer of Aerogels® SpaceLoft® blanket sandwiched between two layers of PTFE or PVC helps boost the R value and can insulate very thin spaces. The thermal efficiency of AeroLite is six times greater than a single-skin fabric system. Soundtex® panels are used in a wide variety of building complexes, airports, and sports facilities to reduce noise and produce clearer acoustics. The product's ultrathin profile can replace a thick mineral mat, which also has its advantages. Lighter weight composites with greater performance are in higher demand as builders are looking for material savings and waste reduction.

In addition to inflatable textiles, frame-supported fabric structures will continue to flourish, providing shade and wind protection in the form of lightweight, low-embodied energy enclosures. Such structures are capable of exhibiting spatial complexity and multiple visual readings appearing opaque during the day and translucent by night when illuminated. Bio-based textiles have been applied to building furnishings and will eventually be scaled to buildings. Tensile fabric structures used for interior walls are constructed for long-term use. These fabric walls need to be torqued (like the twist of a propeller blade) to introduce stiffness to the warp and weft of the fabric weave. The construction of these textile walls is accomplished by the introduction of double curvature or the "anticlastic" (i.e., surfaces have double curvature in diametrically opposite directions, like a saddle), curving of fabric in two directions at every point on the surface of the fabric.

Fiber wall is a fully biodegradable textile-based panel system that consists of plant fibers and plant-based resin. Designed to combine properties such as high structural stiffness, light transmittance, and the appearance of natural fiber, it functions as a self-bearing, translucent space divider. It consists of three shapes of double-curved composite panels made from sisal fiber, linen textile, and soy-protein resin that may be combined to extend the surface in multiple directions. Circular cutouts create multiple possibilities in transparency and light filtering. Fabric will not only harvest energy, but also utilize energy to transmit light and visual effects efficiently. Delight Cloth from LumenCo. Ltd., Tokyo, Japan, is a light-emitting textile made of thousands of fiber-optic strands. With a diameter of only 0.25–0.5 mm, the optical fibers are woven into a large translucent tapestry that can be hung vertically or horizontally, providing a low-energy light source. Developed by Japan-based Tsuya Textile Co. Ltd. in cooperation with the University of Fukui Engineering Center, the material is currently used for wall or ceiling treatments as well as banner signage or clothing, and may be used to emit a wide variety of colors of illumination. Future applications include the integration of optical fiber textiles into fabric structures that emit their own light.

Textiles will also be used to convey moving images and information. Fabcell is a chameleon-like fabric that changes color when conducting an electric charge. Developed by Dr. Akira Wakita's Information Design Laboratory at Keio University in Japan, Fabcell is a flexible fabric made of fibers dyed with liquid crystal ink and conductive yarns. These materials are connected to electronic components and woven into a square textile. When a low voltage is applied, the temperature of the fabric increases, changing the color of the fabric. When arranged in structural matrices, Fabcells can display transforming images within the complex curvature formed by flexible textiles.

Apart from interior applications, textiles are perceived by the construction industry today as mainly for temporary structures. Besides tapestry and curtains, textiles are used in roofing, insulation, and cladding; in sun, water, wind, fire, and noise protection; in floor and concrete reinforcement; and in UV and electromagnetic shielding.

9.5 Nanotextiles in Civil Engineering Applications

Nanotechnologies have already found application in construction. Nanotechnology plays a significant role in the improvement of functional performance of the building by increasing mechanical, chemical, photochemical, and biological properties. In civil engineering, the use of nanofibers and nanotubes can help to make lighter and stronger concrete materials that last longer and able to resist strong shocks generated by earthquakes.

Textiles coated with nanomaterials can provide a variety of attractive features such as enhanced thermal/acoustic insulation, light transmission/reflection, UV and electromagnetic shielding, hydrophilic/hydrophobic, fire resistance, self-cleaning characteristics, and aesthetic finishing for building exteriors.

Seismic wallpaper composite concept is based on a reinforced textile composite system, combining different materials like multiaxial warp-knitted glass and polymer fibers, nanoparticle-enhanced coatings for the textile fabric, and nanoparticles-enhanced mortar to bond the textile to the structure.

Bibliography

Agnew, W., Erosion control product selection, Geotechnical Fabrics Report, IFAI, Roseville, MN, April, 1991.

Armijos, S. J., The economics of fabric structure, *Specialty Fabrics Review*, July 2010, http://specialtyfabricsreview.com/articles/0710_f1_economics.html (accessed May 29, 2013).

Armijos, S. J., The cost of building fabric structures, *Fabric Architecture*, 2004 , http://fabricarchitecturemag.com/articles/0110_f1_costs.html (accessed May 29, 2013).

AS 3706.4-2001. *Determination of Burst Strength of Geotextiles—California Bearing Ratio (CBR)—Plunger Method*, Standards Australia International Ltd, Sydney, New South Wales, Australia, 2001.

AS 3706.5-2000. *Determination of Impact Strength (Dynamic Puncture Strength) of Geosynthetics by Falling Cone Method*, Standards Australia International Ltd, Sydney, New South Wales, Australia, 2000.

American Society for Testing and Materials (ASTM) D-4439, Standard terminology for geosynthetics.

Aziz, M. A., Paramasivam, P., Lee, S. L. Prospects for natural fibre reinforced concretes in construction, *Cement Composite Lightweight Concrete* 1981, 3(2), 123–132.

Bathurst, R. J., *Geosynthetics—Classification*, International Geosynthetics Society (IGS), http://geosyntheticssociety.org/Resources (accessed May 29, 2013).

Bathurst, R. J., *Geosynthetics—Functions*, International Geosynthetics Society (IGS), http://geosyntheticssociety.org/Resources (accessed May 29, 2013).

Beaudoin, J. J., *Handbook of Fiber-Reinforced Concrete: Principles, Properties, Developments and Applications*, Park Ridge, NJ, Noyes, 1990.

Berger, H., Fabric structures, *Architecture Structure Magazine*, November 2004, p. 26.

Brownell, B., Driving the future of fabric structures, *Specialty Fabrics Review*, http://specialtyfabricsreview.com/articles/0611_f1_fabric_structures.html (accessed May 29, 2013).

Engineering use of geotextiles, UFC 3-220-08FA, January 16, 2004.

Ernster, B., Potential builds for fabrics in construction, *Specialty Fabrics Review*, May 2010, http://specialtyfabricsreview.com/articles/0510_f2_architectural.html (accessed May 29, 2013).

Gourc, J. P. and E. M. Palmeira, *Geosynthetics—Drainage and Filtration*, International Geosynthetics Society (IGS), http://geosyntheticssociety.org/Resources (accessed May 29, 2013).

Handbook of Geosynthetics, Geosynthetic Materials Association (GMA), IFAI, USA.

Holtz, R. D., Geosynthetics for soil reinforcement, The Ninth Spencer J. Buchanan Lecture, 2001.

Huang, H.-Y. and Gao, X., Geotextiles, http://www.engr.utk.edu/mse/Textiles/Geotextiles.htm (accessed May 29, 2013).

Ingold, T. S. and K. S. Miller, *Geotextiles Handbook*, Thomas Telford, London, U.K., 1990.

Kazarnovsky, V. D. and B. P. Brantman, Geotextile reinforcement of a temporary road, Smolenskaya region, USSR, in *Geosynthetics Case Histories*, G. P. Raymond and J. P. Giroud (eds.), on behalf of ISSMFE Technical Committee TC9, Geotextiles and Geosynthetics, 1993.

Koerner, R. M., *Designing with Geosynthetics*, 5th Edition, Pearson–Prentice Hall, Upper Saddle River, NJ, 2005, 296 pp.

Lawson, C., Retention criteria and geotextile filter performance, Geotechnical Fabrics Report, IFAI, Roseville, MN, August 1998.

Observatory nano briefing No.:23, Nano-enabled textiles in construction and engineering, March 2012.

Plaggenborg, B. and S. Weiland, Textile-reinforced concrete with high-performance carbon fibre grids, *JEC Composite Magazine*, 44 (October 2008).

Sarsby, R. W., *Geosynthetics in Civil Engineering*, Woodhead Publication, Cambridge, U.K., 2007.

Shukla, S. K., *Geosynthetics and Their Applications*, Thomas Telford, London, U.K., 2002.

Shukla, S. K. and J.-H. Yin, *Fundamentals of Geosynthetic Engineering*, Taylor & Francis Group, London, U.K., 2006.

10

Textiles in Automobiles

10.1 Introduction

The process of urbanization is increasing, and population growth continues at a rapid pace. Mobility is a basic requirement of most of the human activity. According to the latest forecast (Hunter, 2011), there will be more than 9 billion people living on our planet in 2050, and approx. 1.2 billion cars will occupy the world's roads. The automobile industry's yearly growth rate is expected to exceed 5.5% from 2010 to 2015, reaching a value of more than $5.1 trillion by 2015, according to a recent survey by MarketLine (2012). In line with this, the automobile industry has burgeoned into one of the most important business sectors in the last century. Automotive textiles is one of the most dynamic and promising sector of the industrial textiles. It is the single largest consumer of the industrial textiles with over one million tons per annum. Automotive textiles account for approximately 22% in the total market of industrial textiles. Automotive textiles are used for filters for air and fuels, vehicle awnings, air bag cushions, seat belts, composites for structural components, molded parts for interiors, interior decoration for molded parts, seat or protective covers, molded parts for seats, car floor coverings, drive belts, and hoses.

10.2 Automobile Industry: Global Scenario

In 1982, the number of vehicles (cars, trucks, and buses) manufactured around the world was 36.2 million. In 2012, the number of vehicles manufactured more than doubled to 84 million. China dominated the automobile industry in 2010 and 2012, producing 18.3 million and 19.3 million vehicles, respectively, and it has been able to increase its production rapidly since the early 1980. In 2012, China was followed by United States with 10.3 million vehicles, then

Japan with 9.9 million, and Germany with 5.6 million. According to estimates provided by the Roland Berger management consultancy company, approximately 18 million Chinese will purchase a new car in 2015 (Berger, 2010).

After a 9% decline in 2009, global car production immediately jumped back the following year with a 22% increase in 2010. In 2012, for the first time in history, more than 60 million passenger cars were produced in a single year. Fluctuations in vehicle manufacturing affected the automotive textile market consequently.

10.2.1 Automotive Textiles: Market Scenario

Automobile industry is growing to a great extent, so that, the share of automotive fabrics is also rising. Within this context, automotive textiles can make a very valuable contribution, because they primarily combine the desirable properties such as light weight, good sound insulation, UV resistance, rigidity, formability, and resistance to wear. As a result, experts are predicting that textile usage will have excellent growth prospects in the vehicles around the globe. The enormous variety of different ways of using automotive textiles in vehicle manufacturing provides huge economic potential. If the proportion of textiles in a medium-sized car still only accounts for 25 kg at the moment, this figure will increase to 30–35 kg in just a few years. Proportionately, 50%–60% will be nonwoven fabrics and felts, while 40%–50% will be other textile fabrics. The use of natural fibers or fiber mixtures in car manufacturing will also continue to grow partly because it is easier to recycle them, but mainly because natural fibers are lighter than chemical fibers and this can lead to weight savings of up to 40% per car (Source: IVGT, 2010).

The usage of spacer fabrics especially in vehicle interiors can facilitate effective climate control. The composite materials that are new or recycled, produced with reinforcement materials such as nonwoven, flocked fibers, or membranes can provide better sound insulation in vehicles. Recent trends in the automotive textiles sector are highly focused in the direction of safety and functionality. Today, safety is the major concern while designing a vehicle along with innovative functions that can be achieved using smart or intelligent textiles. About 70% of all technical innovations depend on the material properties—that is, the need for new materials is spurring on research and development.

In the present business scenario, the operations are almost standardized by the vehicle manufacturers in order to customize the vehicles to meet environmental and legal requirements, as well as the preferences of their end users. As the latest technological advancements are complemented with a touch of eco-concern, the automobile industry will experience an added push in the near future, doubling its results, marking an tremendous change. On an average, 165 tons of fabrics are used in car manufacturing process every year. The insistence for comfort, concern for safety while driving, and an increasing focus toward reducing fuel consumption and CO_2 emissions have all augmented the usage of specialty textiles in automobiles, especially cars. In a midsized car, 20 kg of textiles were used generally. The rise of necessity for sophistication, and a stringent urge for using protective textiles, the use of textile materials for the same has currently reached 26 kg and is even predicted to increase by 35 kg by the end of 2020. Two-thirds of the automotive textiles are used in trims, seat covers, roof, door liners, and carpets. Other fabric applications include tires, hoses, safety belts, and air bags.

Air bags, trims, and truck covers account for 28%, proving to be the largest share of the coated fabric demand. The U.S. demand for coated fabrics is forecast to increase 3.5% annually to 635 million square yards in 2016. Sales for the same increased 3.4% annually and reached $3 billion by 2012. The automotive textile sector takes the credit for this rise in sales. The rapid increase in automobile sales will motivate the automotive textile sector to further develop new products. While nonrubber-coated fabrics will also have a good market, rubber-coated fabric materials will be in high growth trajectory. Rubber-coated fabrics used in automobiles are expected to post enormous gains by 2013. Knitted and woven fabrics will have a dominating share in the global automobile fabric arena, closely followed by composites and nonwoven. Circular knitted fabrics are used in the interiors of cars for seat covers, door panels, headliners, headrests, boot covers, sunroofs, and parcel shelves. Having the virtues of high flexibility, comfort while traveling, stretchability, and high-grade visual quality, these fabrics will have good potential. Woven fabrics will see a profitable market in the making of door panels, seat covers, side door paneling, and headrests.

Lighter cars prove to be more economical, consume less fuel, and also involve less manufacturing costs. The application of natural-fiber composites have seen a tremendous growth in the automobile industry in recent years, due to its concept being benefited by the manufacturers, consumers, and the environment. Innovative materials and product development in this sector will further enhance its growth. The latest is the creative combination of seat belts and air bags. The belt cum air bag is made with round and smooth edges to give added comfort to its wearer. This process uses high-volume production of cold gas inflator, making the seat belt cum air bag to swell almost five times larger than the width of the normal seat belt. The swelling process is initiated through a deployment signal from a sensor system. This technique will minimize the extent of damage, as it distributes the force across the wearer's body. The interior fabrics of the car are made with a special kind of microfiber, which is ultralight, flameproof, and is also resistant to abrasion. This is believed to be 100% eco-friendly as well.

10.3 Major Components of Automotive Textiles

10.3.1 Seat Belts

According to World Health Organization, road accidents are a worldwide problem, and now over 1.2 million people die every year from road crashes all around the globe. Occupant restraints are the best solution such as seat belts and air bags, which are highly effective in preventing death and injury from traffic collisions. Seat belts are easy to use, effective, and inexpensive means of protection in an accident. Seat belts are expected to reduce the overall risk for serious injuries in crashes by 60%–70% and the risk for fatalities by about 45%.

Seat belt plays a crucial role in the safety of passenger during sudden accident. The wearing of both front and back seat belts is now compulsory in many countries of the world, and all new cars made contain at least four diagonal and lap seat belts each made from about 250 g of woven fabric. Due to ever-increasing concern over safety of passenger during accidents, seat belts show better holding position in industrial textile market.

10.3.1.1 Seat Belts: Market Scenario

The narrow fabric industry's revenue for 2011 was reported at $9 billion. A recent study carried out in six European cities shows that advanced seat belt reminders can increase seat belt use among drivers in urban areas up to rates of 93%–100% (European Transport Safety Council, 2006).

10.3.1.2 Seat Belts: A Lifesaving Guard

The seat belt is an energy-absorbing device that is designed to keep the load imposed on a victim's body during a crash-down to survivable limits. Fundamentally, it is designed to deliver nonrecoverable extension to reduce the deceleration forces, which the body encounters in a crash. Seat belts function as safety harnesses that secure the passengers in a vehicle against harmful movements during collision or similar incidents.

10.3.1.3 Seat Belts: Classification

Use of belt depends upon the weight of passenger; as per passenger's weight, belt width is specified by British standards (Table 10.1).

10.3.1.4 Seat Belts: Dynamics

a. *Work–energy principle*
   The change in the kinetic energy of an object is equal to the net work done on the object. For a straight-line collision, the net work

TABLE 10.1

Classification of Seat Belt

BS Standard	Application	Shoulder	Lap
BS 3254 Part 1 1988	"Restraining devices for adults" width [mm]	Min. 35	Min. 46
BS 3254 Part 2 1991	"Restraining devices for children" width [mm]	Weight 9–18 kg Weight 18–36 kg	Min. 25 Min. 38

done is equal to the average force of impact times the distance traveled during the impact.

$$W_{net} = \frac{1}{2}mv^2_{final} - \frac{1}{2}mv^2_{initial}$$

Average impact force × Distance traveled = Change in kinetic energy

If a moving object is stopped by a collision, extending the stopping distance will reduce the average impact force.

b. *Forces acting on passenger during accident with and without seat belt*
According to Newton's first law of motion, an object at rest tends to stay at rest and an object in motion tends to stay in motion with the same speed and in the same direction unless acted upon by an unbalanced force. Thus, driver continues in motion, sliding forward along the seat. A driver in motion tends to stay in motion with the same speed and in the same direction, unless acted upon by the unbalanced force of a seat belt. Seat belts are used to provide safety for passengers whose motion is governed by Newton's laws. The seat belt provides the unbalanced force, which brings driver from a state of motion to a state of rest.

The task of the seat belt is to stop the driver with the car so that the stopping distance is probably four or five times greater than if the driver is with no seat belt. A crash that stops the car and the driver must take away all its kinetic energy, and the work–energy principle then dictates that a longer stopping distance decreases the impact force (Figure 10.1).

Assume car speed is 13.41 m/s, driver's weight is 50 kg, then what will be the impact force acting on the car driver?

1. *If driver is wearing a nonstretchable seat belt, stopping distance is 0.304 m.*

$$\text{Impact force on driver} = \frac{(0.5 \times 50 \times 13.41 \times 13.41)}{0.304}$$

$$= 14788.5 \text{ N} = 1.5 \text{ tons}$$

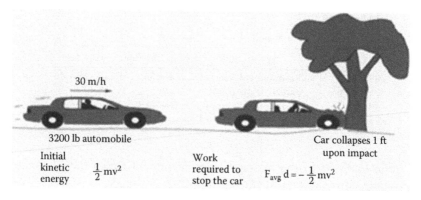

30 m/h

3200 lb automobile

Initial kinetic energy $\frac{1}{2}mv^2$

Work required to stop the car

$F_{avg}\, d = -\frac{1}{2}mv^2$

Car collapses 1 ft upon impact

FIGURE 10.1
Forces acting during car crash.

$$\text{Deceleration of driver} = \frac{14788.5}{50} = 295.77 \text{ m/s}^2$$

With no seat belt to stop the driver with the car, the driver flies free until stopped suddenly by impact on the steering column, windshield, etc. The stopping distance is estimated to be about one-fifth of that with a seat belt, causing the average impact force to be about five times as great. The work done to stop the driver is equal to the average impact force on the driver times the distance traveled in stopping. A crash that stops the car and the driver must take away all its kinetic energy, and the work–energy principle then dictates that a shorter stopping distance increases the impact force (Figures 10.2 and 10.3).

2. *If driver is without seat belt, suppose stopping distance is 0.0608 m*

$$\text{Impact force on driver} = \frac{(0.5 \times 50 \times 13.41 \times 13.41)}{0.0608}$$

$$= 73942.5 \text{ N} = 7.5 \text{ tons}$$

$$\text{Deceleration of driver} = \frac{73942.5}{50} = 1478.8 \text{ m/s}^2$$

3. *If driver is wearing a stretchable seat belt, stopping distance is 0.456 m*

$$\text{Force on driver} = \frac{(0.5 \times 50 \times 13.41.13.41)}{0.456} = 9467.65 \text{ N} = 0.96 \text{ ton}$$

$$\text{Deceleration of driver} = \frac{9467.65}{50} = 189.353 \text{ m/s}^2$$

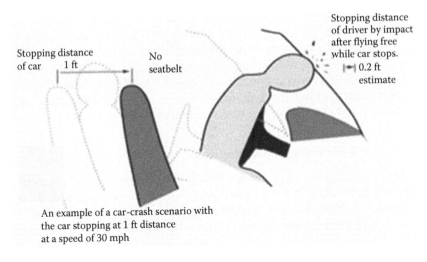

Stopping distance
of car   1 ft

No
seatbelt

Stopping distance
of driver by impact
after flying free
while car stops.
0.2 ft
estimate

An example of a car-crash scenario with
the car stopping at 1 ft distance
at a speed of 30 mph

FIGURE 10.2
Forces acting without seat belt.

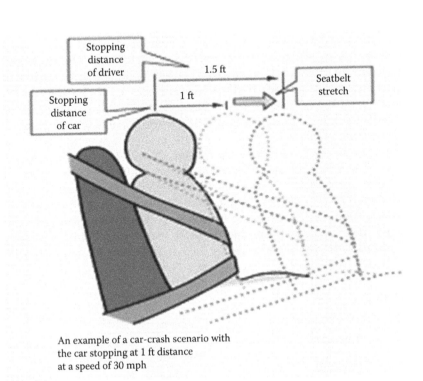

Stopping
distance
of driver

1.5 ft

Seatbelt
stretch

Stopping
distance
of car

1 ft

An example of a car-crash scenario with
the car stopping at 1 ft distance
at a speed of 30 mph

FIGURE 10.3
Forces acting with seat belt.

A moderate amount of stretch in a seat belt harness can extend the stopping distance and reduce the average impact force on the driver compared to a nonstretching harness. If the belt is stretched for 0.125 m, it would reduce the deceleration to 189.353 m/s² and the average impact force to 0.96 ton compared to 295.77 m/s² and 1.5 tons for a nonstretching seat belt. Either a stretching or a nonstretching seat belt reduces the impact force compared to no seat belt.

10.3.1.5 Seat Belts: Critical Characteristics

The critical characteristics of the seat belt are abrasion resistance, resistance to light and heat, capable of being removed and put back in place easily, and good retraction behavior. The load-bearing capacity of seat belt is 1500 kg. The surface of the seat belt (generally smooth) is of particular significance because its structure and properties decisively influence the retraction behavior.

10.3.1.6 Seat Belts: Fibers and Fabric Structure

The widely used fibers for manufacturing seat-belt webbings are nylon and polyester. Nylon was utilized in some early seat belts, but due to its higher UV degradation resistance, polyester is now widely used worldwide. Polyester dominates over nylon because of its lower extensibility and higher stiffness.

The structure of seat belt can be single layer or double layer with weave patterns such as plain weave, twill weave, satin weave being widely adopted and manufactured on needle looms. Normally 2/2-twill structure is used. The needle loom used presently is shuttleless and capable of delivering over 1000 picks/min (Table 10.2).

Filament yarn made of nylon or polyester are woven to produce seat-belt webbing. The linear density of synthetic yarns should be between 100 and 3000 decitex, preferably 550–1800 decitex. The filament linear density should be between 5 and 30 decitex, preferably 8–20 decitex. A typical seat belt is made of 320 ends of 1100 decitex polyester each. Most weft yarns made from polyester are 550 decitex. Commonly used seat belt webbing is a narrow fabric measuring 46 mm wide. This structure permits highest yarn packing

TABLE 10.2

Seat Belt: Fabric Structure

Material	Body Warp Yarn	Selvedge Warp Yarn	Weft Yarn	Binder Thread	Locking Thread
PA/PET	1260 D/108 (fil.)	250 D/24 (fil.)	500 D/48 (fil.)	100 D/18 (fil.)	250 D/24 (fil.)
PA/PET	846/2 D	846 D (mono)	423 D (multi)	220 D	150 D
			396 D (mono)		

within a given area for highest strength and sometimes-coarser yarns are used for good abrasion resistance.

10.3.1.7 Seat Belts: Manufacturing Method

Seat belt is manufactured in a needle loom in which weft is inserted through the warp sheet and a selvedge is formed. The other end of webbing is held by an auxiliary needle, which manipulates a binder and a lock thread. Once these are combined with the weft yarn, a run-proof selvedge is created. The selvedge construction is done with utmost care to ensure that it has superior abrasion resistance. It is equally important to ensure that the selvedge is soft and comfortable to wear.

Needle is used in the loom to insert monofilament and multifilament weft yarn across the width of the web in single shed. High tension is applied on monofilament compared to multifilament yarn so that it does not protrude beyond selvedge. Monofilament weft yarns are rigid yarn that provides good stiffness across web and high resiliency. Monofilament weft yarns are under high tension which maintains webbing softness and round shape together with catch cord yarns and finer warp yarns in selvedge than body of the webbing. Low longitudinal stiffness and thinner web give good winding performance with the use of smooth monofilament weft yarns. Selvedge warps are having about 66% lower thermal shrinkage than body warp threads. Weft threads are having 16% higher dry heat shrinkage than selvedge yarns. Thus woven ribbon is heated after dyeing. This gives soft, round shape selvedge.

10.3.1.8 Seat Belts: Finishing

The woven seat belt webbing is processed under tension in dyeing and finishing machines. The grey webbing is subjected to dyeing and heat setting, while webbing made up of spun-dyed yarn is heat set only. Heat setting is carried out to impart precise extension to the webbing and suppress its recovery in the event of crash. This is achieved by subjecting the webbing to compressive shrinkage treatment in a controlled manner, which increases the weight of webbing from 50 g/m to approximately 60 g/m. Seat belt webbing can be colored by two methods: either by incorporating spun-dyed yarns or by piece dyeing. The appropriate technique for piece dyeing polyester webbing is by thermosol process.

10.3.1.9 Seat Belts: Quality Requirements

The seat belt is required to have the following properties:

- Static load-bearing capacity up to 1500 kg and extensibility up to 25%–30%
- Abrasion resistance

- Heat and light resistance
- Lightweight
- Flexibility for use

10.3.1.10 Performance Tests and Standards for Seat Belt Webbing

Tests are carried out for ascertaining the mechanical performance of the webbing as per BSI and the SAE.

- Width
- Thickness
- Breaking strength and elongation for dry and wet webbing
- Abrasion resistance at various environmental conditions and in contact with buckles and fittings
- Influence of extreme environmental conditions and temperatures
- Rubbing fastness
- Microbial resistance

10.3.1.11 Seat Belts: Various Defects

Seat belts are designed to withstand tremendous loads, but if there is any damage to a seat belt, its load-bearing ability is significantly reduced.

1. *Fraying*
   Fraying is one of the most noticeable defects in seat belts. Fraying is where fibers break in the weaving. This results in reduced strength. Seat belts should be repaired or replaced at any sign of fraying. Another form of fraying that is encountered appears as a fuzzy layer on the seat belt surface. This fuzziness may appear harmless, but it is indicative that fibers are breaking in the weave. The seat belt should be replaced or repaired.
2. *UV damage*
   Over time, UV rays will damage seat belts. UV damage is more frequent in helicopters and aircraft.

10.3.1.12 Inflatable Belts

Ford® introduces the auto industry's first-ever production of inflatable seat belts, which are designed to provide additional protection for rear seat occupants, often children and older passengers who can be more vulnerable to head, chest, and neck injuries. The inflatable seat belt is an amalgamation of the air bag and the seat belt, which is held by weak stitches that burst open when the belt is inflated. Under impact, the belt gives 450% more surface area than the normal flat belt. This inflatable rear seat belts spread crash forces over five times more area of the body than conventional seat belts; this helps reduce pressure on the chest and helps control head and neck motion for rear seat passengers. These belts could be fitted in the rear seats of the automobile to replace the use of air bags in that compartment.

A folded envelop of 3" is covered by 2" face fabric, which is stitched. A folded envelope is rubber coated. Today's envelopes are made from urethane or silicone to make thin, gas leakage proof, high strength, heat resistant, and when crash occurs, a gas fills and inflates the envelope. The face fabric (web) is made of a resin such as PET. Today's cover fabric uses warp-knitted structure that has good strength and good stretching characteristics and is comfortable. Warp knit uses yarn of 3000 denier or below. By using such belts, kinetic energy of passenger is distributed over a larger area; therefore, load experienced by a passenger is small and passenger is protected very effectively.

10.3.2 Air Bags

An air bag is an occupant restraint system consisting of a flexible envelope designed to inflate rapidly during a vehicle collision. Air bags are gas-inflated cushions built into the steering wheel, dashboard, door, roof, or seat of the car that use a crash sensor to trigger a rapid expansion to protect you from the impact of an accident. Air bags are a type of automobile safety restraint like seat belts. There are several kinds of air bags—driver bag, front passenger bag, thorax bag, curtain bag, rear bag, and knee bag.

In most countries, air bags are mandatory for all passenger cars due to stringency in legislation. According to a recent survey, air bag system has contributed up to 20% reduction in fertilities resulting from front collision. Earlier air bags were considered as a substitute to seat belts and were limited only to high-speed sports cars. But today, air bags are working in coordination with seat belts. In fact, air bags cushion an occupant in an event of crash. This helps in avoiding the heat on collision.

Three decades ago, Mercedes-Benz® introduced the air bag in Germany as an option on its high-end S-Class (W126). In this system, the sensors would automatically pretension the seat belts to reduce occupant's motion on impact and then deploy the air bag on impact. This integrated the seat belt and air bag into a restraint system, rather than the air bag being considered an alternative to the seat belt. In 1987, the Porsche® 944 turbo became the first car in the world to have driver and passenger air bags as standard equipment.

In Europe, air bags were almost entirely absent from family cars until the early 1990s. The first European Ford to feature an air bag was the face-lifted EscortMK5b in 1992; within a year, the entire Ford range had at least one air bag as standard. By the mid-1990s, European market leaders such as

Vauxhall/Opel, Rover, Peugeot, Renault, and Fiat had included air bags as at least optional equipment across their model ranges. During the 2000s, side air bags were commonplace on even budget cars, such as the smaller-engined versions of the Ford Fiesta and Peugeot 206, and curtain air bags were also becoming regular features on mass market cars.

10.3.2.1 Air Bag: Market Scenario

In 2010, the world market for automotive air bags stood at 258 million units, but by 2017, this figure is predicted to rise to 446 million units, representing an increase of 73.2%.

Front air bags are expected to remain the biggest category of air bags in 2017. However, forecasts for the 7 years to 2017 suggest that they will form the slowest growing category whereas sales of side-impact air bags, inflatable curtains, and knee air bags will grow considerably faster.

As a result, the share of front air bags in the total market for air bags will drop from 42.2% to 37.9% between 2010 and 2017 although sales of front air bags are still forecast to increase by 55.2%. The fastest increase, at 539.0%, will be in the sales of knee air bags. However, these will continue to represent by far the smallest category with sales of just 33 million units in 2017, representing 7.4% of the market. Sales of side-impact air bags are set to rise by 74.7% to 128 million units, representing a 28.6% share of the market, while sales of curtain air bags will increase by 65.2% and account for a 26.1% share.

10.3.2.2 Air Bag: Principle of Operation

Air bags inflate at a very quick rate instantly after collision, the rate which is faster than the blink of an eye. During the first 15–20 ms immediately after the collision, air bag sensors detect the crash and then send a electrical signal to ignite the air bags. Typically a squib, which is a small explosive device, ignites a propellant, usually sodium azide. The azide burns with higher intensity, liberating nitrogen, which inflates the air bags. The air bag is supposed to be fully inflated within 45–55 ms. Within 75–80 ms, the air bag is deflated and the event is over (Figure 10.4).

The proper functioning of air bag dramatically reduces the chance of occupants' death or serious injury. However, the speed with which air bags inflate generates tremendous forces. Passengers in the way of an improperly designed air bag can be killed or significantly injured. Unnecessary injuries also occur when air bags inflate in relatively minor crashes when they're not needed.

10.3.2.3 Laws of Motion

The object which tends to move have momentum (the product of the mass and the velocity of an object). The object will continue to move at its existing

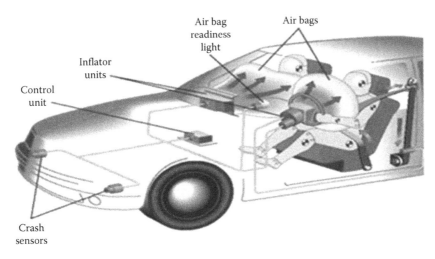

FIGURE 10.4
Principle of air bag operation.

speed and direction, unless an outside force acts on it. A car consists of several objects, such as vehicle itself, loose objects in the car, and passengers. These objects will continuously move at the speed as whatever speed the car is moving at, even if the car is stopped by collision (Figure 10.5).

The force acting over a period of time is required to stop an object's momentum. If a car collides violently with another vehicle, the force required to stop an object is very high, because the momentum of the car has changed immediately while the passengers' has not, as there is not much time to work with. The purpose of fitting any occupant-restraint system in a vehicle is to prevent the passenger from death and major injury.

The main intention of fitting air bags in the car is to slow down the passenger's speed to zero so that he/she can survive with little or no injury. The air bag inflates in a fraction of a second in the small space between the passenger and the steering wheel.

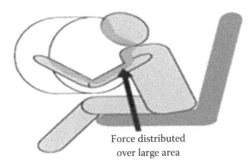

FIGURE 10.5
Force distribution on body during collision.

10.3.2.4 Air Bag Inflation

The air bag is intended to counteract the passenger's forward motion as uniformly as possible in milliseconds. The three main components of an air bag system responsible for achieving this task are the bag, a sensor, and a mechanical switch.

1. The air bag is a thin bag made up of nylon fabric which is folded inside the steering wheel or dash board or seat or door.
2. The crash sensor passes the signal to initiate the ignition, which inflates the bag. Inflation occurs as soon as the vehicle collides.
3. A mechanical switch is turned over suddenly when there is a mass shift that closes an electrical contact, signaling to the sensors that a crash has occurred. The sensors receive information through an accelerometer built into a microchip (Figure 10.6).

An air bag contains a mixture of inorganic compound sodium azide (NaN_3), KNO_3, and SiO_2. The driver side air bag normally contains approximately 50–80 g of NaN_3, with the larger passenger-side air bag containing about 250 g. Within about 40 ms of impact, all these components react in three separate reactions that liberates nitrogen gas. The reactions, in order, are as follows:

$$2\,NaN_3 \rightarrow 2\,Na + 3\,N_2(g)$$

$$10\,Na + 2\,KNO_3 \rightarrow K_2O + 5\,Na_2O + N_2\,(g)$$

$$K_2O + Na_2O + 2\,SiO_2 \rightarrow K_2O_3Si + Na_2O_3Si\,(silicate\ glass)$$

Hot blasts of nitrogen gas inflate the air bag as a result of chemical reaction.

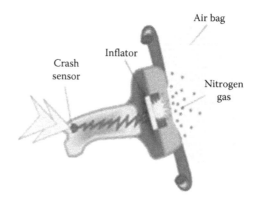

FIGURE 10.6
The air bag and inflation system configured in the steering wheel.

Air bag inflation device

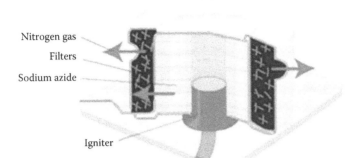

Nitrogen gas

Filters

Sodium azide

Igniter

FIGURE 10.7
The inflation system uses a solid propellant and an igniter.

Researchers made an early attempt to incorporate air bags in cars which got bumped due to prohibitive prices and technical obstacles involving the storage and release of compressed gas.

The igniter in the air bag system helps to ignite the chemical mixture, which burns rapidly to liberate huge volume of nitrogen gas to inflate the bag. The bag then immediately bursts at a rate of 200 mph which is faster than the an eye blink. After a second of successful inflation of the air bag, the gas dissipates through the tiny holes in the bag in a quick manner,facilitating the deflation of the bag so that the passenger can move.

The entire process of inflation and deflation happens in 1/25th of a second, the remaining time is sufficient to treat the passenger for injury.

10.3.2.5 Types of Air Bags

1. *Air bags for frontal injury*
   Air bags are specially designed to deploy in frontal and near-frontal collisions, which are comparable to hitting a solid barrier at approximately 13–23 km/h (8–14 mph). The barrier collision at the rate of 14 mph is roughly equivalent to hitting a parked car of same size across full front of each vehicle at about 29 mph. This is because the parked car absorbs some of the energy of the crash and is pushed by the striking vehicle (Figure 10.8).

2. *Side air bags for cars*
   The side air bags, a relatively new technology designed to protect the head and/or torso (chest and abdomen) in side-impact collisions, are becoming increasingly common in automobiles (Figure 10.9).

FIGURE 10.8
Air bag system for frontal collision.

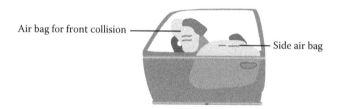

FIGURE 10.9
Side air bags for cars.

3. *Air bags for motorcycle*

Honda Motor Co., Ltd., announced that it has succeeded in developing the world's first production motorcycle air bag system. The motorcycle air bag system comprises the air bag module, which includes the air bag and the inflator; crash sensors, which monitor acceleration changes; and an ECU, which performs calculations to instantly determine when a collision is occurring (Figure 10.10).

4. *Pedestrian protective air bags*

The air bag device for pedestrian protection according to the present invention is able to prevent a pedestrian mounted on the hood panel from falling down on road surface. The air bag device for pedestrian protection according to the present invention is mountable on a motor vehicle with a hood panel and includes an air bag inflatable in front of the vehicle.

10.3.2.6 Raw Materials (Fibers) for Air Bags

Air bag fabrics are made of nylon 6,6 multifilament yarns with counts from 235 to 940 Tex. The number of monofilaments in the multifilament yarn

FIGURE 10.10
Air bags in motorcycle.

ranges from 70 to 220. Air bag fabrics are normally dense in construction. The critical properties considered during the designing of air bag are tensile strength, elongation, tear strength, and weight requirements. Air permeability should be uniform across the whole width of the air bag fabric. There are currently two principal material types that are used in the manufacture of air bags. They are uncoated nylon (polyamide 66) and coated nylon. Two types of commonly used coatings are silicone and neoprene. In general, coated materials are used for driver's side air bags and side impact bags, while passenger-side air bags are made from uncoated nylon materials. The advantages of the coated type are better nongas permeation, easier bag-pressure control, and greater heat resistance to burning particles. The noncoated type is lighter in weight, thinner, more flexible, and less expensive (Table 10.3).

Although nylon 6,6 and polyester have similar melting points, the large difference in specific heat capacity causes the amount of energy required to melt polyester to be about 30% less than that required to melt nylon 6,6. Hence in any inflation event that uses a pyrotechnic or pyrotechnic-containing

TABLE 10.3

Comparison of Fiber Properties of Air Bag

Properties	Nylon 6,6	Polyester
Density (kg/m³)	1140	1390
Specific heat capacity (kJ/kg/k)	1.67	1.3
Melting point (°C)	260	258
Softening point (°C)	220	220
Energy to melt (kJ/Kg)	589	427

inflator, cushions made from polyester yarn are far more susceptible to burn or melt-through in the body of the cushion or at the seam.

The second advantage of nylon 6,6 is its lower density. Lower mass has key advantages—reducing the mass of the cushion lowers the kinetic energy of impact on the occupant in out-of-position situations, thus enhancing safety, while allowing the overall weight of the vehicle to be reduced.

10.3.2.7 Manufacturing of Air Bag Fabrics

Air bags are made from a fabric that is coated on at least one of the surfaces. The air bag fabric may be coated to increase the strength of the fabric so that the fabric is reinforced. A coated air bag base fabric made of a textile fabric that has an excellent air barrier property, high heat resistance, improved mountability, and compactness and excellent adhesion to a resin film is characterized in that at least one side of the textile fabric is coated with resin, at least parts of the single yarns of the fabric are surrounded by the resin, and at least parts of the single yarns of the fabric are not surrounded by the resin. An air bag is characterized by using such a coated air bag base fabric. A method for manufacturing the coated air bag base fabric is characterized by applying a resin solution having a viscosity of 5–20 Pas (5,000– 20,000 cP) to the textile fabric using a knife coater with a sharp-edged coating knife at the contact pressure between the coating knife and the fabric of 1–15 N/cm.

10.3.3 Automotive Filters

The capability of an automobile is boosted extensively by the usage of appropriate filter media in the engine compartment and in various other places of the automobile. The incorporation of filter media in an automobile is intended to filter and retain the contaminant present in the filtrate. The main aim of the filter is to optimize the filtration performance with the required cleanliness level.

10.3.3.1 Materials and Functions of Filters

The filter media range from mesh screens to depth-style media such as threads or chopped paper to 100% natural cellulose to 100% man-made fibers to almost any conceivable combination in between. Table 10.4 outlines an overview of filtration applications in the automotive industry.

It is a difficult task to choose the right type of filter medium for an engine system due to availability of a wide range of filter media. Cellulosic filters are basically used to retain only larger contaminants. As the size of contamination to be removed gets smaller and smaller, the type of media changes from more complex cellulose to blended media where cellulose and man-made fibers are blended together in various configurations. The filter media made from synthetic fibers is recommended over cellulosic media, for

TABLE 10.4

Automotive Filtration Applications and Filter Media

Application	Type of Filter Media
Carburetor air filters	Mainly nonwoven (wet, dry, needled, or spun bonded)
Engine oil filters	Resin-impregnated wet-laid nonwoven (paper)
Fuel tank filters	Activated carbon
Cabin interior filters	Electrostatically charged fiber media, nonwovens, activated carbon, specialty paper
Diesel/soot filters	Ceramic materials
ABS wheel/brake filters	Metal or fiber woven screens
Power steering filters	Mainly screen fabrics
Transmission filters	Woven fabrics or needle felts
Wiper washer screen filters	Woven fabrics
Air-conditioning recirculation filters	Nonwoven/activated carbon
Crank case breathe filters	Nonwovens

retaining extremely small contamination. The reason attributed for choosing man-made fibers is its level of achievable fineness. Filter media made of finer fibers have higher surface area which helps to achieve higher filtration efficiency. Recently, nanofiber materials are becoming popular.

10.3.3.2 Air Filters

Air is essential to vehicles' engine which gets mixed with fuel, ignited, and with the resulting controlled explosion provides power to vehicle. It takes between 10,000 and 12,000 gal of air for each gallon of fuel. The only way for air to enter the engine is through the air intake after passing through the air filter (Figure 10.11), which is essential in removing contaminants such as dirt particles, dust, and debris from the air penetrating the engine, which can cause damage to engine cylinders, wall, pistons, and piston rings while allowing high volume of air to pass through. If the air filter is clogged up, engine performance is reduced, engine power decreases, and engine wear is increased.

An established site for the air filter in the engine of an automobile is in the carburetor. The filter stationed here is normally a wet-laid nonwoven material. Commonly this medium is a mixture of cellulose fibers that are derived from the wood pulp, with small amount of synthetic fiber. The fiber scaffold is imparted with stiffness by a resin binder. Air filters can consist of 1–20 layers, which are normally pleated to increase the surface area (Figure 10.12). In multilayer, the densest layer resides on the exit plane and a less dense layer covers the entry. This arrangement confers a gradient density capability whereby gradually smaller particles are held in the deeper layers of the filter. Another version of this medium is available in which the fibers are more intermingled (Figure 10.12).

FIGURE 10.11
Air filters.

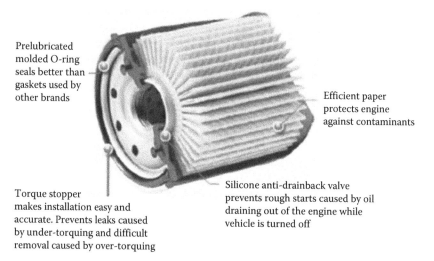

Prelubricated molded O-ring seals better than gaskets used by other brands

Efficient paper protects engine against contaminants

Torque stopper makes installation easy and accurate. Prevents leaks caused by under-torquing and difficult removal caused by over-torquing

Silicone anti-drainback valve prevents rough starts caused by oil draining out of the engine while vehicle is turned off

FIGURE 10.12
Air filter technology.

Apart from these two techniques, it is also possible to manufacture the filter by dry laying. Generally, this kind of filter consists of a fleecy top web, an open-structured inner layer, and a dense lower layer. This type of filter is composed of a blend of synthetic and cellulosic fibers. At present, most development on the carburetor air filter is being conducted in Japan, where Toyota has designed a more expensive medium that is made from needled felt and spun-bonded material.

Filtration through an engine-mounted air filter is accomplished through four mechanisms. The first mechanism is where foreign particles adhere to the media via the impact impingement mechanism. Here, dirt strikes the filter and attaches itself permanently. In the second mechanism, dirt can collect as cluster aggregates, which are collection of particles, created by the triboelectric nature of fibers or the electrostatic charge that is generated during fluid flow. In the third, dirt can be trapped between the pores of the filter media, particularly when they become clogged and reduced in size. The last opportunity for a filter to capture dust arises from the interaction between fluid flow and pressure drop. When all these mechanisms are employed, the total void volume of an air filter can be filled to 90% by the end of its useful life.

10.3.3.3 Oil Filters

Dirt is one of the major causes of engine wear. Dirt particles are carried by the oil into the high-precision clearances between bearings and other moving parts, which abrades and reduces the life of the parts. These dirt particles grind and groove the surfaces of the working parts, modifying clearances, and generating more debris which are abrasive in nature. The repetition of this abrasive action makes the precision components become weak and fatigued, which reduces its life. Apart from deteriorating engine components, the oil that provides vital engine lubrication also get dirty due to these contaminants. Dirty carbon particles generated during combustion can be forced past piston rings and into the oil. These particles by their very nature act like tiny sponges, absorbing critical additives, thus shortening oil life.

The by-products of the combustion process chemically react in the presence of moisture to liberate corrosive and rust-producing acids. Typically about 80% of engine wear is due to contaminants <10 μm. The function of oil filter is to remove soot, rust particles, and other solid contaminants from the oil, providing maximum protection and safety to engine (Figure 10.13). Figure 10.14 shows the oil filter technology. The earliest incarnations of these filters were made from woven, mesh, and paper media that caught dirt by surface phenomenon. Resin-bonded polyester needle felts are capable of trapping smaller contaminants within its structure. In an important development, nanofiber oil filters are made with premium advanced synthetic media technology that results in fibers that have a controlled size, shape, and smaller fiber diameter. The modern filter-media-manufacturing process aims to produce media with low density, greater surface area, and small pore size that is highly possible with the nanofiber oil filters, which deliver both higher dirt-holding capacity at differential pressure and higher filtration efficiency.

Oil filters made up of cellulosic fibers have high variations in size and shape, permitting more contaminants to pass through, resulting in pressure drop due to higher restriction and lower throughput. Synthetic filter media offer better durability with usage. Oil filters based on resin-bonded cellulosic fibers are prone to resin degradation when they react with hot oil.

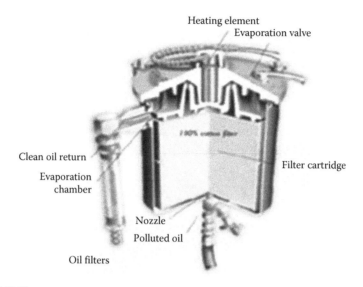

Heating element
Evaporation valve

Clean oil return
Evaporation
chamber

Filter cartridge

Nozzle
Polluted oil

Oil filters

FIGURE 10.13
Oil filter.

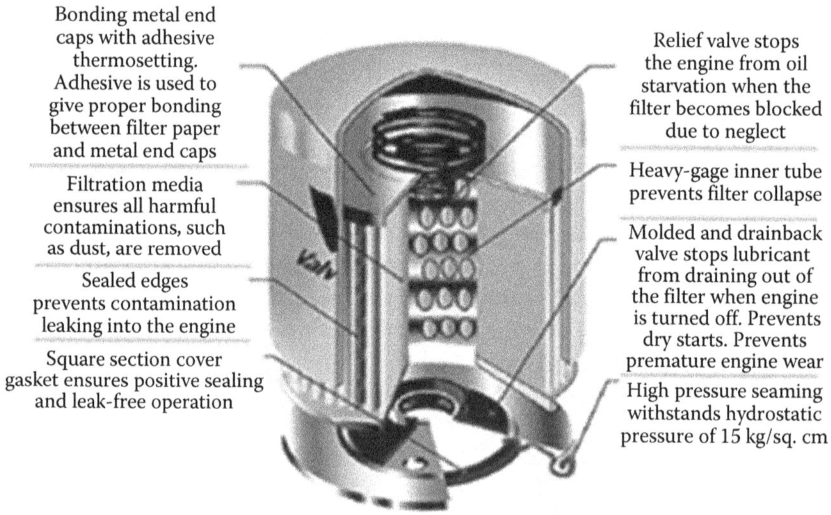

Bonding metal end
caps with adhesive
thermosetting.
Adhesive is used to
give proper bonding
between filter paper
and metal end caps

Filtration media
ensures all harmful
contaminations, such
as dust, are removed

Sealed edges
prevents contamination
leaking into the engine

Square section cover
gasket ensures positive sealing
and leak-free operation

Relief valve stops
the engine from oil
starvation when the
filter becomes blocked
due to neglect

Heavy-gage inner tube
prevents filter collapse

Molded and drainback
valve stops lubricant
from draining out of
the filter when engine
is turned off. Prevents
dry starts. Prevents
premature engine wear

High pressure seaming
withstands hydrostatic
pressure of 15 kg/sq. cm

FIGURE 10.14
Oil filter technology.

The synthetic media technology uses a wire screen backing pleated with the media, resulting in superior strength. Nanofiber oil filters offer extended service intervals, greater engine protection to prolong engine and equipment life, improved lubricant flow, improved cold start performance, and lower operating costs.

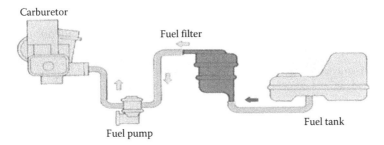

FIGURE 10.15
Passage of fuel in fuel filter.

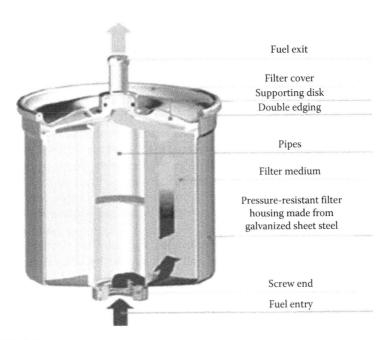

FIGURE 10.16
Fuel-filtration technology.

10.3.3.4 Fuel Filters

Fuel is a volatile petroleum liquid that is used to power the engine. The fuel that enters the engine must first pass through the fuel filter as shown in Figures 10.15 and 10.16, which is essential in helping to protect fuel system components such as plug fuel injectors, carburetors, etc., from contaminants that may be present in the fuel. In an extremely cold climate, they are needed to prevent engine trouble caused by water freezing in the fuel (see Table 10.5).

An engine will exhibit greater performance with higher power, greater fuel economy, and lower emissions, if it receives fuel with less contaminants.

TABLE 10.5

Types of Contaminants and Their Associated Problems in Automobiles

Contaminants	Problems
Particulate and debris	Enters when fuel is transferred between storage tanks. Particulate in the fuel can disrupt engine combustion and cause wear on injectors
Water	Water in the fuel causes corrosion and reduces the lubricity of fuel. It can negatively affect the combustion process and consequently damage system components. Water enters the fuel from storage tanks
Wax/paraffin	Dropout of fuel in cold weather conditions
Microbes (bacteria)	Can grow in the water at the fuel interface
Fuel degradation products	Fuel by-products result from the thermal and oxidative instability of fuel prior to combustion
Asphaltenes	Found naturally in crude oil and can be found in refined fuel
Air	Enters the system from leaks in the fuel line or system connections

Contaminations can originate from dirty and rusty service station storage tanks and, as the vehicle ages, from corrosion within the fuel system components. The major contaminants present in the fuel are fungus, bacteria, wax, water, asphaltite, sediment, and other solid matters. Water is the greatest concern because it is the most common form of contaminant that is likely to be introduced into the fuel during fueling due to poor climatic control and housekeeping practices in fuel storage tanks. Water can cause a tip to blow off an injector or reduce the lubricity of the fuel, which can cause seizure of close tolerance assemblies such as plungers. Further the microorganisms (also called humbugs) live in water and feed on the hydrocarbons found in fuel, which cause plugging of a fuel filter through multiplying colonies spreading throughout a fuel system.

Bacteria may be any color, but is usually black, green, or brown. Draining the system will reduce microbial activity, but will not eliminate it. The possible way to avoid microbial growth once it has started is to clean the system and have a biocide treatment. Temperatures below a fuel's cloud point will result in wax precipitation and filter plugging. To prevent plugged filters due to wax formation, the cloud point of fuel must be at least +12°C (+22°F) below the lowest outside temperature.

Asphaltites are components of asphalt that are generally insoluble and are generally present to some extent in all fuel. These black, tarry asphaltites are hard and brittle, and are made up of long molecules. Fuel with a high percentage of asphaltites will drastically shorten the life of a fuel filter. Sediment and other solids often get into fuel tanks and cause problems. Most of the sediment can be removed by settling or filtration. Fuel filters designed for specific applications will remove these harmful contaminants before they cause further system wear and damage.

Pleated cellulosic base materials are widely used in fuel-filter media. The developed fuel filter has undergone compatibility tests with a variety of diesel fuels including biodiesel. The filter media is designed in such away that larger contaminants get trapped on the outer layer and finer contaminants are held deeper in the media. Cellulose media is usually treated with silicone for effective water separation. This cellulosic filter media is typically fitted on the suction side of the fuel system for the removal of harmful water and coarser contaminants. Water coalesces on media and drains to the bottom of a can/bowl. The third-generation fuel filter water separator media uses both cellulose and melt-blown synthetic layers to achieve the highest levels of fuel filtration performance. This double-layered media increases particulate holding capacity and is a high-performance water separator. It has the ability for high-efficiency emulsified water separation and can be used in both suction and pressure sides of fuel systems. The polyester layer improves water separation and dirt-holding capacity performance.

The first petrol filters were made from wire mesh, and although they were efficient, they were unable to separate water from the passing fuel. New filters based on vinylidene chloride polymers (Saran® monofilament) are produced by various companies under different trade names. The filaments are made by melt spinning and are then stretched. Some of their significant properties are resistance to water, fire, light, and bacterial attack. Saran® is a prime material for petrol filters and is resistant to automotive fuel, delivers high mechanical strength and recovery, is a flame retardant, and does not absorb water. In addition, the Saran® filter is able to prevent the ingress of air in to the fuel tank. The success of the filter is also due to its wicking capability. To facilitate wicking, the surface tension of the Saran® filter fiber is greater than the critical surface tension of the fuel. Wicking ensures that the filter is constantly seeped into the fuel in the presence of air and fuel vapor. Indeed, even when the fuel tank is nearly empty, a fine film of petrol collects over the filter's surface and prevents air from penetrating its structure. Overall, the wicking function is governed by the filter medium and its construction, chiefly by the choice of fiber, finish, and pore distribution. Most Saran filters are fitted into vehicles that employ a carburetor system. Here, when the engine is running, the flow of fuel is variable and intermittent. The role of the Saran filter is to intercept any contamination and prevent it entering its structure or passing into the fuel tank. Particles cannot adhere to its surface as filter cloth is exceptionally smooth. Instead, the filter performs a self-cleaning operation called "back washing," whereby impurities are shed from its surface when car stops.

Filter units are sized by flow rate, expressed in gallons per minute or hour. Since a four-cycle diesel returns two to four times as much fuel as it burns and two-cycle diesel returns five or six times, a handy rule of thumb is to multiply peak engine fuel consumption by 3.5 or 4 and then divide by 60 to get the fuel flow rate in gallons per minute for a four-cycle diesel.

10.3.3.5 Cabin Filters

The quality of the air inside the vehicle is a major factor in protecting the health and safety of the driver and passengers. The level of air-borne pollution ranges from 0.05 to 0.5 mg/m³. The particles that are larger than 3 µm can be trapped in human breathing channels and are responsible for instigating allergic reactions like hay fever and asthma. The cabin filters fitted in an automobile is responsible to filter and retain dust, bacteria, pollen, and exhaust gases that may pass into a automobile ventilation system, keeping the car interior a healthier and comfortable place.

Cabin filters are generally located under the hood of vehicles, inside the glove compartment, or under the dashboard, and most are within easy reach for quick replacement. When the vehicle is in motion, or when the ventilation system is in use, all impurities that are present in the outside air are sucked into the cabin like a vacuum. The cabin filter is normally a pleated paper filter which is incorporated in the outside-air intake for the automobile's passenger compartment. The cabin air filters also shields the heating and air-conditioning system from pollution, guarantee a clear view, and thus contribute to safer driving.

Cabin filters are classified as pollen filter, combination filter, and particulate filter. The type of cabin filter used is decided by environment conditions and vehicle type. A pollen filter removes spores, pollen, particulates, bacteria, fungi, and road dust. Larger particles (>3 µm) are mechanically separated from the airflow by the dense weave of the filter fabric, although extremely tiny particles such as diesel soot may not be removed (Figure 10.17).

A composite filter with a laminated construction consisting of particle filter fabric, activated carbon layer, and substrate medium can absorb both particles like the pollen filter and odors and harmful gases. Activated carbon has an extremely higher surface area that can be subdivided into different pore sizes. A filter of this type can noticeably reduce and even eliminate hydrocarbon molecules, benzene, odors, and other gases. Ozone decomposes almost completely into harmless oxygen when it contacts activated carbon.

Particulate filter is a multilayer design composed of a prefilter, an electrostatically charged microfiber layer, and a cover layer. The prefilter is made from coarse polyester fibers that are supported by binder. To impart the prefilter with certain functional properties, the binder is formulated with a

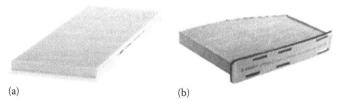

(a)   (b)

FIGURE 10.17
(a) Pollen filter. (b) Activated carbon filter.

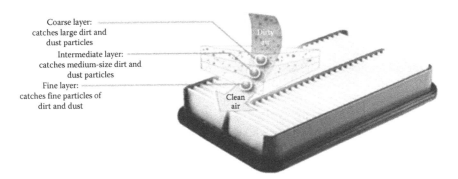

Coarse layer: catches large dirt and dust particles

Intermediate layer: catches medium-size dirt and dust particles

Fine layer: catches fine particles of dirt and dust

Dirty air

Clean air

FIGURE 10.18
Cabin filter: particulate filter.

mixture of antibacterial, water-repellent, and flame-retardant agents. Prefilters are designed to capture larger contaminating particles, including pollen and mold spores. The electrostatically charged microfiber layers, made up of melt-blown polypropylene fibers, attract and hold smaller-size particles from smoke, bacteria, and other contaminants. The cover layer is constructed of a nonwoven filter medium that adds stability and protects the microfiber layer from damage. It also provides an additional barrier against harmful contaminants. The activated charcoal filter has all of the features of the particulate filter, plus an activated charcoal layer that absorbs harmful gases and their odors (Figure 10.18).

10.3.4 Tire Cords

Tires are highly engineered structural composites whose performance can be designed to meet the vehicle manufacturers' ride, handling, and traction criteria, plus the quality and performance expectations of the customer. The tires of a mid-sized vehicle roll about 800 revolutions for every mile. Hence, in 50,000 miles, every tire component experiences more than 40 million loading–unloading cycles, an impressive endurance requirement. The first practical pneumatic tire was made by John Boyd Dunlop in 1887. Dunlop is credited with "realizing rubber could withstand the wear and tear of being a tire while retaining its resilience." Today, over 1 billion pneumatic tires are manufactured in around 450 tire factories in the world.

The tire is a complex technical component and must perform a variety of functions. The reinforcing material gives the product the shape and resistance to dynamic stresses, and it also determines the service life, loading capacity, abrasion resistance, and many other properties of the product. Most important of all, however, it must be capable of transmitting strong longitudinal and lateral forces (during braking, accelerating, and cornering) in order to assure optimal and reliable road-holding quality. Textiles used in tire reinforcement are specially prepared and processed fabrics called tire cord fabrics. Tires being made of rubber are not dimensionally stable by itself on application of load. Also with the advent of pneumatic tire, the dimensional stability requirement increased as tires became a pressure vessel. The tire also acts as a contact between the vehicle and the road and steers the vehicle; the reinforcing textile cord thus monitors directional stability of the tire.

Vehicles with powerful engines require, for example, good grip—particularly on wet and flooded roads. On the other hand, a corresponding improvement in the tread compound can affect tire life, rolling resistance, and ride comfort. One point, however, has absolute priority over all other tire design objectives, and that is safety. In addition, tire plays an important role in the overall energy consumption of an automobile, for example, in a running car, tires uses about 5%–6% of fuel on average.

10.3.4.1 Functions of Tires

1. Vehicle to road interface
2. Support vehicle load
3. Road surface friction
4. Absorb road irregularities

10.3.4.2 Tire: Parts

Depending upon the alignment of tire cords, tires are classified as bias, belted bias, and radial (Figures 10.19 and 10.20).

    a. *Diagonal (bias) tires*
       In bias tires, body ply cords are laid at angles substantially less than 90° to the tread centerline, extending from bead to bead. These types

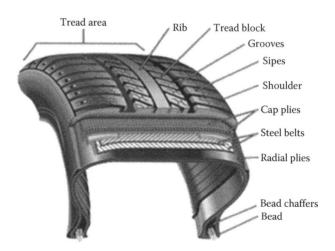

FIGURE 10.19
Tire: cross section.

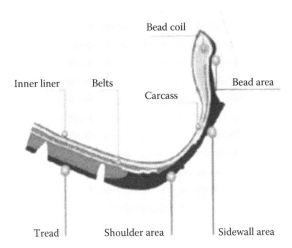

FIGURE 10.20
Tire parts.

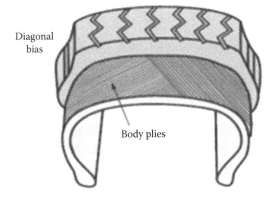

FIGURE 10.21
Diagonal tire.

of tires are used in trucks, trailers, and farm implements. Simple construction and ease of manufacture are advantageous. As the tire deflects, shear occurs between body plies, which generate heat. Tread motion also results in poor wear characteristics (Figure 10.21).

b. Belted bias tires

Belted bias tires, as the name implies, are bias tires with belts (also known as breaker plies) added in the tread region. Belts restrict expansion of the body carcass in the circumferential direction, strengthening and stabilizing the tread region (Figure 10.22).

Improved wear and handling due to added stiffness in the tread area are advantageous. Body ply shear during deflection generates

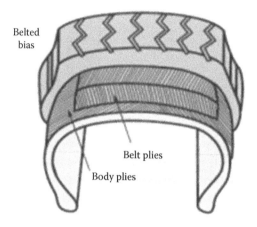

FIGURE 10.22
Belted bias tire.

heat. The drawback associated with this tire type are higher material and manufacturing cost.

c. *Radial tires*

Radial tires have body ply cords that are laid radially from bead to bead, nominally at 90° to the centerline of the tread. Two or more belts are laid diagonally in the tread region to add strength and stability. Radial body cords deflect more easily under load, thus they generate less heat and give lower rolling resistance and better high-speed performance. Increased tread stiffness from the belt significantly improves wear and handling. Complex construction increases material and manufacturing costs (Figure 10.23).

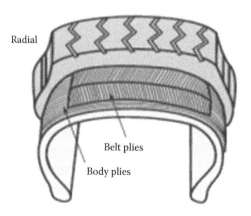

FIGURE 10.23
Radial tire.

TABLE 10.6

Reinforcement Fibers: Advantages and Disadvantages

Fibers Used in Tire Reinforcement	Advantages	Disadvantages
Nylon 6 and nylon 6,6	Good heat resistance and strength; less sensitive to moisture	Heat set occurs during cooling (flat spotting); long-term service growth
Polyester	High strength with low shrinkage and low service growth; low heat set; low cost.	Not as heat resistant as nylon or rayon
Rayon	Stable dimensions; heat resistant; good handling characteristics	Expensive; more sensitive to moisture; environmental manufacturing issues
Aramid	Very high strength and stiffness; heat resistant	Cost; processing constraints (difficult to cut)
Steel cord	High belt strength and belt stiffness improves wear and handling	Requires special processing; more sensitive to moisture

10.3.4.3 Reinforcement Materials

The tire cord and bead wire are the backbone of the tire. The tire cord and bead wire are the predominant load-carrying members of the cord–rubber composite, which determines the load-carrying capacity, the comfort of the tire, and the dynamic behavior. Some tires, when parked, can develop a temporary "set" in the rubber compounds and reinforcement cords, referred to as a "flat spot." Five materials currently make up the major tire textile usage—rayon, nylon, polyester, aramid, and steel (Table 10.6).

a. *Nylon*

Nylon 6 and nylon 6,6 tire cords are synthetic long-chain polymers produced by continuous polymerization/spinning or melt spinning. Its low modulus and low glass transition temperature make it unacceptable as a belt material or for applications where aesthetics, ride, and handling are important, that is, in passenger tires. Nylon is preferred in uses requiring carcass toughness, bruise and impact resistance, high strength, and low heat generation, for example, in tires for medium- and heavy-duty trucks, off-road equipment, and aircraft. In these applications, nylon can be used in the bias-ply tire carcass or in radial tire carcasses with steel or aramid belts (Table 10.7).

b. *Polyester*

Polyester tire cords are also synthetic long-chain polymers produced by continuous polymerization/spinning or melt spinning. The most common usage is in radial body plies with some limited applications as belt plies. Polyester is superior to nylon tire cords in some respects, for example, less thermal shrinkage, less flat spotting tendency, but it suffers lack of bonding with rubber. Polyester cord is not

TABLE 10.7

Typical Properties of Fibers Used in Tire Cord

	Rayon	Nylon 6	Nylon 6,6	Polyester	Aramid	Steel
Tenacity (cN/Tex)	50	80	85	80	190	35
% Elong at break	6	19	16	13	4	2.5
Modulus (cN/Tex)	800	300	500	850	4000	1500
Shrinkage (% at 150C)	<0.1	6.0	5.0	2.0	<0.1	<0.1
Moisture regain (% at RT)	13	4.5	4.5	0.5	<2.0	<0.1
Specific gravity	1.52	1.14	1.14	1.38	1.44	7.85
Melting temperature (C)	>210	225	250	250	>500	—
Glass transition temperature (C)	—	55	55	80	—	—
Heat resistance (C)	150	180	180	180	250	—
Approximate relative cost per unit weight (PET = 1.00)	1.33	—	1.13	100	5.00	—

recommended for use in high-load/high-speed/high-temperature applications, as in truck, aircraft, and racing tires, because of rapid loss in properties at tire temperatures above about 120°C.

c. *Rayon*

Rayon tire cords are made from cellulose produced by wet spinning. It is often used in Europe and in some run-flat tires as body ply material. The low-shrink, high-modulus, good-adhesion properties of rayon make it an excellent choice for use in passenger tires. However, rayon has lost market share to polyester due to higher cost and environmental concerns with production facilities. Rayon is used for racing tires and has gained renewed interest in the development of an extended-mobility self-supporting passenger tire (Table 10.8).

d. *Aramid*

Aramid is a synthetic, high-tenacity organic fiber produced by solvent spinning. Aramid cords have very high strength, high modulus, and low elongation. It is two to three times stronger than polyester and nylon. It can be used for belt or stabilizer ply material as a lightweight alternative to steel cord. It is particularly suited where weight is important, such as in the belts of radial aircraft tires or in overlay plies for premium high-speed tires. In a multiply carcass construction, aramid's low elongation will prevent the outer ply from adjusting to the average curvature, thus placing the inner plies into compression.

e. *Steel cord*

Steel cord is carbon steel wire coated with brass that has been drawn, plated, twisted, and wound into multiple-filament bundles. It is the principal belt ply material used in radial passenger tires.

TABLE 10.8

Comparison of Properties of Reinforcement Fibers

Dimensional stability (carcass)	
Uniformity in curing	Rayon > advanced polyester > nylon
Appearance (sidewall indentations)	Rayon > advanced polyester > nylon
Dynamic stiffness (steering)	Rayon > advanced polyester > nylon
Flat spotting	Rayon > advanced polyester > nylon
Durability (carcass)	
Fatigue resistance/heat generation	Nylon > advanced polyester > rayon
Impact resistance (toughness)	Nylon > advanced polyester > rayon
High-speed/run-flat tires	Rayon > advanced polyester
Strength	Aramid > steel > nylon > polyester > rayon
Modulus (stiffness)	Aramid > steel > rayon > polyester > nylon
Elongation	Nylon > polyester > rayon > steel > aramid
Compression fatigue	Nylon > polyester > rayon > steel > aramid
Chemical resistance	Aramid > nylon > rayon > polyester > steel

Bead wire is a carbon steel wire coated with bronze that has been produced by drawing and plating. Filaments are wound into two hoops, one on each side of the tire, in various configurations that serve to anchor the inflated tire to the rim.

10.3.4.4 Tire Cords

Tire cords are built up from yarns that in turn come from filaments. Filaments from production spinnerets are gathered together, slightly twisted, and placed on "beams" for further processing. The filaments are twisted "Z" into yarns, and the yarns are back-twisted "S" to form a cord. The size of a tire filament, yarn, or cord is measured by its linear density or "denier" or "decitex." Thus a 940/2 8 × 8 nylon cord is formed from 2 to 940 decitex yarns twisted separately at 8 turns/in. and then back-twisted together at 8 turns/in. to form the cord. A 1650/3 10 × 10 rayon cord would comprise 3–1650 denier yarns twisted at 10 turns/in. separately and back-twisted together at 10 tpi.

As with all tire components, choice of a textile cord for a given tire application may require compromises involving cost, intended market segment, and end-use application. The tire engineer has a number of choices for a tire textile including:

1. Chemical composition of textile
2. Cost per unit length and weight (cost in tire)
3. Denier—filament size and strength
4. Cord construction—number of yarn plies

TABLE 10.9

Tire Performance and Tire Cord Property

Tire Performance	Related Property of Tire Cord
Bursting strength	Tensile strength
Tire endurance	Adhesion with rubber
Power loss	Viscoelastic properties
Tread wear	Modulus
Tire size and shape	Modulus
Tire groove cracking	Creep
Flat spotting	Thermal shrinkage
High-speed endurance	Heat resistance

5. Cord twist

6. Number of cords per unit length in ply

7. Number of plies in the tire (Table 10.9)

Twist levels are important for tire cord performance. Higher twists allow a cord to behave like a spring that will not open up under compression, while lower twists allow a cord to behave as a rod, maximizing the strength. As twist increases, the tenacity decreases, fatigue in compression improves (the main reason for higher twists), the cord cost per tire increases (because cords become shorter as they are twisted), and shrinkage during processing and cure increases. Tenacity and fatigue resistance are sometimes reduced with increasing twist. The three major systems used to form the cord are ring twisting, direct cabling, and two for one twisting (Tables 10.10 through 10.13).

10.3.4.5 Weaving of Tire Fabrics

Tire fabrics have changed in response to the constant demand for better tire performance. The cords are woven, with pick cords to maintain spacing, into a wide sheet of fabric. The function of the pick is to maintain a uniform warp cord spacing during the downstream operations, such as shipping, adhesive

Table 10.10

Influence of Twist on the Property of Cord

Cord Property Changes with Increasing Twist Levels
Reduction in strength
Reduction in initial modulus
Reduction in cyclic tension fatigue resistance
Increase in elongation to break
Increase in rupture energy
Increase in cyclic compression fatigue resistance
Increase in cord cost per tire

TABLE 10.11

Functions of Tire Cord

Maintains durability against bruise and impact

Supports inertial load and contain inflating gas

Provides tire rigidity for acceleration, cornering, and braking

Provides dimensional stability for uniformity, ride, and handling

TABLE 10.12

Cord Requirements

Large length-to-diameter ratio, e.g., long filaments

High axial orientation for axial stiffness and strength

Good lateral flexibility (low bending stiffness)

Twist to allow filaments to exert axial strength in concert with other filaments in the bundle

Twist and tire design to prevent cord from operating in compression

TABLE 10.13

Ideal Cord Properties for a Tire Carcass

Dimensional stability—low shrink during cure, no flat spotting, no long-term growth

High tensile strength

High tensile modulus

Low bending modulus

High durability—fatigue resistance, low heat generation on flexing, high adhesion to rubber, chemical and oxidation resistance, heat resistance

High toughness—impact and abuse resistance

Low hysteresis loss at high speeds

dipping and heat treating, calendering, tire building, and lifting. Usually rapier and shuttle looms are used for weaving cord fabric; nevertheless, air jet machines from different companies have been specially designed for tire cord weaving. Uniform cord distribution in the finished tire is essential for tire uniformity and performance. The fabric is then passed through a four-roll calender where a thin sheet of rubber (body ply skim) is pressed onto both sides and squeezed between the cords of the fabric. The calendered fabric is wound into rolls, with a polypropylene liner inserted to keep the fabric from sticking to itself, and then sent to a stock-cutting process.

10.3.4.6 Heat Treatment

The heat treatment is mainly applied to the woven fabrics made of thermo-plastic synthetic fibers, such as nylon and polyester, to be incorporated in tires. The heat treatment occurs under controlled time (exposure), tension, and temperature conditions. It helps to optimize the fabric properties, especially improving the dimensional stability of the fabric and the overall setting effect.

10.3.5 Automotive Interiors

Automotive interiors typically comprise traditional textiles, which are now increasingly being replaced by more functional, new-generation industrial textiles. Global automotive interiors market is estimated to rise to $210 billion by 2015.

The fabrics used worldwide as surface materials for car interiors can be woven, knitted (both circular and warp-knitting), or nonwovens. Woven fabrics represent the dominant application area in seat covers, headrests, and door panels. Fabrics are characterized by a large variety in design, stable shape retention, and high mechanical resistance. Another crucial function of textiles in the auto interior is sound control. The micro denier fabrics, usually valued for their aesthetic and tactile properties, are favored for efficient sound absorption of their greater surface area. Being lightweight, they reduce the weight of the vehicle and improve mileage.

10.3.5.1 Automotive Seats

  a. *Seat comfort*

  During a journey, vehicle occupants are subjected to both mental and physical stresses when exposed to road vibrations, dense traffic, noise, and different weather conditions. Human exposure to mechanical vibrations causes fatigue and discomfort. The magnitude of the effect depends on the intensity, duration, and directions of the excitation. While defining comfort in a vehicle, the seat is considered as a primary aspect compared to other components. Comfort in an automotive seat is governed by a combination of static and dynamic factors. Term "comfort" is used to define the short-term effect of a seat on a human body, that is, the sensation that commonly occurs from sitting on a seat for a short period of time. In contrast, the term "fatigue" defines the physical effect caused by exposure to the seat dynamics for a long period of time. Comfort is subjective, and it is difficult to define this term objectively in order to determine the design specification of a seat that will provide this attribute to an occupant (Pywell, 1993). Seating comfort is strongly related to physical comfort of an occupant. Physical comfort can be defined as the physiological and psychological states perceived during the autonomic process of relieving physical discomfort and achieving corporeal homeostasis. There are three modes of comfort identified: static, transient, and dynamic comfort (Shen and Vértiz, 1997). Homeostasis is a state of equilibrium between different but interrelated functions or elements, as in organism or group.

b. *Seat materials*

The construction of the automotive seat and the cushioning material employed to provide the occupant a comfortable place to sit while operating a motor vehicle have continued to evolve. Most early autos had large, padded seats essentially similar to furniture seating. Some of the materials employed in the 1920s in auto upholstery are springs, hair and its substitutes, cloth, duck, sheeting, cambric, muslin, buckram, webbing, cotton wadding and battings, down, feathers, top materials, slip cover materials, coated ducks, and cords and bindings. In order to ensure that the hair fibers did not disintegrate or loosen from the seats, latex-bonded seating interiors were developed. A light impregnation with natural or synthetic rubber latex bonds the fibers together where they touch. Rubberized hair pads have been described by Pole in 1959 (Figure 10.24).

However, Mercedes® used fiber/latex seating for many years since it apparently gives their seats the desired performance of breathability and good lateral support cornering especially at high speeds. Rubberized coconut fiber seating was used by VW in Brazil up until 1994. Latex foam seating (without any reinforcing fibers) has been used for decades. Latex foam was considered a much superior, comfortable product to previous types of seating, for example, horsehair, wadding, springs. Latex foams were considered the comfort yardstick. They were replaced by polyurethane (PU) foam types that for

FIGURE 10.24
Rubberized hair automotive pads.

many years did not equal latex foam in comfort. They can be considered for two major reasons.

1. The urethane process is an easy one to master and run in production to obtain consistent product, easily trimmed into car seats.

2. The cost of producing urethane cushioning was lower than the more complicated latex processes.

One of the major advantages of PU foams over latex foam is their resistance to bottoming-out when stressed. Latex foam has inferior UV resistance compared with PU foam, and latex may show the early formation of a heavy crust whereas PU foam may exhibit some surface crumbling and/or friability after long UV exposure. DuPont® was able to make successful foam using their "Teracol®" (polybutylene ether) products. It had the extra resiliency required without the bad creep of the current polyester-based flexible PU foams. Its major drawback was that it was too expensive so other new materials had to be found. These can be generally classified as polyether polyols such as

a. Polypropylene glycols (UCC, Dow)

b. Polypropylene–polyethylene glycols (UCC—UCONS)

c. Pluronics—Wyandotte (now BASF)

High-resilience (HR) PU molded foams are key components of automobile interiors and contribute to passenger comfort especially in seats. They have superior vibration damping capability over a wide range of frequencies at low material density. HR foam had a latex rubber-like feel, higher support ratio, lower hysteresis loss, good flex fatigue, and improved inherent flammability resistance. Some special PET nonwovens have recently been developed such as a fiber mass stabilized with elastomer fiber-fused bonding, a folded web and stitch-bonded web, and a PET three-dimensional (3-D) knit fabric with super water absorbency. One of the most important advantages of this fibrous material over urethane foam is their good moisture permeability, which keeps passenger's physiological comfort (Table 10.14).

The traditional seat covering used in the vehicles in the early days was leather, and the use of a textile fabric was rare and constituted a surcharge. In today's modern vehicles, the role of textile fabrics is more than leather coverings. Leather was used for vehicle interiors because of its durability and strength regarding wear, despite the cost involved. The use of textile fabric was considered, but the fabrics produced would have been basic and inadequate for the task of an interior covering for a vehicle. Today's involvement of the textile industry in the automotive industry is considerable. The essential property requirement of seat cover laminates are aesthetic effect,

TABLE 10.14

Important Requirements of Car Interior

Good management of heat and water vapor transfer
Antistain/easy-to-clean characteristics
Antimicrobial/antibacterial properties
Antiallergic trimming
Flame resistance
Antistatic properties
Improved acoustic performance

FIGURE 10.25
Spacer fabric structure.

physiological conformability, strength, color fastness, flame retardancy, heat resistance, and nonvolatile substance content. For some high-end cars, customers expect it to have special functions such as antibacterial, antistatic, and stain-resistant properties.

Polyester fiber is predominantly used for making seat covers and four-some seat cushion material. It is also used for making door trim material because polyester fiber has a higher modulus, good heat stability, excellent resistance to color change, and high durability for sunlight degradation and is less expensive. Nonwoven polyester fabrics made from recycled fibers and novel knitted structure such as spacer fabrics, kunit, multiknit, and struto have been considered as substitutes for PU foam in the cover laminate. Spacer fabric is a knitted textile with threads running perpendicular to the plane of the fabric with a knitted layers each side (Figure 10.25).

Multiknit comprises two stitch layers with the pile in between, which makes fabrics from fibrous webs using Malimo knitting techniques. Kunit consists of a stitch layer with a pile on the top.

The seat covering fabrics are produced with fabric structures such as pile weave, tricot with raising, pile double raschel knit, and pile circular knit. Pile fabrics are usually related to an increase in the values of tactile and visual aesthetic effects. General trends are toward an increase in knit fabrics (tricot, double raschel, and circular knit) that are less expensive and have more formability than weave fabrics. Recently, a suede fabric using PET microfiber nonwoven as base material has been introduced. Seat cover containing phase change material has also been developed, which can increase the microclimate comfortability of the seat.

10.3.5.2 Door Trims, Roof Trims, and Floor Coverings

Door trims are usually made up of plastics such as vinyl chloride and poly-olefin. But textile materials are also used for higher-class cars. In most cases, the textile material is the same as the seat cover fabric. However, the fabric needs to have high enough formability to be made into the complicated shape of a door trim. Its lower end is usually covered by the same carpet as the floor because the door often gets kicked. The fabrics used for door trims should have high degree of stretchiness. This can be obtained by incorporating elastomeric yarns into the structure of fabrics. Now a layer of foam is added to the door panel fabrics to get a more luxurious touch.

Coconut fibers have been used in cars for more than 60 years in such applications as interior trim and seat cushioning. Unlike plastic foam, the coconut fibers have a good "breathing" property, which is a distinct advantage for vehicle seats being used in countries where the climate is hot, as is the case in Brazil. Coconut fibers are also naturally resistant to fungi and mites, and the remains of the fibers also make an effective natural fertilizer at the end of their lifetime. In 1994, Daimler Chrysler started using flax and sisal fibers in the interior trim components of its vehicles. They continued investing in their application of natural fibers, and in 1996, jute was being utilized in the door panels of the Mercedes Benz E-class vehicles. German car manufacturers are aiming to make every component of their vehicles either recyclable or biodegradable.

PET nonwoven and tricot are used as a roof trim. The use of needle-punched nonwoven in particular has increased, and especially velour-patterned nonwoven is usually formed into roof trim by integration with base material. Trims need to have good resistance to color fastness for sunlight, heat resistance, mechanical durability, light durability, formability, nonvolatile substance content, and strain resistance in addition to lightness. Sound-absorbing capability and heat insulation are also needed. The base material for roof trim can be mainly classified as polyester foam sheets and fiber-reinforced porous polymer sheets. Glass fiber is usually used for reinforcing these sheets.

Floor coverings are made up of tufted cut-pile or tufted loop-pile or needle felt. The uses of nonwoven carpet have significantly increased because it is more economical and it has better formability. Road noise is considered as an environmental pollution in few countries. Carpets by providing thermal and acoustic protection thus directly contribute to safety. The absorption potential of a tufted carpet as well as the pile density and the yarn is to be taken into account besides the thickness of the absorbing layer.

10.3.6 Other Textile Applications in Automobiles

a. Headliners

The headliner plays a number of roles within the car's design and engineering; its first role is to cover the inside of the metal panel of the roof. As well as covering the metal, the unit also hides wiring and curtain air bags. They also act as carriers for other components such as storage boxes and overhead lighting. Important requirements of headliners are lightweight; thin profile but rigid without any tendency to buckle, flex, or vibrate; directional stability; aesthetically pleasing; and soft in touch.

Headliner consists of two core materials: the substrate (which can be something as basic as cardboard, but it is normally PU or another polymer impregnated with glass fibers for strength and stiffness) and the facing fabric. In European vehicles, PU-based headliners are used while in North America, PU is joined by polyester (PET) and a range of new materials. As well as covering the roof panel and acting as a carrier for components with the overhead system, the headliner assembly has also an important acoustic function: either built in to its makeup or design or through acting as a cover for acoustic pads or filler materials that are fixed to the inside of the car's roof.

b. *Trunk liners*

Trunk liners give a most luxurious appearance and also serve as protection to both exterior walls and trunk content such as luggage perfect to date. The liners must be decorative and functional, yet have relative cost. These are usually made from waste fiber that are needled and then naturated with elastomeric materials. However, even spun-bonded polypropylene is also used as a substrate for a foamed rubber material that is used as trunk liner.

c. *Parcel shelves*

Parcel shelves, also referred to as package trays, are now almost invariably covered with needle-punched nonwoven mainly in polypropylene or polyester.

d. *Dashboard*

The dashboard, probably the hottest area in the car interior, offers some opportunity for textiles. The dashboard shape being highly curved and also complex, only knitted fabric and 3-D knitted fabrics would be eminently suitable. The performance requirements of textiles used for dashboard are low gloss (no glare or reflections on the windscreen), soft touch, pleasant aesthetics, nonfogging, nonodorous, UV stability, resistance to heat aging, resistance to low temperature, and high abrasion resistance.

e. *Sun visors*

Sun visors are produced from raised warp-knit fabric and polyvinyl chloride. Injection molding produces some sun visors; others are composed of metal frames, and rigid foam or cardboard is also used. The article is close to the windscreen, and UV light and heat resistance must be of the highest standard.

10.4 Nonwovens in Automotive Applications

Nonwoven fabrics are increasingly used in the automotive industry, while the use of woven and knitted fabrics in automotive applications shows only minimal growth. Nonwoven fabrics offer a better price/performance ratio than textiles made out of yarns for many applications. Nonwoven fabrics are likely to become most widely used in the automotive sector as the trend toward manufacturing lighter weight cars with lower fuel consumption continues. Higher productivity and lower production cost of nonwoven production are the reasons attributed to this. To meet the increasing demand for automotive nonwovens, enhancement of the production capacity of high-quality and versatile nonwoven materials is being intensively focused on by nonwoven manufacturers.

10.4.1 Nonwovens Used in Cars

1. Door lining: edge trim, door mirror, armrest, and lower part (door pocket)
2. Sun visors
3. ABC-pillar covering (covering of seat belt)
4. Headliner (molded roof): roof insulation, sun roof (cover), hood, and hood padding
5. Parcel shelves: speaker covering
6. Boot lining: floor mat, sides (wheel casings), rear cover, back seat wall, and spare wheel case
7. Filters: air filter, cabin filter, fuel filter, oil filter during car manufacture, and lacquering
8. Engine housing: bumper felts, bonnet lining, rear side, dashboard, battery separators, and other insulation points
9. Instrument panel: insulation and instrument panel (lower part)
10. Dashboard mat
11. Seats: lining for backs of seats, laminated padding for seat covers and bottom of seats, upholstered wadding, upholstery cover, reverse sides, headrest cushioning, seat subpadding, foam reinforcement, and padding for center armrest
12. Floor mats with tunnel: cladding and subupholstery (insulating material, stuffing)
13. Interior rear wall lining: floor of the car body and under the back seats (exterior wheel case)
14. Estate cars and convertibles: side wall covering (lining for the wheel case), boot floor, lining for the hood-case, and cover for the hood-case

10.5 Natural/Biodegradable Fibers in Automotive Textiles

The recent increase in consumer environmental awareness, along with increased commercial desire to use "greener" materials, has led to new innovations. A study conducted in 1999 indicated that up to 20 kg of natural fibers could be used in each of the 53 million vehicles being produced globally each year. This means that for each new model of car, there would be a requirement of between 1,000 and 3,000 ton of natural fibers per annum, with some 15,000 tons of flax being used in 1999 in the European automotive industry alone. A study by the Nova Institute in 2000 reviewed market possibilities for the use of short hemp and flax fibers in Europe. In this study, a survey of German flax and hemp producers showed that 45% of hemp fiber production went into automotive composites in 1999. One of the attractions of hemp, as compared with flax, is the ability to grow the crop without pesticide application. The potential for fiber yield is also higher with hemp. The type of natural fiber selected for manufacture is influenced by the proximity to the source of fiber, thus panels from India and Asia contain jute, ramie, and kenaf; panels produced in Europe tend to use flax or hemp fibers; and panels from South America tend to use sisal and ramie.

Virtually all of the major car manufacturers in Germany (i.e., Daimler–Chrysler, Mercedes, Volkswagen Audi Group, BMW, Ford, and Opel) now use natural fiber composites in automotive applications. Ford uses from 5 to 13 kg (these weights include wool and cotton). The car manufacturer, BMW, has been using natural materials since the early 1990s in the 3, 5, and 7 series models with up to 24 kg of renewable materials being utilized. In 2001, BMW used 4000 ton of natural fibers in the 3 series alone. Here, the combination is a blend of 80% flax with 20% sisal for increased strength and impact resistance. The main application is in interior door linings and paneling. Wood fibers are also used to enclose the rear side of seat backrests, and cotton fibers are utilized as a sound-proofing material. In 2000, Audi launched the A2 mid-range car, which was the first mass-produced vehicle with an all-aluminum body. To supplement the weight reduction afforded by all-aluminum body, door trim panels were made of PU reinforced with a mixed flax/sisal mat. This resulted in extremely low mass per unit volume, and the panels also exhibited high dimensional stability.

Recently, in the last few years, Volvo has started to use Soya-based foam fillings in their seats along with natural fibers. They have also produced a cellulose-based cargo floor tray—replacing the traditional flax and polyester combination used previously, which resulted in improved noise reduction (Table 10.15).

The two most important factors now driving the use of natural fibers by the automotive industry are cost and weight, but ease of vehicle component recycling is also an ever-increasing consideration to meet the requirements of the end-of-life vehicle directive.

TABLE 10.15

Natural Fiber Usage in
Automotive Components

Component	Weight (kg)
Front door liners	1.2–1.8
Rear door liners	0.8–1.5
Boot liners	1.5–2.5
Parcel shelves	<2
Seat backs	1.6–2
Sunroof shields	<0.4
Headrests	2.5

10.6 Nanotechnology in Automotive Textiles

Nanotechnology has a very wide range of potential applications in many scientific fields. Nanotechnology is defined as the precise manipulation of individual atoms and molecules to create layered structures. The basics of nanotechnology remain in the fact that properties of substances suddenly change when their size is reduced to the nanometer range. The conversion of bulk material into tiny particles of nanosize display hybrid properties synergistically derived from the bulk and tiny components. The incorporation of nanomaterials into a textile can impart functional properties such as strength, shrinkage, electrical conductivity, and flammability. Nanofibers can be defined as fibers with a diameter in the range of 1000 nm. Nanofibers are characterized as having a high surface area to volume ratio and a small pore size in fabric form. The dispersion of nanosize fillers into the polymer matrix produces nanocomposite fibers. The properties of the particles themselves (size, shape, distribution) can profoundly change the characteristics of a polymer system. Nanofillers can be distributed in a polymer matrix through either a mechanical or a chemical process. Nanomodification creates improved fiber characteristics, for example, mechanical strength, thermal stability, the enhancement of barrier properties, fire resistance, optical properties, for use in different application fields.

Tailoring and controlling of properties of material on a nanoscale level are considered to be key factors for the development of high-performance materials in various fields. Nanotechnology plays a significant role in the automobile sector by improving the performance characteristics of various components of automobiles. The automobile industry is potentially a major beneficiary of nanotechnology developments that promise enhancements and benefits at various levels providing lighter, stronger, harder materials, improved engine efficiency, reduced fuel consumption, improved safety, reduced environmental impact, and comfort. Textile materials have an essential role here, for their use spans from interior door panels, seats materials and paddings, dashboard, to cabin roof and boot carpets, headliner, seat belts, air bags, various filters, tire cord, and trimmings. Traditional textiles used for automotive interiors face several major challenges such as protection from dust and dirt, ventilation, durability and wear, and fire resistance, all of which call for new high-tech textiles providing enhanced functionalities. Nanoenabled textiles may have novel solutions and address multiple functional requirements.

10.6.1 Applications of Nanotechnology in Automotive Textiles

Polymer nanocomposites find use in many automotive applications to achieve specific functional requirements. Toyota Motors used N6/clay nanocomposites for making the toothed belt cover that exhibited good rigidity and excellent thermal stability. The weight saving was up to 25% due to the lower amount of clays used. N6 nanocomposites have also been used as engine covers, oil reservoir tank, and fuel hoses in the automotive industry because of their remarkable increase in heat distortion temperature and enhanced barrier properties together with their mechanical properties. To reduce noise within the car, a nanofibrous layer of nonwoven polyvinyl alcohol layered on fibrous underlay materials may be used. Compared to conventional materials, this offers improved noise reduction, while also providing good heat insulation and weight reduction. Nanoclay/polypropylene nanocomposites are used for seat back in Honda Acura. Thermoplastic polyolefin nanocomposites doors were developed by Chevrolet Impalas. The weight reduction of polymer nanocomposites can have a significant impact on environmental protection and material recycling.

Polycarbonates have many characteristic advantages such as being lighter, safer, and a more flexible material to design, but the drawbacks such as its poor scratch resistance and low UV shielding restricted its widespread usage. The coating of transparent silica nanoparticles imparts superior scratch resistance to polycarbonates.

10.6.2 Future Scope of Nanotechnology in Automotive Textiles

Many researches are going on in the automotive field in bringing nanoenabled textiles for various applications in automobiles. Coating textiles with a hydrophilic coating of TiO_2 nanoparticles or by plasma treatment may provide good moisture wicking and transpiration absorption that are valuable for comfortable driving. Nanocomposites containing organically modified clay dispersed in selected polymer matrices have attracted considerable attention for imparting flame resistance. Novel filters using nanofibers have the potential to offer superior capabilities compared to conventional products such as a large surface area while maintaining porosity to remove harmful substances and greatly improve air quality and safety. These nanofilters can be used in air filters and cabin filters in automobiles, which will provide high efficiency in particulate filtration. Car interior application will mostly deal with comfort issues—dirt-repellent and antimicrobial textiles and surfaces, and nanoparticulate air filters. The application of silver (Ag) nanoparticles to the textile surface will impart antimicrobial or antibacterial properties that could improve hygiene within the car interior and contribute to eliminating unpleasant odors.

Synthetic fibers possess poor antistatic properties, but fabrics containing electrically conductive nanosized materials, like titanium dioxide (TiO_2), zinc oxide (ZnO), and antimony-doped tin oxide (SnO), have been proved to be effective in dissipating accumulated static charge. Nanoparticles such as SiO_2, Al_2O_3, ZnO, and CNT are the most widely used to improve tear/wear resistance of the textiles. They can be mixed with many fiber precursor polymers such as polystyrene, polypropylene, or polyvinyl alcohol before spinning or, alternatively, applied to fabrics by spray or dip coating.

The textiles with the NanoSphere® finish produced by Schoeller, presenting a naturally self-cleaning effect, and the Tencel™ material based on nanofibrils of cellulose produced by Lenzing, which combines a good moisture management, reduced energy consumption, reduced growth of bacteria, antistatic behavior, and heat-absorbing properties compared to common polyester, make the material a good candidate for seat car covers. The nanomodified microfiber Evolon® by Freundenberg, used for headliner, dashboard, carpet backing, doors, and the underbody shield, allows a weight and thickness reduction and better noise absorption. Elmarco's NanoSpider acoustic web offers similar sound reduction in addition to heat insulation and weight reduction.

Bibliography

Adanur, S., *Wellington Sears Handbook of Industrial Textiles*, Technomic, Basel, Switzerland, 1995, pp. 500–506.

Bauxton, A., Tech. Text. Market, October, 1992, 34.

Berger, R., Roland Berger Strategy Consultants' Outlook For The Chinese Automotive Supplier Market: How Long Will The Party Last, Germany 2010 http://www. theautochannel.com/news/2010/12/30/512725.html (accessed May 29, 2013).

Brody, H., ed., *Synthetic Fibre Materials, Polymer Science and Technology Series*, Longman Scientific and Technical, John Wiley and Sons, New York, 1994.

Blair, G. R., J. I. Reynolds, and M. D. Weierstall, *Automotive Cushioning through The Ages—A Review*, The Molded Polyurethane Foam Industry Panel, September 2008.

Borgers, W., *Technical Textiles with Benefits for Weight, Function and Design*, Johann Borgers GmbH & Co.KG, Bocholt, Germany.

Buist, J. M. and Woods, G., Molding of flexible urethane foams, Rapra Code 43C6-6124-83, *Transactions of the Institute of the Rubber Industry*, 41(1), 1965.

Dupont, www2.dupont.com/Automotive/en_us (accessed May 29, 2013).

Du Preez, W. B., Damm, O. F. R. A., Trollip, N. G., and John, M. J., Advanced materials for application in the aerospace and automotive industries, CSIR Conference Proceeding, pp. 1–5, 2008, http://researchspace.csir.co.za/dspace/bitstream/10204/2637/1/Damm_2008.pdf (accessed May 29, 2013).

European Transport Safety Council, Seat belt reminders increasingly standard in Europe—but not in all countries, NEWS RELEASE Brussels, embargoed until 25 October 2006, http://www.etsc.eu/documents/Press%20release%20PIN%20Flash%203.pdf (accessed May 29, 2013).

Fibre2fashion.com, Booming market for automobile textiles: Forecasts for 2012, http://www.fibre2fashion.com/industry-article/29/2837/booming-market-for-automobile-textiles-forecasts-for-20121.asp (accessed May 29, 2013).

Ford, Ford introduces industry's first inflatable seat belts to enhance rear seat safety, http://media.ford.com/article_display.cfm?article_id=31360 (accessed May 29, 2013).

Gemeinhardt, P.G. et al., Molding flexible polyether PU foams, *SPE Journal*, October, p. 1117, 1960.

Global Automotive Retail, World Market Automotive Retail, MarketLine, U.K., pp. 2–7, May 2012.

Gupta, S., Automotive textiles in the driver's seat, *ATA Journal for Asia on Textile & Apparel*, December 2008.

Hufenbach, W., Bohm, R., Thieme, M., Winkler, A., Mader, E., Rausch, J., and Schade, M., Polypropylene/glass fibre 3D-textile reinforced composites for automotive applications, *Materials & Design*, 32(3), 1468–1476, 2011.

Hunter, B., mtex press release, Mobile textiles—More valuable than ever, November 22, 2011, www.mtex-chemnitz.de (accessed May 29, 2013).

International Organization of Motor Vehicle Manufacturers (OICA), http://www.worldometers.info/cars (accessed May 29, 2013).

International Organization of Motor Vehicle Manufacturers, 2012, Production statistics, http://oica.net/category/production-statistics/(accessed May 29, 2013).

Johnson, T., *Fuel Filters: Choose Well and Prosper*, Pacific Fishing, University of Alaska Sea Grant, Marine Advisory Program, Anchorage, AK, August 1998.

Kihlander, I. and Ritzen, S., Compatibility before completeness—Identifying intrinsic conflicts in concept decision making for technical systems, *Technovation*, 32(2), 79–89, 2012.

Kovac, F. J., Tire technology, Goodyear Tire and Rubber Co., 1970.

Lewin M., Sello S. B., eds, *Handbook of Fiber Science and Technology*, Marcel Dekker, Inc., New York, 1989.

Mantovani, E. and Zappelli, P., Nanoenabled automotive textiles, Observatory nano Briefing No. 24, December 2011, http://www.observatorynano.eu/project/filesystem/files/Briefing%20No.24%20Nano-enabled%20automotive%20textiles.pdf (accessed May 29, 2013).

Marshall Brain, How airbag works, http://auto.howstuffworks.com/car-driving-safety/safety-regulatory-devices/airbag.htm (accessed May 29, 2013).

Mukhopadhyay, S. K. and J. F. Patridge, Automotive Textiles, *Textile Progress*, 29(1/2), 3–25, 1997.

Mustafa, Y. and Rahman, A., Roslan, Development of an automotive seat for ride comfort, Project Report, 2004, http://eprints.utm.my/2640 (accessed May 29, 2013).

Nave, C. R., Seat belt physics, Hyper physics, Department of Physics and Astronomy, Georgia State University, http://hyperphysics.phy-astr.gsu.edu/%E2%80%8Chbase/seatb.html#cc1 (accessed May 29, 2013).

Nissan, Fuel filter manual, nissan.com.sg/ForOwners/fuel_filters.pdf (accessed May 29, 2013).

Pelmar Engineering Ltd., Tire fundamentals, http://www.pelmar.com/english/Article.aspx?item=1009 (accessed May 29, 2013).

Pywell, J., Automotive seat design affecting comfort and safety, SAE Technical Paper 930108, 1993.

Rajendran, S., Scelsi, L., Hodzic, A., Soutis, C., and Al-Maadeed, M.A., Environmental impact assessment of composites containing recycled plastics resources, *Conservation and Recycling*, 60, 131–139, 2012.

Reed, M. P., L. W. Schneider, and L. L. Ricci, *Survey of Auto Seat Design Recommendations for Improved Comfort-Technical Report*, University of Michigan Transportation Research Institute, Ann Arbor, MI, 1994.

Science Serving Society, http://www.scienceservingsociety.com/ts/text.htm (accessed May 29, 2013).

Senthil Kumar, R., Seat belt, *Asian Textile Journal*, July, 27–29, 2010.

Senthilkumar, R., Acoustic textiles—Sound absorption. *Melliand Textilberichte*, March 2011.

Shen, W. and Vértiz, A., Redefining seat comfort, SAE Technical Paper 970597, 1997.

Singha, K., Strategies for in automobile: Strategies for using automotive textiles-manufacturing techniques and applications, *Journal of Safety Engineering*, 1(1), 7–16 (2012).

Skolnik, L., Tire Cords, Kirk-Othmer, *Encyclopedia of Chemical Technology*, 2nd edn, Vol. 20, p. 328, John Wiley and Sons New York, 1969.

Smith, W., Automotives—A major textile market, *Textile World*, September 1994.

Soutter, W., Nanotechnology in the automotive industry, *Micrometrics*, July 2012, http://www.azonano.com/article.aspx?ArticleID=3031 (accessed May 29, 2013).

Supplier Business Ltd., The Headliners, Soft Trim and Acoustics Report, 2009.

Tanner, D., 3.1 Tires, *High Technology Fibers (Part B), Handbook of Fiber Science and Technology III*, Lewin, M. and Preston, J., eds., Marcel Dekker, Inc. New York, 1989.

U.S. Department of Transportation, The Pneumatic Tire, DOT HS 810 561, February 2006.

Vigano, F., Consonni, S., Grosso, M., and Rigamonti, L., Material and energy recovery from Automotive Shredded Residues (ASR) via sequential gasification and combustion, *Waste Management*, 30(1), 145–153, 2010.

Viju, S. and Mukhopadhyay, A., Automotive filters, *Asian Textile Journal*, 15(5), 49–55, 2006.

Volland, R. and Rothermel, H. M., Advantages of a new molding technology, SAE paper—83 04 86.

WebBikeWorld, http://www.webbikeworld.com/motorcycle-news/honda-motorcycle-airbag.htm (accessed May 29, 2013).

Yang, H. H., *Aromatic High Strength Fibers*, John Wiley and Sons, New York, 1989, p. 228.

11

Miscellaneous Applications in Industrial Textiles

11.1 Bolting Cloth

Bolting cloth is woven of extremely smooth, durable stainless steel with a plain square mesh pattern fabric. It features high capacity and strength and widely used for accurate dry or wet sifting and separating. The primary fields of application of bolting cloth are flour and grain milling, food processing, and general industry. Most bolting cloth sizes are stocked in 40″, 48″, and 60″ widths. The woven wire cloth in bolting grade is extremely smooth, highly durable wire to speed up the bolting action and increase the capacity of vibratory screening machines. The enhanced strength and toughness of stainless steel bolting cloth make it most durable and ideal for industrial use light materials requiring high throughput allowed by enhanced useable open area.

11.2 Membrane Fabric

Membrane fabric today plays a vital role in the filtration purpose. Figure 11.1 shows the hollow fiber membrane fabric used in reverse osmosis process for purification of effluent or drinking water in commercial or industrial purposes. The hollow fiber is planer membrane wound in spiral form, in which nonwoven is used for the substrate of membrane. The hollow fiber, whose membrane is nonporous, by applying higher pressure water permeates through the membrane into its hollow part, and substance materials such as salt, virus, and pyrogen can be rejected at the surface of active separation layer in membrane.

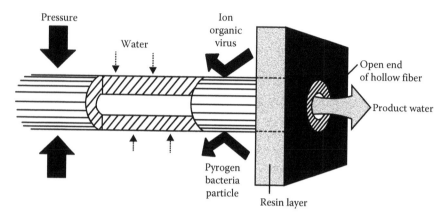

FIGURE 11.1
Membrane made out of hollow fiber in reverse osmosis.

a. *Reuse and recycling of water:* In order to efficiently utilize water, several technological systems for the reuse of wastewater or water recycling have intensively developed. In these systems, RO–MF-membrane hollow fibers have been utilized according to water-quality requirement. Activated carbon fiber is often used in combination with the RO–MF-membrane hollow fiber as a major component of water-purifying equipment for drinking water. The main role of activated carbon fiber in the equipment is to remove such substance as chlorinated organic chemical and smell constituents.

b. *Photocatalyst fibers:* A water purification system using conical nonwoven fabric filter units made of photocatalyst fiber has been developed. The constituent of the fiber is gradually varied from silica at the core part to TiO_2 (photocatalyst) at the surface of the filter. In the system, an ultraviolet (UV) lamp is located at the axial center of the conical unit line. Contaminants in the water are oxidized and degraded by the photocatalyst under the UV light during the passage of water through the conical unit line. The system is going to be widely applied for purifying several kinds of water in recycled water usages such as public bath, spring bath, swimming pool, and industrial-use water.

c. *Ion-exchangeable fibers:* There are some kinds of ion-exchangeable fibers. They can be effective to remove toxic ingredients of heavy metal from water. The fiber made of ion-exchangeable polystyrene resin is used for purification of recycled water from atomic power plants.

11.3 Filter System

a. *Bag filter:* Bag filters are widely used for cleaning the exhaust gas from several kinds of incinerators. They are usually made of glass fiber-woven fabric or synthetic fiber needle-felt nonwoven or its combination with woven fabric. Polyphenylene sulfide, *m*-aramid, polyimide, and polytetrafluorocarbon are typically used as the synthetic fiber material. The dust contained in exhaust gas is accumulated on the surface of a bag filter whose shape is tubular. When the inlet pressure increases up to a settled value by the layer-up of accumulated dust, it is dropped off from the bag surface by counter pulse jet or mechanical vibration. Acid gas and some other harmful gas substances can also be removed by introducing such materials as slaked lime and activated carbon in the bag (Figure 11.2).

b. *Air filter:* Air filters used for the removal of dust for clean room and office contribute to energy saving by the recycling of conditioned air. The filter medium is usually nonwoven fabric, in which the finer fibers are used for improved filter efficiency (Figure 11.3).

Removal efficiency is the lowest at 0.05–0.1 µm of dust particle size, because inertial collision effect for catching dusts within air flow by fiber within the medium becomes more significant for larger particles with particle size above 0.1 µm, and diffusion effect is more effective for small particles, below 0.05 µm. Filter medium is usually pleated in the filter of high removal efficiency, because pleating increases the area of filtration and can reduce the pressure drop of the filter by lowering the velocity of air passing through the filter medium.

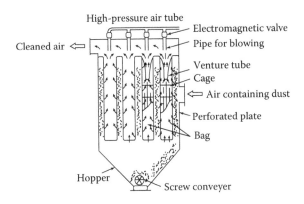

FIGURE 11.2
Bag filter system.

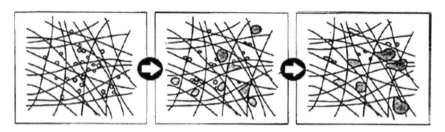

FIGURE 11.3
Effect of fiber density in filtration efficiency in air filter system.

c. *Particle filters:* Most particle filter systems are equipped with woven or nonwoven felt for filtering the particles. The woven fabrics that are used as filter fabrics have the functionality imposed for the filtration process. Structure and properties of the woven filter fabrics are adequately differentiated, for the principles of the filtering process. The paper defines the structural and functional elements that are specific to the filtering woven fabrics, which have a structure that is simple, balanced, and unbalanced in yarn count and thread density. The methods of filter design for the fabrics with simple structure are based on the specific geometry of the structure element fabrics. Filter fabric with uniform pore (size and shape) is recommended for use and weaves balanced with equal segments and uniform distribution (Figure 11.4).

11.4 Tissue Engineering

Artificial blood vessel: The structural material of an artificial blood vessel of middle and large diameters is usually polyethylene terephthalate (PET) fabric or polytetrafluoroethylene (PTFE) membrane. The tubular vessel is corrugated like bellows, to be flexible. In the case of PET fabrics, some of them have velour surface and/or are made of microfiber. Tissue culturing to such a tube is easier because cell has a tendency to be more easily cultured along fine fiber. To avoid the leakage of blood, some vessels have usually a barrier layer made by filling such materials as gelatin or made of a biocompatible elastic polymer.

11.5 Acoustic Uses

a. *Speaker diaphragm:* Material of speaker diaphragm is required to have high tensile modulus, high internal friction loss, and low specific gravity for getting good vibration characteristics.

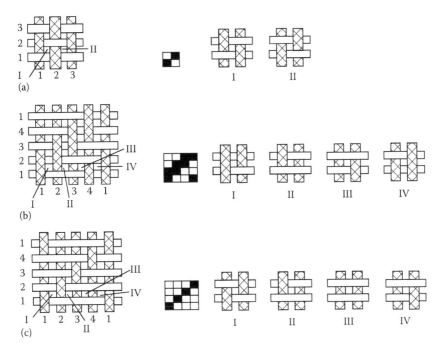

FIGURE 11.4
Pore size variations in filter cloth structure: (a) plain weave, (b) 2/2 twill weave, and (c) 1/3 twill weave.

Therefore, a fiber such as carbon fiber, organic fiber having high modulus, and glass fiber is used for its reinforcement. The examples of high-quality diaphragm are (1) composites sheet composed of carbon fiber and thermoset resin, (2) glass fiber-reinforced thermoplastic board made by injection molding, and (3) paper blended with such a high-performance fiber as those mentioned earlier. Screen integrated with speaker is made of PEN woven fabric, and it can be an image display medium and speaker diaphragm at the same time. The high modulus and reasonable price of PEN fiber make the material suitable for this application. Its sonic features are (1) simple system, (2) clear-cut sound quality by its high-vibration damping, and (3) no sound directivity with higher sound pressure.

b. *Musical instrument parts:* Nonwoven felt used for a musical instrument such as a piano is usually made of wool whose functions are (1) providing cushion, (2) damping of vibration, and (3) absorption of sound. But in some electronic pianos, synthetic fiber and/or fiber combined with elastic foam are used for its felt part. For the stringed instrument, several kinds of materials such as silk yarn, gut, synthetic fiber, and steel wire are also used.

11.6 Decatizing Cloth

Decatizing, also known as *crabbing*, *blowing*, and *decating*, is the process of making permanent finish on textile cloth, so that it does not shrink during garment washing. It is used mainly for wool, also applied to processes on fabrics of other fibers, such as cotton, linen, or polyester. The decatizing clothes are made out of woven or nonwoven fabric, made out of wool/cotton-blended fabric or high-performance fiber nonwoven felt. Crabbing and blowing are minor variations on the general process for wool, which is to roll the cloth onto a roller and blow steam through it. The decatized wool fabric is interleaved with a cotton, polyester/cotton, or polyester fabric and rolled up onto a perforated decatizing drum under controlled tension. The fabric is steamed for up to 10 min and then cooled down by drawing ambient air through the fabric roll. The piece is then reversed and steamed again in order to ensure that an even treatment is achieved. There are several quite different types of wool decatizing machines including batch decatizing machines, continuous decatizing machines, wet decatizing machines, and dry decatizing machines.

11.7 Printed Circuit Board

Printed electronic circuit: The electronic circuit board is made by etching copper foil laminated on the board and/or by metal plating or by electric conductive paste printing on the board. The circuit is usually patterned by screen printing and/or photo resist method. In many cases, the circuit on one side of the board is connected to the circuit on another side through holes covered by metal plating. Recently, the board built up by circuits in multiple integrated layers has been developed. The board is usually made of rigid paper or glass cloth impregnated by thermoset resin. In some cases, it is reinforced by woven fabric or wet-formed nonwoven of organic high-performance fiber (Figure 11.5).

11.8 Screen for Electronic Printing

Fine screen (mostly plain woven fabric made of monofilament) is one of the key materials of screen printing for electric devices. In many cases, the monofilament is made of sheath–core bicomponent fiber. Typical examples

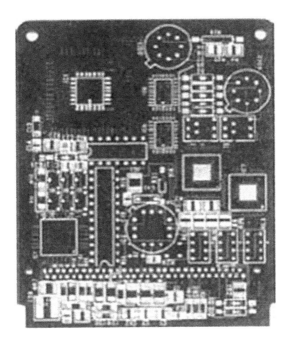

FIGURE 11.5
Printed circuit board.

of sheath/core are (1) nylon/polyester, (2) toughened PET/PP, and (3) wholly aromatic polyester-reinforced flexible polymer. The screen made of (3) is the most feasible to precision screen print, because it has high resistance for mechanical deformation. Screen printing using such materials is used for patterning of printed circuit, printing of fluorescent pigment for plasma display, as shown in Figure 11.6, printing dielectric paste for condenser, and printing of dam wall for liquid crystal display.

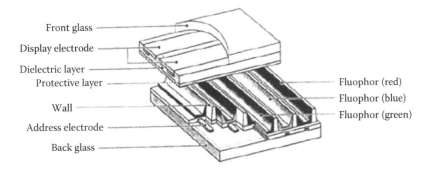

FIGURE 11.6
Structure of plasma display.

Circuit systems for E-textiles: E-textiles are (1) the circuit material must be flexible enough, (2) it should have enough mechanical toughness, and (3) it should have enough washability. It is also desirable that it is stretchable and can be colored. The following are wiring methods for the circuit: (1) weaving or knitting of electric (or photo) conductive yarn into the cloth, (2) embroidering or stitching of such a yarn on the cloth, (3) printing of conductive material on the cloth, and (4) laminating of conductive tape on the cloth. There are electric conductive yarns, which are mixed with ultrafine metal fiber and organic electric conductive fiber. Nowadays electric conductive fiber is made of conventional materials such as PET and PVA. Electric conductive and stretchable yarn whose core consists of elastomeric yarn spirally covered by such an electric conductive fiber is developed. Such yarns must be useful because they can be stretchable and dyeable and have enough toughness as textile products.

11.9 Battery and Fuel Cell

a. *Electrode:* The electrode of fuel cell must have high gas permeability with high electric conductivity. Carbon fiber sheet is useful as the material for this electrode. Activated carbon fiber is also promising as the electrode material of battery for the storage of night electric power. Carbon nanofiber and nanotube are very useful as electrode material for several kinds of battery because of their good electric conductivity and high specific surface area.

b. *Separator:* Wet-formed nonwoven felt made of such fiber material as polypropylene (PP) is used for the separation sheet of nickel/hydrogen battery. The sheet is composed of fine fibers and is finished by hydrophilic treatment, which enhances its working life, its electric power, and its ability to suppress self-discharge.

11.10 Smart Woven Fabrics: Renewable Energy

The polymer-based piezoelectric fibers can be used as either weft or warp into the woven structure. The conductive polymeric fibers can be used as negative and positive electrodes for charge transfer, so that the resultant fabric can produce energy for micropowered electronics. The main advantage of the use of polymer-based piezoelectric material in this application is its flexibility and the fact that it can easily be incorporated in the woven structures without causing any problem. It is impossible to integrate existing

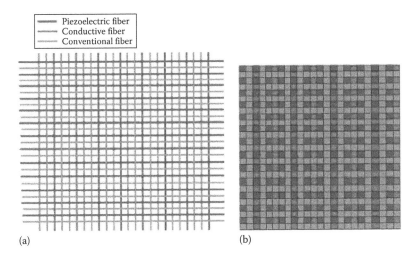

FIGURE 11.7
Smart fabric made out of conductive polymeric fibers: (a) yarn pattern, (b) fabric simulation.

ceramic-based piezoelectric fibers into similar structures because these fibers are rigid and brittle thus can cause major problems in the weaving process. Figure 11.7 shows that the polymer-based piezoelectric fibers used as either weft or warp into the woven structure and conductive fibers can be used as negative and positive electrodes for charge transfer; therefore, the resultant fabric can produce energy to power small electronic devices.

The smart piezoelectric woven fabrics can be used where they can be subjected to mechanical strain/stress or vibrations. Depending on the application and energy need, smart piezoelectric woven fabrics can be used to produce whole textile structure or only a part of it. For instance, tents, awnings, and umbrellas can be wholly made of smart piezoelectric fabrics and produce electricity under rain as well as wind. However, waterproof finishing is needed if the fabric will be used for outdoor applications. The best weaving technique for smart piezoelectric fabric is plain weaving and its derivatives. Depending on energy need, different fabric designs can be made with less or more interlacing. Other existing fabric-making methods (other than weaving) such as embroidery can also be used to produce smart fabrics.

11.11 Composites

In recent years, the uses of textile structures made from high-performance fibers are finding increasing applications in composites. High-performance textile structures may be defined as materials that are highly engineered fibrous structure having high specific strength, high specific modules, and designed to perform at high temperature and high pressure (loads) under corrosive and extreme environmental conditions. Significant developments have taken place in fibers, matrix polymers, and composites manufacturing techniques. The textile-manufacturing processes are less complex than injection molding and laminating, and they have advantages in greater control of fiber placement and in ease of handling preforms. These textile structures may be planar 2-D fabrics, for example, nonwoven, knitted, or nonwoven materials or 3-D woven, braided, nonwoven, or knitted fabrics. Making use of the unique combination of lightweight, flexibility, strength, and toughness, textile structure has long been recognized as an attractive reinforcement form for many composite applications.

The advantages of textile techniques are homogeneous distribution of matrix, reinforcing fibers, high drapeability, free of solvents, and low financial expenditure. The use of thermoset matrices is widespread at present. The resin is applied to the textile preform at the consolidation stage. Polyesters or epoxy matrices are applied by resin transfer molding process. Faster cure is possible with other resin formulation, mainly polyurethanes, suitable for reaction injection molding process. The textile industry now has access to multifilament with high tensile strength and modulus and high resistance to chemical, heat, and hydrolysis for use in high-performance applications and in aggressive environment. The use of renewable cellulosic natural fibers as reinforcing fillers in fiber composites or adding a fiber blend in technical textile products is appealing because of the properties of the resultant composites and environment. The advantages of biofibers as low-cost, renewable biodegradable raw materials can be utilized in some technical textile products to a much greater extent than it is being done today.

11.12 Battery Separators

Many advances have been made in battery technology in recent years, both through continued improvement of specific electrochemical systems and through the development and introduction of new battery chemistries. A separator is a porous membrane placed between electrodes of opposite polarity, permeable to ionic flow but preventing electric contact of the electrodes. A variety of separators have been used in batteries over the years. Separators have been manufactured from cellulosic papers and cellophane to nonwoven fabrics, foams, ion exchange membranes, and microporous flat sheet membranes made from polymeric materials. As batteries have become more sophisticated, separator function has also become more demanding and complex. Separators play a key role in all batteries. Their main function is to keep the positive and negative electrodes apart to prevent electrical short circuits and at the same time allow rapid transport of ionic charge carriers that are needed to complete the circuit during the passage of current in an electrochemical cell. They should be very good electronic insulators and have the capability of conducting ions by either intrinsic ionic conductor or by soaking electrolyte.

a. *Separator requirements:* A number of factors must be considered in selecting the best separator for a particular battery and application. The characteristics of each available separator must be weighed against the requirements and one selected that best fulfills these needs. A wide variety of properties are required of separators used in batteries. The considerations that are important and influence the selection of the separator include the following:

- Electronic insulator
- Minimal electrolyte (ionic) resistance
- Mechanical and dimensional stability
- Sufficient physical strength to allow easy handling
- Chemical resistance to degradation by electrolyte, impurities, and electrode reactants and products
- Effective in preventing migration of particles or colloidal or soluble species between the two electrodes
- Readily wetted by electrolyte
- Uniform in thickness and other properties

b. *Separator types:* Separators for batteries can be divided into different types, depending on their physical and chemical characteristics. They can be molded, woven, nonwoven, microporous, bonded, papers, or laminates. In recent years, there has been a trend to develop solid and gelled electrolytes that combine the electrolyte and separator into a single component. In most batteries, the separators are made of either nonwoven fabrics or microporous polymeric films. Batteries that operate near ambient temperatures usually use separators fabricated from organic materials such as cellulosic papers, polymers, and other fabrics, as well as inorganic materials such as asbestos, glass wool, and SiO_2. In alkaline batteries, the separators used are either regenerated cellulose or microporous polymer films. The lithium batteries with organic electrolytes mostly use microporous films.

c. *Microporous separators:* They are fabricated from a variety of inorganic, organic, and naturally occurring materials. Nonwovens made out of polymeric fibers (e.g., nylon, cotton, polyesters, glass), polymer films (e.g., polyethylene (PE), PP, PTFE, poly(vinyl chloride)), and naturally occurring substances (e.g., rubber, asbestos, wood) have been used for microporous separators in batteries that operate at ambient and low temperatures (<100°C). The microporous polyolefins (PP, PE, or laminates of PP and PE) are widely used in lithium-based nonaqueous batteries and filled PE separators in lead–acid batteries respectively.

d. *Separator properties*

  i. *Thickness:* The lithium-ion cells used in consumer applications are made out of thin microporous separators. The thicker the separator, the greater the mechanical strength and the lower the probability of punctures during cell assembly but the smaller the amount of active materials. The thinner separators take up less space and permit the use of longer electrodes. This increased both capacity and, by increasing the interfacial area, rate capability.

  ii. *Permeability:* The separators should not limit the electrical performance of the battery under normal conditions. Typically, the presence of separator increases the effective resistivity of the electrolyte by a factor of 6–7. The ratio of the resistivity of the separator filled with electrolyte to the resistivity of the electrolyte alone is called MacMullin number. MacMullin numbers that are as high as 10–12 have been used in consumer cells.

  iii. *Gurley (air permeability):* Air permeability is proportional to electrical resistivity, for a given separator morphology. It can be used in place of electrical resistance measurements once the relationship between Gurley and electric resistance is established. The separator should have low Gurley values for good electrical performance.

  iv. *Porosity:* It is implicit in the permeability requirement; typically lithium-ion battery separators have a porosity of 40%. Control of porosity is very important for battery separators. Specification of percent porosity is commonly an integral part of separator acceptance criteria.

  v. *Wettability:* The separators should wet out quickly and completely in typical battery electrolytes. A separator should be able to absorb and retain electrolyte. Electrolyte absorption is needed for ion transport. The microporous membranes usually do not swell on electrolyte absorption.

  vi. *Chemical stability:* The separators should be stable in the battery for a long period of time. They should be inert to both strong reducing and strong oxidizing conditions and should not degrade or lose mechanical strength or produce impurities, which can interfere with the function of the battery. The separator must be able to withstand the strong oxidizing positive electrode and the corrosive nature of the electrolyte at temperatures as high as 75°C. The greater the oxidation resistance, the longer the separator will survive in a cell. Polyolefins (e.g., PP, PE) exhibit high resistance to most of the conventional chemicals, good mechanical properties, and a moderate temperature range for application, making them ideal polymers for lithium-ion battery separators. PP separators exhibit better oxidation resistance properties when in contact with the positive electrode in a lithium-ion cell. Thus, the oxidation resistance properties of trilayer (PP/PE/PP) separators with PP as the outside layer and PE as inner layer are superior.

vii. *Dimensional stability:* The separator should lay flat and should not curl at the edges when unrolled, as this can greatly complicate cell assembly. The separator should also not shrink when exposed to electrolyte. The cell winding should not affect the porous structure in any adverse way.

viii. *Puncture strength:* The separators used in wound cells require a high puncture strength to avoid penetration of electrode material through the separator. If particulate material from the electrodes penetrates the separator, an electrical short will result and the battery will be rejected. The separators used in lithium-ion batteries require more strength than the one used in lithium primary batteries. The primary lithium batteries have only one rough electrode, and thus it requires less strength. As empirically observed, for most applications, the puncture strength should be at least 300 g/mil for separators used in lithium-ion cells. Mix penetration strength is a better measure of separator strength in a battery compared to puncture strength.

ix. *Thermal stability:* Lithium-ion batteries can be poisoned by water, and so materials going into the cell are typically dried at 80°C under vacuum. Under these conditions, the separator must not shrink significantly and definitely must not wrinkle. Each battery manufacturer has specific drying procedures.

x. *Pore size:* A key requirement of separators for lithium batteries is that their pores be small enough to prevent dendritic lithium penetration through them. Membranes with submicrometer pore sizes have proven adequate for lithium batteries.

xi. *Tensile strength:* The separator is wound with the electrodes under tension. The separator must not elongate significantly under tension in order to avoid contraction of the width. A tensile strength specification is sometimes given, but the key parameter is Young's modulus in the machine direction. Since Young's modulus is difficult to measure, a 2% offset yield is a good measure; less than 2% offset at 1000 psi is acceptable for most winding machines.

11.13 Cigarette Filter

The plug wrap paper helps to hold the cigarette filter in cylindrical form; cellulose acetate tow is plasticized but requires time to cure. The plug wrap keeps the fibers in contact allowing interfiber bonding to occur, which allows the rod to harden. Plug wrap is made in two basic types: nonporous (standard) and porous. The type of plug wrap used determines how the cigarette filter is likely to be ventilated: the greater the amount of air that can be drawn through a plug wrap, the lower the delivery of the cigarette. The standard plug wrap is used when the customer either does not ventilate the filter or has online laser perforation capability during cigarette manufacture.

- *Porous plug wrap:* Porous plug wraps are used when the customer uses pre-perforated tipping. It has an interlocking network of cellulose fibers interspersed with spaces (pores); these spaces allow air to pass through the paper (Figures 11.8 and 11.9).
- *Stiff plug wrap:* Stiff plug wraps are heavier, approximately 100 gsm, and are used to provide rigidity to the end of the cigarette. Porous plug wrap filter gives an alternative end appearance, typically used on premium brand cigarettes, and an option to use colored plug wraps (Figure 11.10).

11.14 Coated Abrasives

A coated abrasive is an abrasive grain bonded to a flexible substrate using adhesives. Common substrates are paper, cloth, vulcanized fiber, and plastic films and come in grit sizes ranging from very coarse (~2 mm) to ultrafine (submicrometer). The international standard for coated abrasives is ISO6344. The sandpaper and emery cloth are coated abrasives for hand use,

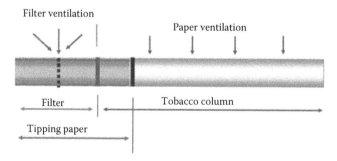

FIGURE 11.8
Cigarette filter ventilation and plug wrap.

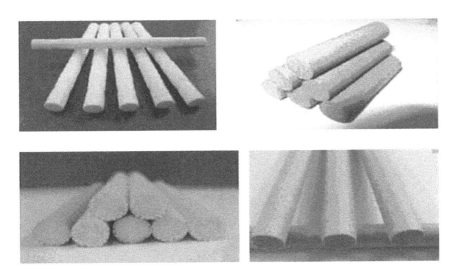

FIGURE 11.9
Cigarette recess filters.

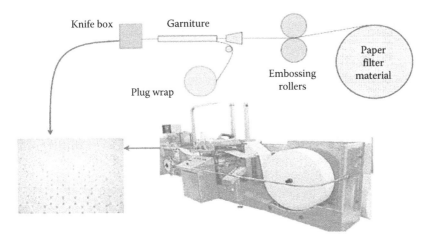

FIGURE 11.10
Paper filters: production.

usually nonprecision. These two terms are used by general public in place of "coated abrasives." Other coated abrasive forms include sanding cords, pads, belts, and disks. Although the most familiar types of coated abrasives are probably the individual sheets of sandpaper with which home woodworkers prepare furniture or crafts for painting, the trade term "coated abrasives" actually encompasses a much wider array of products for both individual and industrial uses (Figure 11.11).

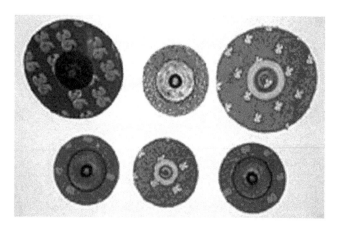

FIGURE 11.11
Coated abrasives.

Today, industrial uses for coated abrasives range from hand polishing with sheets of coated abrasive to grinding steel with large machines that use 300-horse power electric motors to drive belts several feet wide.

a. *Raw materials:* The "sandpaper" is actually a misnomer, as most coated adhesive products contain neither sand nor paper. Generally, they consist of some type of abrasive mineral, which can be organic or synthetic; flexible backings; and adhesives. The abrasive grain, the key part of coated abrasive products, may be either a natural or a synthetic mineral. Due to their extreme hardness, natural minerals such as garnet or emery (corundum with iron impurities) find limited use in products for wood-related applications, while crocus mineral (natural iron oxide) is limited to use as a polishing agent because of its softness. However, such natural minerals comprise less than 1% of the abrasives market.

b. The use of a particular coated abrasive product determines the mineral that will be used in that product. Aluminum oxide is the most common abrasive, followed by silicon carbide. Because silicon carbide is harder and sharper, it is used for applications involving glass and other nonmetal materials. Aluminum oxide, which is the tougher abrasive, is used for metalworking applications where high forces are common. Minerals containing zirconium alumina and alumina are typically used where extremely rugged abrasives are needed, such as in foundries. Expensive and extremely hard minerals such as diamond or cubic boron nitride are restricted to special polishing processes.

c. *Backings:* The backing is the flexible platform to which the abrasive mineral is attached. The development of coated abrasives as a versatile manufacturing tool can in part be attributed to improvements in backing materials. Without a strong and flexible backing, coated abrasives could not survive rough handling or the effects of liquids that are often used as grinding aids.

d. Backings come in four basic materials, each with unique attributes. Paper is the lightest of the backing materials and also the weakest. Although its lack of material strength limits paper's usefulness for hand applications, its flexibility makes it ideal for applications in which the coated abrasive must fit closely to the contour of a work piece. Graded on a scale that increases with the physical weight of a ream, paper backings come in weights rated A to F. Backings made from woven fibers come in progressively heavier weight designations of J, X, Y, M, and H and are typically made of cotton, polyester, or rayon. The two other types of backing are fiber backing and film backing.

e. The pattern of weave in the backing varies from fibers woven at 90° angles to fibers overlaid at 90° angles and stitched together. A less common mesh or screen pattern is used for backings in materials needed in wet, low-pressure applications. Fiber backings are made of multiple layers of resin-impregnated cloth fibers that are used in some dry, high-pressure applications. Film backings, a recent development, have improved the effectiveness of coated abrasives in precision finishing. Uniformly thick synthetic film can be used with special micrometer-sized minerals to produce highly reflective finishing and precision dimensions on parts.

11.15 Nonwoven Abrasives

Specially designed for their unique combination of aggressive action for buffing, blending, surface conditioning, fine polishing, cleaning, and scratch removal, nonwoven materials are finding their way into an ever-increasing number of products and applications (Figure 11.12). Applications using non-woven materials have become almost as varied as those for coated abrasive products. Manufactured from polyester fibers, nonwoven and synthetic steel wool pads would not rust, shred, or splinter. Abrasive grain is uniformly distributed throughout and resin bonded to the nonwoven fibers for consistent finishes and long life. Synthetic steel wool and nonwoven abrasive pads can be cut or folded, and can be used with or without chemicals. The properties of nonwoven abrasives are the following:

- Cleaning glass, fiberglass, porcelain, ceramic, chrome, copper, and stainless steel
- Highlighting and rubbing wood

FIGURE 11.12
Nonwoven abrasives.

- Use with liquid detergent
- Light cleaning steel, nonsteel
- Denibbing plastic
- Defuzzing wood
- Finishing areas for blending

Bibliography

Arora, P. and Z. Zhang, Battery separators, *Chemical Review*, 104, 4419–4462 (2004).
Borkowski, J., *Uses of Abrasives and Abrasive Tools*, Prentice Hall, Englewood Cliffs, NJ, 1992.
Compass Wire Cloth Corporation, Vineland, NJ, www.compasswire.com/bolting_cloth.html (accessed May 25, 2013).
King, R. I. and R. S. Hahn, *Handbook of Modern Grinding Technology*, Chapman and Hall, New York, 1984.
Matsuo, T., Advanced technical textile products, *Textile Progress*, 40(3), 123–181 (2008).
McKee, R. L., *Machining with Abrasives*, Van Nostrand Reinhold, New York, 1982.
O.E. Meyer Co., Sandusky, OH, www.oemeyer.com/industrial/abrasives (accessed May 25, 2013).

12

Testing of Industrial Textiles

12.1 Testing of Filter Fabrics

The filter fabrics are made out of natural or synthetic materials using various woven or nonwoven fabric structures. These fabrics are used in various industrial applications in automobile, medical, air filtration, water filtration, etc. According to the end-user requirements, the test requirements will vary. Some of the important tests required for the filter fabrics are as follows:

1. Filter fabrics (Woven type)
   a. Mass per square meter (ASTM D: 5261/ASTM D: 3776/ISO: 9864/IS: 14716)
   b. Thickness (ASTM D: 5199)
   c. Thread density (ASTM D: 3775/IS: 1963)
   d. Yarn number (IS: 3442/ASTM D: 3883)
   e. Tear resistance (trapezoid strength) (ASTM D: 4533/IS: 14293)
   f. Grab strength (ASTM D: 4632/ISO: 13934/2)
   g. Water permeability (ASTM D: 4491/BS: 6906/3/IS: 14324/ISO: 11058)
   h. Air permeability (ASTM D: 737)
   i. Pore size by porometer (ASTM D: 6767)
   j. Apparent opening size (ASTM D: 4751)
   k. Bursting strength (ASTM D: 3786/IS: 1966)
   l. Breaking strength and elongation (ASTM D: 5035/DIN EN ISO: 13934/1)
2. Filter fabrics (Nonwoven type)
   m. Mass per square meter (ASTM D: 5261/ASTM D: 3776/ISO: 9864/IS: 14716)
   n. Thickness (ASTM D: 5199)

o. Tear resistance (trapezoid strength) (ASTM D: 4533)

p. Breaking strength/grab strength (ASTM D: 4632)

q. Water permeability (ASTM D: 4491/BS 6906/3/IS: 14324/ISO: 11058)

r. Air permeability (ASTM D: 737)

s. Pore size by porometer (ASTM D: 6767)

t. Apparent opening size (ASTM D: 4751)

u. Bursting strength (ASTM D: 3786/IS: 1966)

12.1.1 Testing of Filter Media

A large and ever-increasing number of standard tests are available for characterizing either filters or their associated media. These standards are established either by national authorities, such as the British Standards Institution and the American Society for Testing Materials (ASTM), by specific industry organizations, such as TAPPI (the American Pulp and Paper Industry), or by regional or international organizations, such as Comit European de Normalization and the International Organization for Standardization (ISO).

12.1.2 Testing Filtration Characteristics

Most of the test procedures designed to characterize a medium with respect to the filtration-specific properties involve "challenging" the medium, either with a suitable clean fluid or with a fluid containing dispersed particles of selected and controlled characteristics. Challenging with a clean fluid permits the evaluation of the permeability or resistance to flow per unit area of medium, such as the flow rate of air or water under a defined pressure and the size of the pores of the medium. The fluid containing dispersed particles permits the determination of the smallest particle that can be retained with 100%, which is the "absolute rating" of a medium. The relationship between particle size and retention efficiency is typically expressed as a grade efficiency curve, i.e., the relationship between the quantity of material filtered and the increasing resistance to flow, and hence the dirt-holding capacity of the medium under the specific operating conditions.

12.1.2.1 Permeability

Permeabilities are generally expressed in two main forms, even if in a considerable variety of units. The more common form, appropriate for sheets of media but effectively treating thickness as a constant, characterizes them in terms of the rate of flow of a specified fluid, usually air, per unit area. A far less widely used form, which is more rigorous fundamentally and takes cognizance of the thickness, characterizes a medium by its permeability coefficient.

12.1.2.1.1 Air Permeability

The most common form for expressing permeability disregards the thickness of the medium, so that the permeability is empirically quantified by the flow rate of air per unit area, under a defined differential pressure. An appropriate example is the Frazier scale widely used internationally in the paper and textile industries; this is based on the flow of air and was formerly specified as cfm/ft² at 0.5 in. WG. Metric versions require care since they may use various combinations of definitions of air volume (liters or cubic meters), time (minutes or seconds), area (square centimeters, square decimeters, or square meters), and differential pressure (mm WG or Pa). Considerably higher pressures (than the 12.5, 20, or 25 mm WG used in air tests), ranging up to 1 bar, may be used where flow rates are relatively low, due either to testing with water instead of air or to the fineness of pores in media such as membranes.

12.1.2.1.2 Measuring Permeability

Suitable measurements can be made with an apparatus of varying degrees of sophistication: a very simple measuring device is the Gurley densometer or air resistance tester, used in the paper industry. With this instrument, pressure is provided by a vertical piston that slides down under its own weight, thus forcing a known volume of air through a standard orifice holding the sample being tested. The number of seconds taken for the predetermined volume of air to flow through the sample provides an empirical definition of its permeability. The Frazier Precision Instrument Company manufactures a differential pressure air permeability machine in two models. The first model, the low pressure machine, with air flow generated by suction up to pressure differentials of 5 kPa (0.05 bar), was developed for measuring the air permeabilities of textile-type materials. Frazier developed a second model, the high-pressure machine: this utilizes the same principle but with pressurized air flow, providing much greater versatility of use due to its higher attainable air flow with differential pressures up to 0.7 bar.

12.1.2.1.3 Pore Size Evaluation Tests

Bubble point testing, also known as *liquid expulsion testing*, utilizes a controlled air pressure to empty through pores that had previously been filled with a wetting liquid. A simple relationship between the pressure, the properties of the liquid, and the diameter of an ideal circular pore permits the calculation of the equivalent pore diameter. This method is normally used for pores in the size range 0.05–0.50 mm, but is, of course, only a secondary test, since it does not actually measure a pore dimension.

Challenge tests determine the effective size of open pores by challenging them with suspensions of particles of known sizes. This method is typically used for pores in the size range 0.005–0.001 m, and this is now a direct measure of through-pore size.

Mercury porosimetry, known also as mercury intrusion, involves filling the pores with mercury under pressures up to 400 MPa. The volume of mercury forced in, which can be measured very accurately, is related to pore size and pressure by the same relationship used in the bubble point test. This method, which is the subject of BS 7591: Part 1:1992, is reported to be suitable for many materials.

Gas adsorption, as described in BS 7591: Part 2:1992, typically involves measuring the quantity of nitrogen adsorbed as its relative pressure is progressively increased at a constant cryogenic temperature. The minimum size of pore that can be studied is restricted by the 0.4 nm size of the nitrogen molecule; the maximum is limited to about 50 nm by the practical difficulty of measuring the amount of nitrogen adsorbed at high relative pressure. The method is therefore most appropriate for pores in the size range 0.0004–0.04 lam.

12.1.2.1.4 Filtration Efficiency

Filtration efficiency is usually stated in terms of the percentage of particles of a certain size that would be stopped and retained by a filter medium. This raises two quite difficult problems: where a test dust comprises particles of a range of sizes, what is the actual size to which the percentage efficiency relates? The numerical differences in percentage efficiency of a wide variety of media are often relatively small; many media being over 95% efficient, therefore, is percentage efficiency a meaningful basis for the comparison of different media? There are two alternative expressions for percentage efficiency: one is percentage penetration and the other is the Beta ratio (fl ratio). Very high efficiency air filters, for which efficiencies range upward from 99.99%, are sometimes characterized in terms of percentage penetration and are in fact classified as ultra-low penetration air (ULPA) filters: thus, European Standard class EU 15 can be described as having an efficiency of 99.9995% or a penetration of 0.0005%. The fl ratio is based on counts of particles of specific sizes.

12.1.2.1.5 Filtration of Liquids

Two different techniques are used for determining the efficiency when filtering liquids, respectively, are identified as the single-pass test and the multipass test. Although these tests have much in common, there is a significant difference in the particle size distribution presented to the filter, which may significantly affect the stated efficiency of the medium. The single-pass test, as its name implies, passes a consistent, unchanging distribution of particles through the test circuit just once. The filter medium, ranging from say a 47 mm diameter disk to a 300 × 300 mm sheet, is held in a leak-free support, with sampling points sited as close to the filter as possible. By contrast, the objective of the multipass test with the pressure filtration is to challenge the filter with a gradually increasing percentage of smaller particles: this is felt to be more representative of real systems in which a fluid is recirculated repeatedly and where larger particles are not only removed by filtration but are also being ground down to smaller dimensions. This test was originally developed for hydraulic oils but has become the basis of standards relating to other fluids such as lubricating fluids and water: e.g., the internal combustion engine lube oil filter standard is ISO 4585. Therefore, the multipass test, as defined by ISO 4572, now specifies online sampling and analysis as mandatory, recognizing that efficiency will tend to change as the filter progressively blocks continuous monitoring with an analyzer and can provide a direct read-out of fl ratio.

12.1.2.1.6 Filtration of Gas Air

All gas-phase filtration tests are of single-pass format, but there is considerable variety both in the nature of the suspension of particles used to challenge a filter and in the analytical methods whereby performance is assessed. Three types of tests can be distinguished, respectively, identified as staining tests, weight arrestance, and particle concentration efficiency.

Atmospheric dust spot efficiency is the first of the *staining tests*. This test is a standard procedure for air filters used in air conditioning and general ventilation. The tests is based on the intensity of staining of a "target" filter paper caused by the flow of a quantity of atmospheric air through it. The staining arises from the natural contaminants in the local atmosphere. The intensity of the staining is monitored by an opacity meter and provides an empirical measurement of the concentration of the contaminants in the air drawn through the target.

Synthetic dust weight arrestance test is a standard procedure for air filters used in air conditioning and general ventilation. It is described in detail in Part 1 of BS 6540. The essence of the procedure is to challenge a filter with a dispersion of test dust, the filtrate passing on through a second or final filter, which collects that part of the dust that penetrates through the filter under test. The dust dispersion is created continuously by a suitable combination of a dust feeder and a compressed air venturi ejector. The weight of dust passing through the filter under test is determined by reweighing the final filter. The full procedure, which is designed for testing complete air filters or filter panels (rather than simply a sample of filter medium), includes feeding a weighed quantity of dust in a series of equal increments. The first increment is being restricted to 30 g, to permit determination of the initial synthetic dust weight arrestance. Between consecutive increments, measurement is made of the weight of dust passing the filter under test, the corresponding pressure loss across the test filter, and its atmospheric dust spot efficiency.

The synthetic dust weight arrestance, A (%), for any particular period is given by

$$A - 100 \times \left(1 - \frac{W_2}{W_1}\right)$$

where
   W_1 is the weight of synthetic dust fed
   W_2 is the weight of synthetic dust passing the filter under test

12.1.2.1.7 Particle Concentration Efficiency

For the various grades of high-efficiency air filters (HEPA, ULPA, etc.), particle concentration efficiencies are measured and expressed in terms of differences between upstream and downstream concentrations of submicrometer particles determined by continuous online monitoring. While the concept is simple, the practical reality tends to be complex because of the sophisticated technique and equipment required both to generate consistently suitable aerosols and to determine the size, size distribution, and concentration of the particles. Aerosol particles are variously solid or liquid and range from almost monosized to heterogeneous mixtures. For example, an aerosol of sodium chloride crystals can be generated by atomizing a 1% solution to produce fine droplets, from which water is removed by evaporation; the particle size is determined by the atomization step. Alternatively, an aerosol of dioctylphthalate (DOP) droplets is formed by the condensation that occurs when warm air containing DOP vapor is quenched by dilution with cold air; the particle size is controlled by the temperature difference between the two air streams.

12.1.2.1.8 Dirt-Holding Capacity

The dirt-holding capacity of a medium can conveniently be assessed as part of either the multipass liquid filtration test or the synthetic dust weight arrestance test for air filters.

12.1.3 Testing of Mechanical Properties

Most filter media manufacturers have their own very specific mechanical property demonstrations. However, there are some generally accepted methods, which are as follows.

12.1.3.1 Strength

The strength of a material is typically characterized by generating stress/strain data using an extensometer, in which a strip of textile is stretched by a suspended weight. A linear relationship (Hooke's law) exists between applied stress and the amount of extension per unit length up to the elastic limit, beyond which stretching accelerates and then rupture occurs. This pattern provides a variety of parameters and definitions by which the material may be characterized, the most widely used being tensile strength; others are breaking, rupture or yield strength, yield point, elastic limit, and ultimate elongation. The extensometer is not designed to test fabric as far as rupture, but only within the range of stress where both stretching and recovery can occur, i.e., over the linear limits of Hooke's law. In practice, with sheet materials such as textiles and paper, it is customary to treat the sheet thickness as a constant and to relate the stress only to the width of the strip, i.e., as kg/cm. The bursting strength is an empirical value that depends on the diameter of the disk tested in accordance with appropriate standards, such as BS 3137:1995 for paper and BS 4768:1991 for textiles.

12.1.3.2 Resistance to Abrasion

Various devices are available whereby the resistance of textiles to abrasion can be quantified. Examples are the Frazier Schiefer Abrasion Tester, and the Martindale and Taber testers available from SDL International; these subject samples to continuous rubbing under a controlled pressure.

12.1.3.3 Thickness, Compressibility, and Resilience

The compressometer permits the evaluation of the thickness, compressibility, and resilience of a wide variety of materials (textiles, rubber, felt, nonwovens, paper, films, etc.) especially where observations are required at a range of compressive loads extending from 0.3 mbar to 1.7 bar. The sample to be tested is placed between the instrument base or anvil and the circular pressure foot that is fastened to the vertical spindle; three sizes of pressure foot are available (diameters 25, 75, and 125 mm). The lower dial indicates the thickness of the specimen, while the upper dial shows the pressure applied by a helical spring in the tube between them, this pressure being set manually using a rack and pinion device to compress or relax the spring.

12.1.3.4 Membranes

The filtration action of micro- and ultrafiltration membranes is very similar in principle to that of other continuous media. The delicacy and very fine pore structure of membranes, however, result in some major differences in test methods and procedures that are based on two areas: structure-related parameters and permeation-related parameters. Certain tests are also used to establish the integrity of membranes in specific applications. The direct measurement of pore statistics is routinely carried out by electron microscopy (by scanning electron microscopy and transmission electron microscopy). It should be noted that the asymmetric structure of most ultrafiltration membranes, with top layer pore sizes in the range 20–1000 A, means that many of the methods of characterization of microfiltration membranes and other continuous media cannot be applied. Bubble point and mercury intrusion methods require high pressures that could damage or destroy the membrane structure. The methods that can be used with ultrafiltration membranes include permeation experiments and test methods such as gas adsorption–desorption, thermoporometry, permporometry, and rejection measurements.

12.1.3.4.1 Some Other Tests for Filter Media

12.1.3.4.1.1 Diffusion Testing A diffusion test is recommended in high-volume systems with final filter surface areas of 0.2 m² or greater. This *test* is based on the *fact* that gas will diffuse through the liquid in the pores of a fully wetted filter. The diffusion rate is proportional to the differential pressure across the membrane and to its surface area. The flow of gas is limited to diffusion through water-filled pores at differential pressures below the bubble point pressure of the material under test. In the diffusion test, pressure is typically applied at 80% of the bubble point pressure of the material. When there is liquid downstream of the filter, the volume of gas flow is determined by measuring the flow rate of displaced water. The rate of diffusion can also be measured by a gas flow meter. In industrial settings, the flow rate is often measured on the upstream side of the filter, which does not require a tap into the sterile downstream side. The measurement technique used by many automated devices is pressure decay, after the gas on the upstream side is pressurized to the desired test pressure.

12.1.3.4.1.2 Mercury Intrusion Method The mercury intrusion test relies upon the penetration of mercury into the membrane pores under pressure. The volume of mercury forced into the membrane is related to the pore size and pressure; the size is inversely proportional to the pressure. As with the bubble point test, a morphology or shape factor must be introduced. In the test, the pressure of mercury is gradually increased, and at the lowest pressure the largest pores will fill with mercury. The increasing pressure progressively fills the smaller and smaller pores, until a maximum intrusion of mercury is achieved. At high pressure, however, erroneous results may be obtained due to the deformation or damage to the membrane material. In addition, the method also measures dead-end pores, which are not active in filtration. The size range of the test covers 5 nm to 10 lam pores, i.e., it covers microfiltration and some ultrafiltration membranes. Overall, it gives pore size and pore size distribution.

12.1.3.4.1.3 Water Integrity Test This test is relevant to sterilizing-grade hydrophobic filters that are used for the sterile filtration of air streams and gases in many pharmaceutical and biological applications. It is based on the same principles as the mercury intrusion test and may be performed in situ after sterilization without any downstream manipulations and can be directly correlated to the bacterial challenge tests. The upstream volume of the housing or filter must be completely flooded with water; pressure is then applied by air on the water volume and the rate of water permeation determined.

12.1.3.4.1.4 Bacterial Challenge Test A bacterial challenge test system for the evaluation of the effectiveness of high-efficiency membrane filters uses a nebulizer adapted for high-pressure operations. The device uses two impinge-type samplers in series upstream and a silt sampler downstream of the test filter. A minimum challenge of 3×108 spores is recommended for filters operating 300 days/year with average flows of 850 dm³/min. The bacterial challenge test is a destructive method, and it therefore must be correlated with practical nondestructive integrity tests, e.g., bubble point and diffusion methods, to ensure filtration reliability. Filters that retain 100% of the challenge organism *Pycnoclavella diminuta* normally have water bubble point values of 3 bar or more.

12.1.3.4.1.5 Latex Sphere Test   Latex spheres make up one of the variety of closely sized inert test dust materials that may be used in challenge tests. The object of these tests is to characterize the pore size and the filtration efficiency of media.

12.1.3.4.1.6 Gas Adsorption–Desorption   The use of gas adsorption–desorption is frequently practiced for the measurement of pore size and size distribution of porous media. Typically, nitrogen is used as the adsorbing medium, the method determining the quantity of gas adsorbed (and desorbed) at a particular pressure up to the saturation pressure. A model is required that relates the pore geometry to the adsorption isotherms. The method is limited generally to more uniform structures. The ceramic membranes have been satisfactorily characterized by this method.

12.1.3.4.1.7 Thermoporometry   Thermoporometry uses the calorific measurement of solid–liquid transition in a porous medium. The method typically uses water as the fluid and is based on the fact that the freezing temperature in the pores of a membrane (i.e., the top layer) depends upon the pore size. The extent of undercooling is inversely proportional to the pore diameter. The method also measures the dead-end pores in the membrane. The material of the medium should have enough elasticity to resist the expansion of water as it freezes.

12.1.3.4.1.8 Permporometry   Permporometry is a method that characterizes only the *active* pores in the membrane. It is based on the blockage of pores by a condensable gas, linked with the measurement of gas flux through the membrane. The pore blockage is based on the same principle of capillary condensation as used in adsorption.

12.1.3.4.1.9 Flow Porometry   A novel method of porometry test has been developed especially for ceramic membranes, although it could, in principle, be applied to other types. In this method, the membrane sample is soaked in a liquid that fills all the (through) pores in the sample spontaneously. One side of the sample is then pressurized with air, which slowly removes the liquid from the pores. The largest pore will become free first, followed by progressively smaller pores, and the air flow rate.

12.2 Testing of Hoses

Hoses used in many industrial applications are for liquid or gas transmission. These hoses are manufactured with polymeric materials and should be capable of withstanding the pressure and impact load. Some of the major tests subjected to hoses are given in the following text.

12.2.1 Hydrostatic Pressure Tests

Hydrostatic pressure tests are classified as follows:

1. Destructive type
    a. Burst test
    b. Hold test
2. Nondestructive type
    a. Proof pressure test
    b. Change in length test (elongation or contraction)
    c. Change in outside diameter or circumference test
    d. Warp test
    e. Rise test
    f. Twist test
    g. Kink test
    h. Volumetric expansion test

12.2.1.1 Destructive Tests

Destructive tests are conducted on short specimens of hose, normally 18 in. (460 mm) to 36 in. (915 mm) in length, and as the name implies, the hose is destroyed in the performance of the test: (a) burst pressure is recorded as the pressure at which actual rupture of a hose occurs and (b) a hold test, when required, is a means of determining whether weakness will develop under a given pressure for a specified period of time.

12.2.1.2 Nondestructive Tests

Nondestructive tests are conducted on a full length of a hose or hose assembly. These tests are for the purpose of eliminating hose with defects that cannot be seen by visual examination or in order to determine certain characteristics of the hose while it is under internal pressure.

1. A proof pressure test is normally applied to hose for a specified period of time. On new hose, the proof pressure is usually 50% of the minimum specified burst except for woven jacket fire hose where the proof pressure is twice the service test pressure marked on the hose (67% of specified minimum burst). Hydrostatic tests performed on fire hose in service should be no higher than the service test pressure referred to earlier. The regulation of these pressures is extremely important so that no deteriorating stresses will be applied, thus weakening a normal hose.

2. With some type of hose, it is useful to know how a hose will act under pressure. All changes in length tests, except when performed on wire braid or wire spiraled hose, are made with original length measurements taken under a pressure of 10 psi (0.069 MPa). The specified pressure, which is normally the proof pressure, is applied, and immediate measurement of the characteristics desired is taken and recorded. Percent length change (elongation or contraction) is the difference between the length at 10 psi (0.069 MPa) (except wire braided or wire spiraled) and that at the proof pressure times 100 divided by the length at 10 psi (0.069 MPa). Elongation occurs if the length of the hose under the proof pressure is greater than that at a pressure of 10 psi (0.069 MPa). Contraction occurs if the length at the proof pressure is less than that at 10 psi (0.069 MPa). In testing wire braided or spiraled hose, the proof pressure is applied and the length recorded. The pressure is then released and, at the end of 30 s, the length is measured; the measurement obtained is termed the "original length."

3. Percent change in outside diameter or circumference is the difference between the outside diameter or circumference at 10 psi (0.069 MPa) and that obtained under the proof pressure times 100 divided by the outside diameter or circumference at 10 psi (0.069 MPa). Expansion occurs if the measurement at the proof pressure is greater than that at 10 psi (0.069 MPa). Contraction occurs if the measurement at the proof pressure is less than that at 10 psi (0.069 MPa).

4. Warp is the deviation from a straight line drawn from fitting to fitting; the maximum deviation from this line is warp. First, a measurement is taken at 10 psi (0.069 MPa) and then again at the proof pressure. The difference between the two, in inches, is the warp. Normally, this is a feature measured on woven jacket fire hose only.

5. Rise is a measure of the height a hose raises from the surface of the test table while under pressure. The difference between the rises at 10 psi (0.069 MPa) and at the proof pressure is reported to the nearest 0.25 in. (6.4 mm). Normally, this is a feature measured on woven jacket fire hose only.

6. Twist is a rotation of the free end of the hose while under pressure. A first reading is taken at 10 psi (0.069 MPa) and a second reading at proof pressure. The difference, in degrees, between the 10 psi (0.069 MPa) base and that at the proof pressure is the twist. Twist is reported as right twist (to tighten couplings) or left twist. Standing at the pressure inlet and looking toward the free end of a hose, a clockwise turning is right twist and counterclockwise is left twist.

7. Kink test is a measure of the ability of woven jacket hose to withstand a momentary pressure while the hose is bent back sharply on itself at a point approximately 18 in. (457 mm) from one end. Test is made at pressures ranging from 62% of the proof pressure on sizes 3 in. (76 mm) and 3.5 in. (89 mm) to 87% on sizes under 3 in. (76 mm). This is a test applied to woven jacket fire hose only.

8. Volumetric expansion test is applicable only to specific types of hose, such as hydraulic or power steering hose, and is a measure of its volumetric expansion under ranges of internal pressure.

12.2.2 Design Considerations

In designing hose, it is customary to develop a design ratio, which is a ratio between the minimum burst and the maximum working pressure. Burst test data are compiled, and the minimum value is established by accepted statistical techniques. This is done as a check on theoretical calculations, based on the strength of reinforcing materials and on the characteristics of the method of fabrication. Minimum burst values are used as one factor in the establishment of a reasonable and safe maximum working pressure. Maximum working pressure is one of the essential operating characteristics that a hose user must know and respect to assure satisfactory service and optimum life. It should be noted that design ratios are dependent on more than the minimum burst. The hose technologist must anticipate natural decay in strength of reinforcing materials, and the accelerated decay induced by the anticipated environments in which the hose will be used and the dynamic situations that a hose might likely encounter in service.

12.3 Testing of Transmission Belt

Transmission belts are used in many applications in industrial end uses for transmitting material from one destination to another destination. The belt drives and their component test for meeting standard requirements are given in the following table.

Standard and/or Project
Belt drives—Flat transmission belts and corresponding pulleys—Dimensions and tolerances
Belt drives—Pulleys—Limiting values for adjustment of centers
Belt drives—Pulleys—Quality, finish, and balance
Belt drives—Pulleys for V-belts (system based on datum width)—Geometrical inspection of grooves
Belt drives—V-belts and V-ribbed belts, and corresponding grooved pulleys—Vocabulary
IBelt drives—Endless wide V-belts for industrial speed-changers and groove profiles for corresponding pulleys
Belt drives—V-ribbed belts, joined V-belts, and V-belts including wide section belts and hexagonal belts—Electrical conductivity of antistatic belts: Characteristics and methods of test
Belt drives—Classical and narrow V-belts—Grooved pulleys (system based on datum width)
Synchronous belt drives—Vocabulary
Synchronous belt drives—Pulleys
Belt drives—Dynamic test to determine pitch zone location—V-ribbed belts
Belt drives—Electrical conductivity of antistatic endless synchronous belts—Characteristics and test method
V-belts—Uniformity of belts—Test method for the determination of center distance variation
Belt drives—Grooved pulleys for V-belts (system based on effective width)—Geometrical inspection of grooves
Belt drive—V-ribbed belts for the automotive industry—Fatigue test
Synchronous belt drives—Automotive belts—Determination of physical properties
Curvilinear toothed synchronous belt drive systems
Installed run-out in curvilinear synchronous belt drive systems

12.4 Testing of Conveyor Belts

The following tests are carried out for conveyor belts that are used in industrial purposes:

a. *Dynamic Splice Testing—Din 22110*

In general, the splice is considered to be the weakest point on an endless conveyor belt. Statically, it is usually easy to achieve a tensile strength that corresponds to that of the belt. For the optimum belt configuration and dependable continuous operation, however, another parameter is vital: the dynamic efficiency of the splice. The fatigue strength of the splice is determined in accordance with Section 3 of the DIN 22110 standard. Four endless belts are placed on the test conveyor, one after the other. The belts are run until they break or at least until the reference number of 10,000 load cycles has been achieved. During each sawtooth-patterned load cycle, the belt travels over the two drums 18 times. Running the test longer would not result in any additional findings. The upper load is specified such that a specimen breaks after achieving a high number of load cycles, and a further specimen runs through at least until it has reached the reference load cycle number. The dynamic efficiency of the splice is calculated from the maximum top load, at which the 10,000 load alternations are achieved. In the DIN 22101 standard, which has been in effect since 2002, the dynamic efficiency is taken directly into account when specifying the necessary belt tensile strength. In general, a high dynamic efficiency of the splice lowers the required belt breaking strength.

b. *Drum Friction Test*—The drum friction test simulates a belt slipping over a jammed pulley or a pulley rotating under a stationary conveyor belt. This measures whether the surface temperature remains under a required maximum after a specific time and under a specific tension. Generally, a rubber conveyor belt can produce a pulley surface temperature of up to 500°C (930°F). The visual appearance of flame or glow is not permitted.

c. *Surface Resistance Test*—An electrostatic charge may build up on the conveyor belt surface and ignite a mixture of flammable gases and air. Therefore, the surface resistance of the conveyor belt covers has to be below 300 MΩ.

d. *High-Energy Propane Burner Test*—Conveyor belts must not propagate fire. In order to determine whether a conveyor belt fulfills this requirement, it is ignited by a propane burner. After the ignition source has been removed, the flames must self-extinguish within a certain time frame or within a certain distance. A typical sample for this test is 2 m long at full width.

e. *Spirit Burner Test*—A small piece of the conveyor belt is held over a spirit burner flame. After a certain time, the burner is removed. The duration of flame and glow has to be within a specified time limit.

f. *Bunsen Burner Test*—A small piece of the belt is held over a Bunsen burner flame, and the duration of the belt's afterglow must be less than a specified maximum time. This test, which was implemented as per the 1969 Federal Coal Mine Health and Safety Act, is similar to a standard that was in force in Europe until the mid-1970s for underground conveyor belts (fire resistant, grade K or S). Later, these conveyor belts have only been allowed to be used in aboveground applications.

12.4.1 Aramid Conveyor Belts

Aramid (two trademarks are "Kevlar" and "Twaron") conveyor belts are used in special applications where a low weight is of paramount importance. Aramid has low fatigue resistance and high compression sensitiveness. Special requirements on the conveyor design have to be incorporated. The (dynamic) splice strength is much lower than that for steel-cord conveyor belts. Aramid is less robust, quickly absorbs moisture, is susceptible to UV radiation, and has a very high transverse thermal expansion and rapid loss of strength due to swatter and humidity. Severe failures of aramid conveyor belts in underground mining have led to their immediate ban some decades ago. Some of the important tests and standards for general and aramid conveyor belts are as follows:

- Maximum breaking strength
- Splice strength, static
- Splice strength, dynamic
- Longevity
- Weight, for the same breaking strength
- Pulley diameters
- Impact resistance
- Rip resistance
- Corrosion resistance

12.4.2 DIN (German)

As per standard, test requirement of technical textile materials in German and their DIN test numbers are given below.

252	Ply separation strength
284	Electrical conductivity; specifications and test method
340	Laboratory-scale flammability characteristics; requirement and test method
505	Method for the determination of the tear propagation resistance of textile conveyor belts
583	Textile conveyor belts; total thickness and thickness of constructional elements
703	Troughability
1120	Determination of strength of mechanical fastenings; static test method
1554	Drum friction test
7716	Rubber products; requirements for storage, cleaning, and maintenance
9856	Determination of elastic and permanent elongation and calculation of elastic modulus
12881-1	Fire simulation flammability testing; propane burner test
12881-2	Fire simulation flammability testing; large-scale fire test
12882	Conveyor belts for general use; electrical and flammability safety requirements
13827	Steel cord conveyor belts; determination of displacement of steel cords

14890	Specification for rubber- or plastics-covered conveyor belts of textile construction for general use
14973	Conveyor belts for underground use; electrical and fire safety requirements
15236-3	Steel cord conveyor belts; special safety requirements for underground use
16851	Textile conveyor belts; determination of endless length
20340	Flammability of conveyor belts; specifications and method of test
22100-1	Synthetic materials for underground use; hygiene requirements
22101	Basics for the design of belt conveyors (superseding the 1982 and 2002 issues)
22102	Textile carcass conveyor belts
22103	Fire-resistant conveyor belts for coal mining
22104	Antistatic conveyor belts; requirements and testing
22109	Textile carcass conveyor belts for underground coal mines ("self-extinguishing" grades)
22110	Testing of splices
22112	Belt conveyors for underground coal mining
22117	Conveyor belts for coal mining; determination of the limiting oxygen index (LOI)
22118	Textile carcass conveyor belts for use in underground coal mines; fire testing
22120	Scraper rubber for underground coal mining
22121	Textile carcass conveyor belts for coal mining; permanent joints
22129	Steel cord conveyor belts for underground coal mining (cross-section)
22131	Steel cord conveyor belts
22721	Textile conveyor belts for underground use
28094	Steel cord conveyor belts; adhesion strength test of core to cover layer
53504	Testing of breaking strength and elongation of rubber
53505	Testing of hardness of rubber (Shore A and D)
53507	Testing of tear strength of rubber
53516	Testing of abrasion resistance of rubber

12.4.3 ISO (World)

As per standard, test requirement of technical textile materials and their DIN test numbers are given below.

251:2003	Conveyor belts with textile carcass—Widths and lengths
252:2007	Conveyor belts—Ply adhesion between constitutive elements—Test method and requirements
282:1992	Conveyor belts—Sampling
283:2007	Conveyor belts—Full thickness tensile strength and elongation—Specifications and method of test
283-1:2001	Textile conveyor belts—Determination of tensile strength, elongation at break, and elongation at the reference load
284:2004	Conveyor belts—Electrical conductivity—Specification and test method
340:2007	Conveyor belts—Flame retardation—Specifications and test method
432:1989	Ply-type conveyor belts—Characteristics of construction
433:1991	Conveyor belts—Marking
505:2000	Conveyor belts—Method for the determination of the tear propagation resistance of textile conveyor belts

583:2007	Textile conveyor belts—Tolerances on total thickness and thickness of covers—Test methods
583-1:2008	Conveyor belts with a textile carcass—Total thickness and thickness of elements—Methods of test
703:2007	Conveyor belts—Transverse flexibility and troughability—Test method
1120:2002	Conveyor belts—Determination of strength of mechanical fastenings—Static test method
1431:2004	Rubber—Resistance to ozone cracking
1554:1999	Conveyor belts—Drum friction test
2148:1974	Continuous handling equipment—Nomenclature
3684:1990	Conveyor belts—Determination of minimum pulley
3870:1976	Conveyor belts (fabric carcass)—Adjustment of take-up device
4195:2007	Conveyor belts—Heat resistance—Requirements and test method
4661-1:1993	Rubber, vulcanized, or thermoplastic—Preparation of samples and test pieces—Physical tests
5048:1989	Belt conveyors; calculation of operating power and tensile forces
5284:1986	Conveyor belts—List of equivalent terms
5285:2004	Conveyor belts—Guide to storage and handling
5293:2004	Conveyor belts—Formula for transition distance on three equal-length idler rolls
5924:1989	Fire tests—Reaction to fire—Smoke generated by building products (dual-chamber test)
7590:2009	Steel cord conveyor belts—Methods for the determination of total thickness and cover
7622-1:1995	Steel cord conveyor belts—Longitudinal traction test—Part 1: Measurement of elongation
7622-2:1995	Steel cord conveyor belts—Longitudinal traction test—Part 2: Measurement of tensile strength
7623:1997	Steel cord conveyor belts—Cord-to-coating bond test—Initial test and after thermal treatment
8094:1984	Steel cord conveyor belts—Adhesion strength test of the cover to the core layer
9856:2004	Conveyor belts—Determination of elastic and permanent elongation and calculation of elastic modulus
10247:1990	Conveyor belts—Characteristics of covers—Classification
10357:1989	Conveyor belts—Formula for transition distance on three equal-length idler rollers (new method)
12881-1:2008	Conveyor belts—Propane burner test
12881-2:2008	Conveyor belts—Large-scale flammability test
12882:2002	Conveyor belts for general use—Electrical and fire safety requirements
13827:2004	Steel cord conveyor belts—Determination of horizontal and vertical position of the steel cords.
14890:2005	Conveyor belts—Specification for conveyor belts of textile construction for general use
14973:2004	Conveyor belts for underground use—Electrical and fire safety requirements
15236:2004	Steel cord conveyor belts (type A) and woven steel cord or steel strand conveyor belts (types B and C)
15236:2004	Steel cord conveyor belts—Vulcanized splices

15236-1:2004	Steel cord conveyor belts for general use—Design, dimensions, and mechanical requirements
15236-2:2004	Steel cord conveyor belts—Preferred types
15236-3:2008	Steel cord conveyor belts; special requirements for underground use
15236-4:2004	Steel cord conveyor belt—Vulcanized splices
16851:2005	Textile conveyor belts—Determination of the endless length
18573:2003	Conveyor belts—Test atmospheres and conditioning
22721:2007	Textile conveyor belts for underground mining

12.5 Testing of Ropes

The following standard tests are recommended for ropes used in industrial applications:

1. *Diameter:* The parameter is measured after applying a load of 10 kg to single ropes, 6 kg to half ropes, and 5 kg to twin ropes. This means that it might be very difficult to accurately test the diameter of your rope in domestic conditions.

2. *Weight:* The parameter is expressed as the weight of the rope per 1 m of its length. The weight of single ropes free of any finishes ranges from 55 to 88 g/m, half ropes weigh ~50 g, and twin ropes ~42 g. The rope core must account for at least 50% of its total weight.

3. *Number of falls:* In the language of the laymen: it gives the number of falls the tested rope has to withstand. The EN 892 standard requires at least five falls for single ropes given a load of 80 kg. Half ropes are tested with a 55 kg load applied. With twin ropes, the two ropes are loaded with 80 kg weights at all times, and the minimum number of falls withstood is 12. The number of falls withstood in the course of the testing is a direct measure of the safety (strength) margin of the rope. New ropes cannot practically get ruptured at shock loading provided that the rope is in satisfactory condition and well handled (protected against sharp edges). The safety of the rope will gradually decrease as a function of the material aging and wear, that is, as a result of the factors detrimental to its strength. Also humidity, which often attacks the polyamide fibers, of which the rope is produced, decreases its strength.

4. *Impact force:* In the course of the testing, the impact force is derived from the testing load at standard falls with the elongating rope gradually absorbing the impact energy before impeding the fall. The lower the impact force, the more comfortable the absorbed fall for the climber. During the testing, each fall increases the impact force in the rope, and the final number of withstood standard loadings is

dependent on the rate at which it increases. The higher the number of standard falls withstood, the higher the rope life. Practical usage of the ropes on the rocks or training walls is different than that in the laboratory conditions. During standard rope tests, the rope ends are tightly fixed, yet in field operation, the belaying tools and systems allow for certain rope slippage, which allows for dynamic withstanding of the falls. As a result of the dynamic belaying, a proportion of the fall energy is dissipated and the impact force diminished. For this reason, it is essential to control and use appropriate dynamic belaying tools. The so-called fall factor is also key to the magnitude of the fall force. The fall factor is a one-dimensional quantity giving the ratio between the length of the fall and that of the rope withstanding the fall. Practically, it is of no importance for the magnitude of the impact force how long the fall is, but rather the magnitude of the fall factor. Falls of 5 m with a fall factor of 1.5 will give rise to a considerably higher impact force than a fall of 6 m with a fall factor of 0.5 m.

5. *Sheath slippage:* The test seeks to quantify the displacement of the sheath with regard to the core when loading the rope surface. A special machine is used for the test and the EN 892 standard specifies that the displacement cannot be higher than 40 mm at the rope length elongation of 1930 mm, that is, ~2%. If, in real practice, the core gets displaced with regard to the sheath, swelling or so-called stockings may occur. If the ends of the ropes have been poorly welded, the core at the rope end may slip out of the sheath or the sheath may slip outside the core. The ends of our ropes are ultrasonically welded into a single indivisible unit, and if the displacement requirements are complied with, the earlier condition will not occur.

6. *Elongation:* The useful static elongation is tested by applying a load of 80 kg onto the rope and may not exceed 8% for single ropes (one-strand ropes) and twin ropes (two strands tested at the same time) and 10% for half ropes (single strand).

7. *Elongation at initial fall:* This parameter gives information on the elongation of the rope in the course of the initial standard fall. The maximum permissible dynamic elongation is 40%, and it reflects the properties of the rope more efficiently than the static value of working elongation. All LANEX ropes with elongation ranging between 30% and 35% comply with the requirement of this (still optional) EN standard by a wide margin.

8. *Knotability:* Excellent flexibility is one of the most important requirements imposed on climbing ropes. How is this parameter quantified? A simple knot is fastened on the tested rope and a 10 kg load is subsequently applied. Then, the inner diameter of the knot is

measured, and the knotability coefficient is derived from the rope diameter. The maximum permissible value amounts to 1.1 multiple of the rope diameter. Poor flexibility essentially impedes both the fastening of knots and the rope's passage through the karabiners of the belaying system, and when withstanding falls, even poor dissipation of the shock forces into the entire rope length. The effects of the weather and insufficient care of the rope are also detrimental to the rope flexibility.

9. *Rope testing at drop tower:* At LANEX, there is a fully equipped testing room where the individual parameters are tested, including the fall tests. Therefore, when newly developed ropes are shipped to European testing rooms for certification, they have already been fully conditioned and furnished with values representing the known technical parameters. LANEX uses a Vienna-based accredited testing room for its products.

12.6 Testing of Composites

The composite materials are both inhomogeneous and of anisotropic nature in comparison to metallic materials. The elastic properties of a composite material are a measure of its stiffness. This property is necessary to determine the deformations that are produced by loads. The elastic properties of the composite are determined by the elastic stretch and recovery properties and volume fraction of the constituents (fiber and matrix) additionally by the orientation of the reinforcement. Elasticity theory can be used to determine the elastic constants given sufficient information; however, the strength of a composite material is affected by a number of parameters including fiber orientation, volume fraction of fibers, the amount of porosity, fiber length, lamination layup, and environmental effects. The properties measured parallel to the loading direction (tensile), that is, in the fiber direction, would be dominated by the strength of the fibers since they are oriented in the load axis. In the transverse direction, however, the fibers take little of the load, and the matrix material dominates the properties. The design of a test specimen to evaluate a material will ensure that the sample fails in the desired mode for the intended application. Differences in fiber layup, fiber length, and orientation may necessitate differences in the geometry of the specimen to achieve the same failure mode. The material properties of composites are affected not only by the chemical makeup (fibers, resins, etc.) but also by the processing route and conditions. To characterize a material completely is an enormous undertaking since the range of properties to describe a material is extensive, for example, strength in shear, tension and compression, density, stiffness, hardness, impact, creep, fatigue, and corrosion. In addition, these properties will change with temperature and perhaps strain rate, particularly for polymeric composites.

Evaluation of composite materials may be further subdivided into destructive and nondestructive. In destructive testing, tests are carried out to the specimen's failure, in order to understand a specimen's structural performance or material behavior under different loads. These tests are generally much easier to carry out, yield more information, and are easier to interpret than nondestructive testing (NDT). NDT is a wide group of analysis techniques used in science and industry to evaluate the properties of a material, component, or system without causing damage.

12.6.1 Types of Loading

12.6.1.1 Tensile

The response of a composite to tensile loads is very dependent on the tensile stiffness and strength properties of the reinforcement fibers, since these are far higher than the resin system on its own, and also mainly depend on the fiber orientation and integrity.

12.6.1.2 Compression

The adhesive and stiffness properties of the resin system are crucial in determining the compression properties, as it is the role of the resin to maintain the fibers as straight columns and to prevent them from buckling.

12.6.1.3 Shear

This load is trying to slide adjacent layers of fibers over each other. Under shear loads, the resin plays the major role, transferring the stresses across the composite. The interlaminar shear strength (ILSS) of a composite is often used to indicate this property in a multilayer composite ("laminate").

12.6.1.4 Flexure

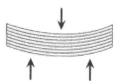

Flexural loads are really a combination of tensile, compression, and shear loads. When loaded as shown, the upper face is put into compression, the lower face into tension, and the central portion of the laminate experiences shear.

12.6.2 Destructive Testing

12.6.2.1 Tensile Testing

The basic physics of most tensile test methods is very similar to that of a specimen with a straight-sided gage section that is gripped at the ends and loaded in uniaxial tension. The tensile test specimens are based on the specimen cross section and the load-introduction method. By changing the specimen configuration, many of the tensile test methods are able to evaluate different material configurations, including unidirectional laminates, woven materials, and general laminates. This test method provides procedures for the evaluation of tensile properties of single-skin laminates, and also these tests are performed in the axial or in-plane orientation. Properties obtained can include tensile strength, tensile modulus, elongation at break (strain to failure), and Poisson's ratio. For most oriented fiber laminates, a rectangular specimen is preferred. Panels fabricated of resin alone (resin casting) or utilizing randomly oriented fibers (such as chopped strand) may be tested using dog-bone (dumbbell)-type specimens. The test axis or orientation must be specified for all oriented fiber laminates.

12.6.2.1.1 In-Plane Tensile Test

12.6.2.1.1.1 *Straight-Sided Specimen Tensile Tests* In the tensile test method, a tensile stress is applied to the specimen through a mechanical shear interface at the ends of the specimen, normally by either wedge or hydraulic grips. Composite materials are usually gripped using some form of "friction grip," where the load is transferred to the specimen through gripping faces that are roughened with serrations or a crosscut pattern. Parallel clamping grips, positively closed by manual or hydraulic means, allow the operator to control the gripping force on the specimen. Bending of the specimen will also occur if the fiber layers are not equally spaced, for instance, as a result of poor consolidation. The material response is measured in the gage section of the specimen by either strain gages or extensometers, and the elastic material properties subsequently determined (Figure 12.1).

Two-axis extensometers are available that measure lateral contraction for Poisson's ratio determination. The important factors that affect tension

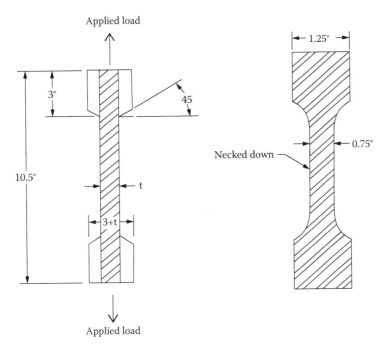

FIGURE 12.1
Tensile test method.

testing results include control of specimen preparation, specimen design tolerances, control of conditioning and moisture content variability, control of test machine-induced misalignment and bending, and consistent measurement of thickness. Fiber alignment, control of specimen taper, and specimen machining (while maintaining alignment) are the most critical steps.

12.6.2.1.1.2 Bow Tie Tension Specimen (for Filament Wound Tubes) Bow tie tension specimen is especially for filament wound tubes. The specimen taper is accomplished by a large cylindrical radius between the wide gripping area at each end and the narrower gage section, resulting in a shape that justifies the specimen nickname of the "dog-bone" specimen. The taper makes the specimen particularly unsuited for the testing of 0° unidirectional materials.

12.6.2.1.2 Out-of-Plane Tensile Test
Out-of-plane testing may be achieved by direct out-of-plane loading by bonding a composite laminate between two fixture blocks or indirect out-of-plane loading using a curved beam. The direct out-of-plane load that uses square or circular loading blocks is pulled out-of-plane using a universal joint to aid alignment. Strength is determined simply by dividing maximum load prior to failure by the specimen cross-sectional area (Figure 12.2).

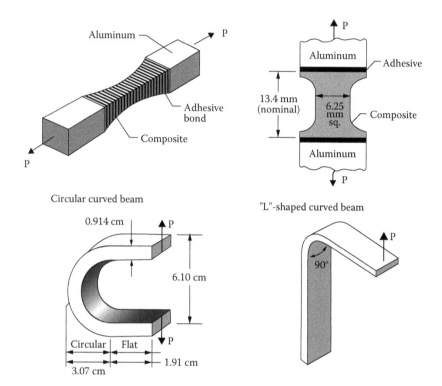

FIGURE 12.2
Out-of-plane tensile test.

In many instances, the material failure will be close to the bonded attachment. A reduced cross-sectional area in the gage length can help force the failure to within the gage length. The curved beam approach to out-of-plane tensile strength (not modulus) takes advantage of the out-of-plane tensile loading induced in the bend of a curved laminate beam subjected to an opening moment. Load is applied using either a four-point bend test fixture or, in a tensile test machine, with suitable loading hinges attached to the beam.

12.6.2.2 Compression Test

Most lightweight structures and substructures include compression members, which may be loaded in direct compression, or under a combination of flexural and compressive loads. The axial stiffness of compression members can be controlled only by the cross-sectional area. The main difficulty in compression testing of composites is prevention of buckling of the specimen. The ratio of compressive to tensile strength is low for the highly anisotropic fibers, but the compressive strength of glass fibers is probably higher than their tensile strength. The parameters found to be significant contributors to the accuracy of the data include fabrication practices, control of fiber alignment, improper and/or inaccurate specimen machining, improper tabbing procedures if tabs are used, poor quality of the test fixture, improper placement of the specimen in the test fixture, improper placement of the fixture in the testing machine, and an improper test procedure. The in-plane compressive test methods are typically used to generate the ultimate compressive strength, strain-at-failure, modulus, and Poisson's ratio of axially or transversely loaded unidirectional composite specimens (Figure 12.3).

There are three basic methods of introducing a compressive load into a specimen: direct loading of the specimen end, loading the specimen by shear, and mixed direct and shear loading. Compressive test methods may also be further classified as having a supported or an unsupported test section to prevent buckling. Unsupported test section method uses an unsupported gage length loaded by shear loading. This test method comprises two fixture types: the conical wedge grips and rectangular wedge grips. These fixtures use tabbed or untabbed specimens and transfer load via wedge-type grips. Another method applies a combination of end loading and shear loading

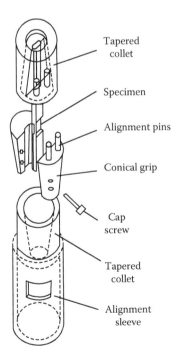

FIGURE 12.3
Compression test: celanese fixture.

to the test specimen. The fixture comprises four blocks clamped in pairs to either end of the test specimen. The surfaces of the fixture blocks in contact with the specimen are roughened to increase the effective coefficient of friction and hence the shear load transfer. The ratio between shear and end loading is adjusted by the torque applied to the clamping bolts. Because of the flexibility of this test method, many different types of composite materials may be tested.

Four-point bending test uses flexure of a sandwich beam to determine compressive properties. The sandwich beam method comprises a honeycomb-core sandwich beam that is loaded in four-point bending, placing the upper face sheet in compression. The upper and lower face sheets are of the same materials and configuration. The upper face sheet is designed to fail in compression when the beam is subjected to four-point bending. The beam is loaded to failure in bending, resulting in the measurement of compressive strength, compressive modulus, and strain-at-failure if strain gages are applied to the upper surface.

Axial compression testing is also useful for the measurement of elastic and compressive fracture properties of brittle materials or low-ductility materials. In any case, the use of specimens having large L/D ratios should be avoided to prevent buckling and shearing modes of deformation. Specimens should have a uniform rectangular cross section, 12 mm (0.5 in.) wide by 140 mm (5.5 in.) long.

12.6.2.3 In-Plane and Interlaminar Shear Properties

Shear testing of composite materials has proven to be one of the most difficult areas of mechanical property testing. While shear modulus measurements are considered accurate, the biggest difficulty is in measuring shear strength. The presence of edges, material coupling, nonpure shear loading, nonlinear behavior, imperfect stress distributions, or the presence of normal stresses make shear strength determination questionable. The most common specimen has a constant rectangular cross section, 25 mm (1 in.) wide and 250 mm (10 in.) long.

The following shear tests are available for in-plane shear measurements.

12.6.2.3.1 ±45° Tensile Test

In-plane shear response tensile measures the shear stress and shear strain properties of a composite material specimen composed of +45° plies and −45° plies. The in-plane shear stress in the gage section is a function of the average applied tensile stress. While this shear stress state may not be pure, it does mimic the stress state within a structural laminate (Figure 12.4).

Measurement of biaxial strain using either extensometers or strain gages will allow shear modulus to be calculated. This test method can be used for unidirectional and woven fabric materials but cannot be used for discontinuous composites. The stress–strain curve from this test is highly nonlinear,

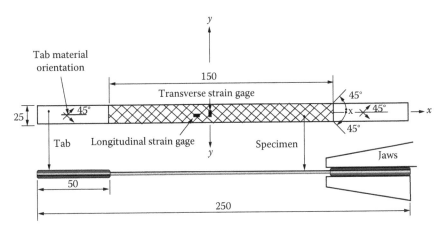

FIGURE 12.4
±45° tensile test.

and the test methods indicate the strain range for modulus calculation and the shear strain that should be used to identify shear strength.

12.6.2.3.2 Iosipescu Shear Test

The Iosipescu or V-notched shear test uses a rectangular beam with symmetrical centrally located V-notches. The beam is loaded by a special fixture applying a shear loading at the V-notch. Either in-plane or out-of-plane shear properties may be evaluated, depending upon the orientation of the material coordinate system relative to the loading axis. The test is not usually used for multidirectional materials but is successfully applied to discontinuous reinforced materials. The notched specimen is loaded by introducing a relative displacement between two halves of the test fixture. Strain gages oriented at ±45° to the loading axis away from the notches and along the loading axis are used to determine the shear response. The specimen notches influence the shear strain along the loading direction, making the shear distribution more uniform than would be the case without the notches (Figure 12.5).

12.6.2.3.3 Rail Shear Test

In rail shear test, load is applied on the laminate by using long rails to apply the shear load while reducing the normal load. While the standard is restricted to in-plane testing, it is capable of testing for either material shear or multidirectional laminate shear properties. The test method is limited to the determination of the modulus or initial shear stress–strain response, because the method cannot sustain the higher loads to failure in high-strength multidirectional laminates. The shear stress state is not uniform in the rail shear test, and failures can be identified outside the gage section, hence reliable strength data cannot always be obtained (Figure 12.6).

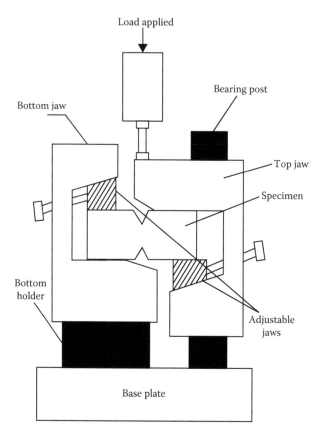

FIGURE 12.5
Iosipescu shear test.

12.6.2.3.4 10° Off-Axis Shear Test

The 10° off-axis shear test represents a simple test method. It uses a straight-sided rectangular specimen where the fibers are unidirectional and oriented 10° off the loading axis. This specimen is not in a state of pure shear so that the test produces results of generally higher modulus and significantly lower strengths than the other shear test methods. It is also not suitable for shear evaluation of multidirectional laminates.

12.6.2.3.5 Short Beam Shear Test

Short beam shear test is used to determine the ILSS of parallel fibers. The specimen is a short, relatively deep, flat laminate loaded in three-point bending with a narrow span. The intent is to minimize flexural stresses while maximizing in-plane shear stresses (Figure 12.7).

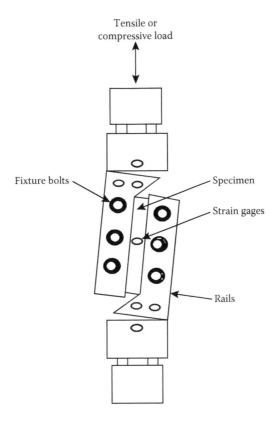

FIGURE 12.6
Rail shear test.

However, the contact stresses induced at the load points greatly interfere with the stress distribution, often leading to failure under the loading nose. However, the V-notched beam is beginning to be the test method of preference to determine interlaminar failure strength. The short beam strength test should be used only for qualitative testing such as material process development:

$$\text{Shear strength}\left(\text{N}/\text{m}^2 \text{ or psi}\right) = \frac{\left[0.75 \times \text{breaking load (N or lb)}\right]}{\left[\text{width (m or in.)} \times \text{thickness (m or in.)}\right]}$$

12.6.2.4 Interlaminar Fracture

Interlaminar fracture, or delamination, continues to be one of the more serious failure modes for laminated composite structures. Delamination arises

FIGURE 12.7
Short beam shear test.

from high out-of-plane loads where there are no fibers to resist these loads. Delamination can occur from tensile, shear loads, or a combination of the two. Several methods to characterize delamination have been developed, which include the following.

12.6.2.4.1 Mode-I Fracture

The double cantilever beam (DCB) specimen has been widely used to measure the mode-I interlaminar fracture toughness of composites. The DCB specimen is a laminate with a nonadhesive insert placed at the mid-plane, at one end prior to curing or consolidation, to simulate a delamination. Generally, unidirectional parallel-sided specimens are used. Typically, loads are applied to the DCB via loading blocks or hinges adhesively bonded to the surface of the DCB. During test, the specimen is subjected to displacement-controlled loading and usually experiences stable delamination growth allowing several values to be determined along the specimen's length. As the delamination grows, fiber bridging usually occurs, increasing the energy required to propagate the delamination further (Figure 12.8).

12.6.2.4.2 Mode-II Fracture

There are several different test configurations for measuring mode-II fracture toughness. The specimen configuration is largely the same; it is the method of applying load to promote a mode-II fracture that varies. The configuration

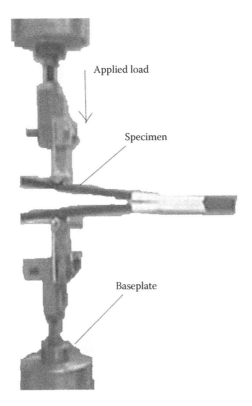

FIGURE 12.8
Mode-I fracture test.

is a parallel, unidirectional beam as in the DCB. The three main methods of loading the specimen are thus as follows:

1. *End-loaded shear*: specimen is loaded as a cantilever in specialized fixturing
2. *End-notched flexure* (ENF): specimen is loaded in three-point bending resulting in unstable delamination growth.
3. *Four-end-notched flexure*: as for the ENF except the specimen is loaded in four-point bending resulting in stable delamination growth (Figure 12.9).

The 4ENF loading has advantages over the other tests. The four-point bending loading fixture uses rollers to support the specimen and to allow it to rotate freely. The upper loading rollers are allowed to rotate about the vertical centerline of the fixture to account for the asymmetric bending of the specimen caused by one half of the beam containing a delamination.

FIGURE 12.9
Four-end-notched flexure load test.

The specimen is loaded in displacement control and generally stable delamination results. Fiber bridging does not always occur for delamination grown in mode II; however, delamination values at delamination initiation are generally used to quote mode-II toughness. However, there is significant friction between the delamination faces with this method.

A compliance calibration technique has been developed for the 4ENF specimen where the mode-II fracture toughness G_{IIC} is determined from

$$G_{IIC} = \frac{mp_c^2}{b}$$

where
  p_c is the critical load at delamination initiation
  b is the specimen width
  m is the slope of the plot of the specimen compliance versus delamination
    length

12.6.2.5 Moisture Diffusivity and Moisture Equilibrium

Polymeric composites absorb moisture, and this may have a deleterious effect. Hence, evaluation of material should be carried out after relevant moisture exposure representing the worst case. The degree of moisture in a composite laminate will depend on the thickness, material, relative humidity, and temperature. Hence, the conditioning procedure should account for the diffusion process and terminate with the moisture content nearly uniform through the thickness. Moisture diffusivity and moisture equilibrium content are obtained using gravimetric tests that expose an initially dry specimen to a humid environment and documents moisture mass gain versus the square root of time. The diffusion process may be accelerated by increasing the exposure temperature. During early exposure, the increase in mass will be linear with root time, and this linear relationship gives the diffusion rate or moisture diffusivity. In time, the mass uptake will reach equilibrium or saturation, and the specimen will gain no more significant weight. The weight percent mass gain at this point is the equilibrium moisture content. If the specimens cannot be tested shortly after conditioning, they should be placed in sealed bags to prevent moisture escape. The mass change at equilibrium is calculated as follows:

$$\text{Mass change} = \frac{W_i - W_b}{W_b} \times 100$$

where
  W_i is the final moisture content in the material
  W_b is the initial moisture content in the material

12.6.2.6 Fiber Volume Fraction

The fiber volume fraction of cured polymer-matrix composites can be obtained by a number of methods including matrix digestion method, ignition loss, and image analysis methods. Image analysis uses digital imaging of a micrograph detailing the fibers and matrix. These methods generally apply to laminates fabricated from most material forms and processes.

12.6.2.7 Void Content of Composites

Void content measures the voids in reinforced polymers and composites. Information on void content is useful because high void contents can significantly reduce the composite strength. Monitoring lot-to-lot void contents can also act as a measure of the consistency of the composites' manufacturing process. For void content, it is necessary to have the theoretical densities of both the resin and the reinforcing material to determine the theoretical density. The individual densities are normally obtained from the supplier of the resin and reinforcing material. After the actual density of the material is determined, the weighed sample (1 in. × 1 in.) is placed into a weighed crucible and burned in a 600°C muffle furnace in air until only the reinforcing material remains. The crucible is cooled and weighed. The resin content (ignition loss) can be calculated as a weight percent from the available data. Void content is calculated by comparing the actual density to the theoretical density:

$$T = \frac{100}{R/D + r/d} \quad V = \frac{T_d - M_d}{T_d}$$

where
  T is the theoretical density
  V is the void content (volume%)
  T_d is the theoretical composite density
  M_d is the measured composite density
  R is the resin weight %
  D is the density of resin
  r is the reinforcement weight %
  d is the density of reinforcement

12.6.2.8 High-Speed Puncture Multiaxial Impact

The high speed impact test is used to determine toughness, load–deflection curves, and total energy absorption of impact events. Since speed can be varied, it can simulate actual impact values at high speeds. This sophisticated impact test provides full force and energy curves during the milliseconds of the impact, using a "Tup," which incorporates an impact head and a load cell. Since many materials (especially thermoplastics) exhibit lower impact strength at reduced temperatures, it is sometimes appropriate to test materials at temperatures that simulate the intended end-use environment. The specimen is clamped onto the testing platform. The crosshead, with the attached Tup, is raised to the appropriate height and is released so it impacts at a specified speed. A load–deflection curve is produced. The instrumented impact tester includes a thermal chamber that allows testing of specimens at elevated and reduced temperatures without the need to transfer the specimens from an oven or freezer just prior to testing. This provides better control and consistency of test results at nonambient temperatures.

12.6.2.9 Accelerated Weathering Testing

Accelerated weathering simulates damaging effects of long-term outdoor exposure of materials and coatings by exposing test samples to varying conditions of the most aggressive components of weathering—ultraviolet (UV) radiation, moisture, and heat. A QUV test chamber uses fluorescent lamps to provide a radiation spectrum centered in the UV wavelengths. Moisture is provided by forced condensation, and temperature is controlled by heaters. No direct correlation can be made between accelerated weathering duration and actual outdoor exposure duration. Up to 20 test samples are mounted in the QUV and subjected to a cycle of exposure to intense UV radiation followed by moisture exposure by condensation. Various cycles are defined depending upon the intended end-use application—for example, a typical cycle for automotive exterior applications would be 8 h UV exposure at 70°C followed by 4 h of condensation at 50°C.

Accelerated weathering provides exposed samples for comparison to unexposed control samples. Often, several exposure times (such as 500, 1000, and 2000 h) will also be compared to each other. Depending upon the performance requirements of concern, such a comparison may involve measurements of haze, transmission, Yellowness Index, color change, and/or physical properties such as impact strength.

12.7 Nondestructive Testing of Composites

NDT is the testing of materials or structures without causing failure of the item being tested. There are three acronyms in general usage: nondestructive testing (NDT), nondestructive evaluation, and nondestructive inspection (NDI) (or nondestructive investigation [NDI]). Various nondestructive techniques are available that outline the testing that can be carried out on (1) incoming materials, such as prepregs, resins, and fibers, to help build quality assurance into the subsequent parts, (2) the types of tests that can be used during fabrication and on the final product, and (3) in-service use of NDT. In-service use of NDT covers the problems related to in-service testing in evaluating such properties as reliability, durability, and life expectancy of a fiber-reinforced composite by nondestructive means. NDT (in some cases nondestructive inspection might be a more appropriate term) can and should cover the complete range of operations from the raw materials through fabrication to final inspection to in-service inspection.

Bibliography

American Society for Testing and Materials, www.astm.org/Standards/D380.htm (accessed May 25, 2013).

BSI, Air filters used in air conditioning and general ventilation. Part 1: Methods of test for atmospheric dust spot efficiency and synthetic dust weight arrestance, BS 6540: Part 1: 1985, British Standards Institution, London, U.K., 1985.

BSI, Porosity and pore size distribution of materials, BS 7591: Part 4: 1993, British Standards Institution, London, U.K., 1993.

Dorman, R. G. and A. S. Ward, *Filtration Principles and Practices Part II*, Marcel Dekker, New York, 1979.

Endo, Y., D.-R. Chen, and D. Y. H. Pui, Collection efficiency of sintered ceramic filters made of submicron spheres, *Filtration and Separation*, 39(2), 43–47 (2002).

Eurovent, Method for testing air filters used in general ventilation, EUROVENT, Staffordshire, U.K., 4/5, 1992.

Hahn, H. T. and R. Y. Kim, Proof testing of composite materials, *Journal of Composite Materials*, 9(3), 297–311 (1975).

Holden, C. and B. Longworth, Improving test methods for polymer melt filters, *Filtration and Separation*, 39(3), 28–29 (2002).

Jena, A. and K. Gupta, A novel technique for characterization of pore structure of ceramic membranes, *Journal of Filtration Society*, 1(4), 23–26 (2001).

Kassapoglou, C., Fatigue model for composites based on the cycle-by-cycle probability of failure: Implications and applications, *Journal of Composite Materials*, 45(3), 261–277 (2011).

Kerschmann, R., Filter media structure in virtual reality, *Filtration and Separation*, 38(7), 26–29 (2001).

NetComposites, UK, www.netcomposites.com (accessed May 25, 2013).

Peuchot, C., IFTS: Past, present and future, *Filtration and Separation*, 37(6), 16–18 (2000).

Purchas, D. B., *Industrial Filtration of Liquids*, Leonard Hill Books, London, U.K., 1967.

Rideal, G., Filter calibration: High precision method, *Filtration and Separation*, 38(2), 26–28 (2001).

Sjoblom, P. O., J. T. Hartness, and T. M. Cordell, On low-velocity impact testing of composite materials, *Journal of Composite Materials*, 22(1), 30–52 (1988).

Smithers Rapra and Smithers Pira Ltd, England, www.rapra.net/composites/mechanical...testing/in-plane-tension.asp (accessed May 25, 2013).

Todd, K., Testing sterile air filter integrity, *Filtration and Separation*, 37(2), 24–25 (2000).

Venkataraman, C. and K. Gupta, Revealing the pore characteristics of membranes, *Filtration and Separation*, 37(6), 20–23 (2000).

13

Textile Composites

13.1 Introduction to Composites

Five decades ago, there has been a huge demand for materials that are stiffer and stronger, yet lighter in aeronautic, energy, civil engineering, and in various structural applications. This necessity and demand accelerated the industry and research organization to develop the concept of combining different materials in an integral composite structure. Composites are the product with the desired properties (e.g., lightweight, strong, corrosion resistant) obtained by combining two or more distinct materials. Composites are produced to synergize constituent material properties to attain desirable properties, for example, mechanical, chemical, and physical properties. Composite materials are quickly attaining ground as recommended materials for many high-performance applications. Composites, especially fiber-reinforced composites, have been used predominantly for decades in many industry sectors such as aerospace, marine, transportation, infrastructure, and consumer goods. The application of composites in structural materials in recent decades in many high technology-oriented projects around the world has instilled confidence among technologists to use it further in many high-performance applications. The science of composites depends significantly on material science, mechanical engineering, in addition to polymers and textiles.

13.2 Composite Materials: Global Scenario

The global composites market represents $108.9 billion in value. The market is growing at an average of 6% per year. In value, the Americas represent 36%, Europe 33%, and Asia Pacific 31%. By 2016, the composite materials industry is predicted that it will reach $27.4 billion—a 7.8% compound annual growth rate. In the same time frame, composite end products will total $78 billion.

During the period 2010–2012, composites have a 3.6% share in transportation based on monetary value, compared to a 68% share in marine. In construction, composites occupy 7% of the market; in aerospace, 10%; and in wind energy, 38%. Population increases, infrastructure needs, and the green movement are only a few of the factors that will drive composite growth to new horizons. In the end, innovations that reduce the cost of composite products by 30% have the potential to triple the composite market.

13.3 Composite Materials: Definition and Classification

Composite material is produced by combining reinforcement and matrix to obtain a unique combination of properties. The earlier definition is more general and can include metals alloys, plastic copolymers, minerals, and wood. The definition of a "composite material," which includes a reference to a physical scale appropriate for present purposes, is as follows (Tuttle, 2004):

> A composite material is a material system consisting of two (or more) materials, which are distinct at a physical scale greater than about 1×10^{-6} m (1 μm) and which are bonded together at the atomic and/or molecular levels.

Composite materials are commonly classified at two distinct levels based on (1) the matrix form and (2) the reinforcement form.

13.3.1 Classification of Composites Based on Matrix

Composite material can be primarily classified with respect to the matrix constituent. The major composite classes include metal matrix composites (MMCs), ceramic matrix composites (CMCs), polymer matrix composites (PMCs), and carbon matrix composites (Figure 13.1).

13.3.1.1 Metal Matrix Composites

Metal-matrix composites consist of two constituent materials, where the matrix is a metal and the reinforcement material may be metal or ceramic or organic compounds. The metals aluminum, copper, and magnesium are widely used as matrix. The characteristic advantages of MMC are fire resistant, operate in wider temperature range, hydrophobic, better conductivity (thermal and electrical), resistance to radiation damage, and do not exhibit outgassing.

13.3.1.2 Ceramic Matrix Composites

Ceramics are generally composed of metallic and nonmetallic elements. CMCs presently in use typically use SiC or inhibited carbon as the matrix.

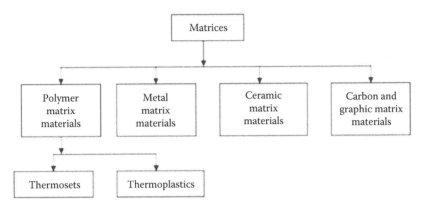

FIGURE 13.1
Composites: classification based on matrix.

The role of the matrix is to provide the required wear and abrasion resistance, or to protect the fiber from oxidation and damage.

13.3.1.3 Polymer Matrix Composites

PMC consists of a polymer(resin) matrix combined with fibrous reinforcement. PMCs are prevalent in many structural applications due to their low cost and simple fabrication methods. Polymers are mostly organic compounds based on carbon, hydrogen, and other nonmetallic elements. PMC have gained the importance among composite groups, and they have found widespread applications. PMC can be easily fabricated into any large complex shape, which is an advantage.

In PMC applications, thermosetting or thermoplastic polymers can be used as the matrix component. PMCs (also termed as reinforced plastics) are, in general, a synergistic combination of high-performance fibers and matrices.

13.3.2 Classification of Composites Based on Reinforcement Forms

Composite material can also be classified based on reinforcement form as fiber-reinforced composites, laminar composites, and particulate composites. Fiber-reinforced composites can be further divided into those containing discontinuous or continuous fibers (Figure 13.2).

13.3.2.1 Fiber-Reinforced Composites

FRCs are composed of fibers embedded in matrix material. FRCs are considered to be short fiber composites if its properties vary with fiber length.

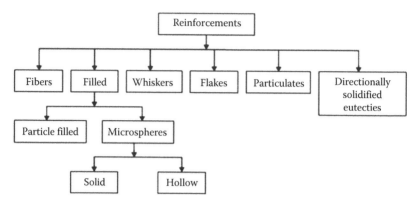

FIGURE 13.2
Composites: classification based on reinforcement forms.

Composites based on continuous fibers can be classified as unidirectional composites, woven composites, and braided composites.

- *Unidirectional composites:* all fibers are aligned in the same direction and embedded within a matrix material.
- *Woven composites* are formed by first weaving continuous fibers into a fabric and then embedding the fabric in a matrix. Hence, a single layer of a woven composite contains fibers in two orthogonal directions.
- *Braided composite* typically contains two or three nonorthogonal fiber directions. Braided composites are then formed by embedding the fabric in a matrix.

13.3.2.2 Whisker-Reinforced Composites

Whiskers are defined as "single crystals grown with nearly zero defects". They are usually made up of materials such as graphite, SiC, copper, and iron and be in short fiber form of different cross sections. Whisker sizes ranging from 2 to 10 are used for producing composites of better property. The ceramic whiskers provide higher specific strength and specific modulus, which are required for lightweight structural composites. The advantages associated with composites based on ceramic whisker are better resistance to temperature, mechanical damage, and oxidation. Metallic whiskers are more dense than ceramic whiskers.

13.3.2.3 Laminar Composites

Laminar composites are composed of layers of materials held together by matrix. Sandwich structures fall under this category. Based on the orientation of the fibers in the lamina, the laminated composites are classified into five types as follows:

1. Unidirectional laminates: they are extremely strong and stiff in the 0° direction but are also very weak in the 90° direction
2. Bidirectional laminates
3. Multidirectional laminates
4. Woven fabric laminates
5. Interplay hybrid laminates

13.3.2.4 Flake Composites

Flakes are often used in place of fibers as can be densely packed. Metal flakes which are in close proximity with each other in polymer matrices can conduct electricity or heat, while mica flakes and glass can resist both. Flakes are economic to produce and usually cost less than fibers. Parallel flake filled composites provide uniform mechanical properties in the same plane as the flakes.

13.3.2.5 Filled Composites

Filled composites result from the incorporation of filler materials to polymer matrices to replace a portion of the matrix, enhance, or modify the properties of the composites. The fillers also enhance strength and reduce weight.

13.3.2.6 Particulate Composites

Particulate composites are composed of particles distributed in a matrix body. The particles may be flakes or in powder form. Concrete and wood particle boards are examples of this category. Particulates are roughly spherical particles with diameters typically ranging from about 1 to 100 µm.

13.4 Reinforcement: Fibers

Reinforcements are important constituents of a composite material and provide required strength and stiffness to the composite. The most common fibers used as reinforcements are glass, carbon, aramid, and boron fibers. Fibers are the important class of reinforcements, as they are the load-bearing members of composites and satisfy the desired conditions influencing and enhancing the material properties as desired. Some of the common types of fiber reinforcements include:

- Continuous carbon tow, glass roving, aramid yarn
- Discontinuous chopped fibers
- Woven fabric
- Multidirectional fabric (stitch bonded for 3-D properties)
- Stapled
- Woven or knitted 3-D preforms

Continuous fibers are used for filament winding, pultrusion, braiding, weaving, and prepregging applications. Continuous fibers are used with most thermoset and thermoplastic resin systems. Chopped fibers are used for making injection molding and compression molding compounds. Chopped fibers are made by cutting the continuous fibers. In spray-up and other processes, continuous fibers are used but are chopped by machine into small pieces before the application. Woven fabrics are used for making prepregs as well as for making laminates for a variety of applications (e.g., boating, marine, and sporting). Preforms are made by braiding and other processes and used as reinforcements for resin transfer molding (RTM) and other molding operations.

Organic and inorganic fibers are used to reinforce composite materials. Almost all organic fibers have low density, flexibility, and elasticity. Inorganic fibers are of high modulus and high thermal stability, and possess greater rigidity than organic fibers and notwithstanding the diverse advantages of organic fibers that render the composites in which they are used.

The following different types of fibers are mainly used: glass fibers, silicon carbide fibers, high silica and quartz fibers, aluminum fibers, metal fibers and wires, graphite fibers, boron fibers, aramid fibers, and multiphase fibers. Among the glass fibers, it is again classified into E-glass, A-glass, and R-glass. There is a greater market and higher degree of commercial movement of organic fibers.

13.4.1 Reinforcement: Fiber Forms

Reinforcement materials are combined with resin systems in a variety of forms to create structural laminates.

13.4.1.1 Discontinuous Fibers

Discontinuous fibers are distributed in a matrix and may be randomly oriented (in which case the composite is isotropic at the macro scale) or may be oriented to some extent (in which case the composite is anisotropic at the macro scale). The orientation of short fibers is usually brought about during the fabrication process used to produce the composite material/structure; fiber alignment often reflects the flow direction during injection molding.

Short-fiber composites are preferred for nonstructural applications where the higher strength and stiffness are not a primary concern. Short-fiber composites can be manufactured at low cost compared to continuous fiber composites. Manufacturing methods such as spray-up, compression and transfer molding, reaction injection molding, and injection molding are much cheaper than the processes used for continuous-fiber composites.

13.4.1.2 Roving and Tow

Continuous fibers are available in various forms ranging from roving to fabric. The most commonly available form of continuous fiber is roving. Fiber roving can also be used in various forms such as chopped, converted to fabric forms such as woven, mats, braids, knitted fabrics, and hybrid fabrics in the composite manufacturing process. Rovings are supplied by weight, with a specified filament diameter. The term "yield" is commonly used to indicate the number of yards in each pound of glass fiber rovings. Rovings are used in many technical applications. During the spray-up process, the fiber roving is chopped with an air-powered gun that forces the chopped fiber strands into the desired mold, while simultaneously applying resin and catalyst in the correct ratio. This process is commonly used for bathtubs, shower stalls, and many marine applications (Figure 13.3).

Similarly, tow is the basic form of carbon fiber. PAN- and pitch-based carbon fibers are available with a moderate (33–35 Msi), intermediate (40–50 Msi), high (50–70 Msi), and ultrahigh (70–140 Msi) modulus. Newer heavy-tow carbon fibers, with filament counts from 48K to 320K, are available at a lower cost than aerospace-grade fibers. Fiber forms made from aligned discontinuous tows are more drapable; that is, they are more pliable and, therefore, conform more easily to curved tool surfaces than fiber forms made from standard tow.

FIGURE 13.3
Tow.

13.4.1.3 Mats

Mats are the nonwoven fabrics made from fibers that are bonded by chemical bonding process. They are available in two forms such as chopped and continuous strand mats. Chopped mats contain short fibers oriented in random fashion and cut to lengths that typically range from 38 to 63.5 mm (1.5 to 2.5 in.). Continuous-strand mat is produced from swirls of continuous fiber strands. The mats are isotropic in structure due to random orientation of fibers which possess same property in all directions. Chopped-strand mats provide low-cost reinforcement primarily in hand layup, continuous laminating, and some closed-molding applications.

13.4.1.4 Woven Fabrics

Fabric can be produced by interlacing or interlocking or interlooping of fibrous strands or by creating adhesion between fibers by chemical or thermal means. Among the various fabric types, woven fabrics are interlaced in looms by varying weight, weaves, and width. Woven fabrics produced by interlacing are bidirectional and have good mechanical properties in both directions of yarn or roving. Impact resistance is enhanced because the fibers are continuously woven.

Woven roving is relatively thick and used for heavy reinforcement, especially in hand layup operations and tooling applications. Due to its relatively coarse weave, woven roving wets out quickly and is relatively inexpensive (Figure 13.4).

Woven rovings are produced by weaving fiberglass rovings into a fabric form. This yields a coarse product that is used in many hand layup and panel molding processes to produce fiber-reinforced polymers. Many weave configurations are available, depending upon the requirements of the laminate. Exceptionally fine woven fiberglass fabrics, however, can be produced for applications such as reinforced printed circuit boards. The various weave structures (see Table 13.1) used as reinforcement in composites are discussed below. The properties of composites are significantly influenced by reinforcement fabric properties and construction particulars especially weave pattern (Table 13.2). All weaves have their advantages and disadvantages, and consideration of the part configuration is necessary during fabric selection.

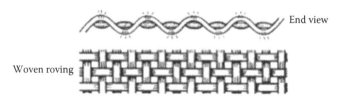

FIGURE 13.4
Woven roving.

TABLE 13.1

Weave Style: Comparison of Properties

Property	Plain	Twill	Satin	Basket	Leno	Mock Leno
Good stability	G	A	P	P	E	A
Good drape	P	G	E	A	VP	P
Low porosity	A	G	E	P	VP	A
Smoothness	P	A	E	P	VP	P
Balance	G	G	P	G	P	G
Symmetrical	E	A	VP	A	VP	G
Low crimp	P	A	E	P	P/E	P

E, excellent; G, good; A, acceptable; P, poor; VP, very poor.

TABLE 13.2

Effect of Weave Pattern on Mechanical Properties of Composites

Fabric Weave Type	Plain	Crowfoot Satin	5-Shaft Satin	8-Shaft Satin
Resin	Polyester	Polyester	Polyester	Polyester
Plies	18	18	12	12
Resin content (wt.%)	37.1	36.7	36.5	37.6
Thickness (mm)	3.15	3.07	3.05	3.05
Flexural strength (MPa)	371	584	435.8	600
Flexural modulus of elasticity (GPa)	26.8	23.5	26.3	22.3
Compressive strength (MPa)	177.2	393	331	443.3
Tensile strength (MPa)	316.5	408.2	404.7	413.7

- *Plain weave:* Each warp yarn passes alternately under and over each weft yarn. The fabric is symmetrical, with good stability and reasonable porosity (Figure 13.5). However, plain weave fabric has poor drape, and the high level of fiber crimp imparts relatively low mechanical properties compared with the other weave styles. This weave style is not to be preferred for producing heavy fabrics while using coarser yarn which develops more crimp.

- *Twill weave:* One or more warp yarns alternately weave over and under two or more weft yarns in a regular repeated manner. Twill weave pattern produces the visual effect of a straight or broken diagonal "rib" to the fabric. Twill weave is characterized by good drape, superior wet-out, and less stability compared to plain weave pattern. Twill fabrics are smoother and have slightly higher mechanical properties at reduced crimp (Figure 13.6).

- *Satin weave:* Satin weaves are basically having longer warp floats to produce fewer intersections of warp and weft. Satin weaves are

FIGURE 13.5
Plain weave.

FIGURE 13.6
Twill weave.

characterized by very flat, good wet out, and a high degree of drape. The low crimp gives good mechanical properties (Figure 13.7). Satin weaves allow yarns to be woven in the closest proximity and can produce fabrics with a close "tight" weave. However, the pattern's low stability and asymmetry need to be considered.

- *Basket weave:* Basket weave is mostly similar to plain weave except that two or more warp yarns alternately interlace with two or more

FIGURE 13.7
Satin weave.

weft yarns. The arrangement of two warps crossing two wefts is designated as 2×2 basket, but the arrangement of yarn need not be symmetrical (Figure 13.8). Basket weave is characterized as flatter, and, though less crimp, stronger than a plain weave, but less stable. It must be used in the production of heavyweight fabrics made with coarser yarns to avoid excessive crimping.

- *Leno weave:* The stability of fabrics having open structure that are produced from fine count yarns are enhanced by leno weaves. It is a form of plain weave in which adjacent warp yarns are twisted

FIGURE 13.8
Basket weave.

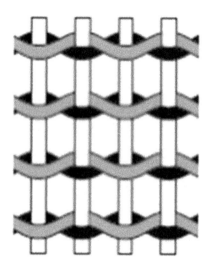

FIGURE 13.9
Leno weave.

around consecutive weft yarns to form a spiral pair, effectively "locking" each weft in place (Figure 13.9). The open structure of leno fabric does not help it to be an effective reinforcement material, so leno fabrics are always used in combination with other weaves as composite reinforcement.

- *Mock Leno weave:* A version of plain weave in which occasional warp yarns, at regular intervals but usually several yarns apart, deviate from the alternate under–over interlacing and instead interlace every two or more yarns (Figure 13.10). This happens with similar frequency in the weft direction, and the overall effect is a fabric with increased thickness, rougher surface, and additional porosity.

FIGURE 13.10
Mock leno.

13.4.1.5 Hybrid Fabrics

Hybrid fabrics can be produced by varying fiber types, yarn compositions, and fabric configurations. Hybrids are used to take advantage of properties or features of each reinforcement type. For example, high-strength strands of S-glass or small-diameter filaments may be used in the warp direction, while less costly strands compose the fill. A hybrid also can be created by stitching woven fabric and nonwoven mat together.

13.4.1.6 Noncrimp Fabrics

Noncrimp fabrics are composed of yarns that are placed parallel to each other and then stitched together using polyester thread. The yarns that are only arranged in warp direction are termed as warp unidirectional yarns that are needed in stiffness-critical applications such as water ski applications where the fabric is laid along the length of the ski to improve stiffness. Noncrimp fabrics provide greater flexibility compared to woven fabrics. Noncrimp fabrics offer greater strength because yarns remain straight, whereas in woven fabrics, yarns bend at interlacing points. Noncrimp fabrics are available in a thick layer and thus an entire laminate could be achieved in a single-layer fabric.

13.4.1.7 Knitted Fabrics

Knytex® is the first knitted reinforcement fabric introduced in 1975 to provide higher strength and stiffness as compared to woven roving reinforcements. The unidirectional reinforcement fabrics are combined and stitched together with the help of polyester filament to produce knitted reinforcement (Figure 13.11).

Mat may also get combined into the construction for some specific applications. Knitted fabrics are most commonly used to reinforce flat sections

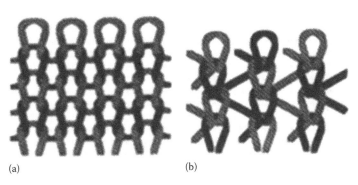

(a)   (b)

FIGURE 13.11
(a) Weft knitting. (b) Warp knitting.

or sheets of composites, but complex 3-D preforms have been created by using prepreg yarn.

13.4.1.8 Multiaxial Fabrics

Multiaxial fabrics are nonwovens that are produced by stacking unidirectional yarn layers in different orientations and bonded by stitching or a chemical binder (Figure 13.12).

The proportion of yarn in any direction can be decided by product requirements. In multiaxial fabrics, the yarn crimp associated with woven fabrics is ignored because the yarns lie on top of each other, rather interlacement. Super-heavyweight nonwovens are available (up to 200 oz/yd²) and can significantly reduce the number of plies required for a lay-up, making fabrication more cost-effective, especially for large industrial structures.

13.4.1.9 Braided Fabrics

Braided fabrics are produced by intertwining of yarns (Figures 13.13 and 13.14).

The braid's strength comes from intertwining three or more yarns without twisting any two yarns around each other. This unique structure offers, typically, greater specific strength than wovens. Braided structure has natural conformability, which facilitates braid suitable for 3-D applications such as sleeve production, preforms because of the ability to fit to any part shape that it is reinforcing. The quasi-isotropic configuration within a single layer of braided fabric can eliminate problems associated with the layering

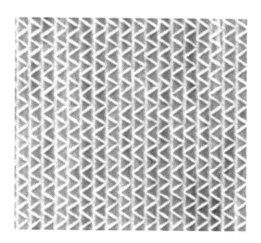

FIGURE 13.12
Multiaxial fabrics.

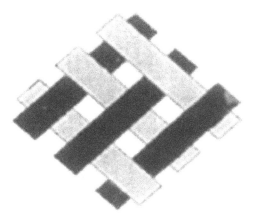

FIGURE 13.13
Braided structure.

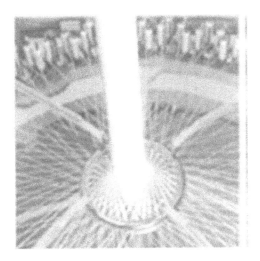

FIGURE 13.14
Braiding.

of multiple 0°, +45°, −45°, and 90° fabrics. Furthermore, the propensity for delamination is reduced dramatically with quasi-isotropic braided fabric. Its 0°, +60°, −60° structure gives the fabric the same mechanical properties in every direction, so the possibility for a mismatch in stiffness between layers is eliminated.

Braid's impact resistance, damage tolerance, and fatigue performance have attracted composite manufacturers in a variety of applications, ranging from hockey sticks to jet engine fan cases. Three-dimensional braiding is a relatively new topic and is mainly developed for industrial composite materials.

13.5 Polymer Matrices or Resins

The primary functions of the resin are to distribute the load experienced to all the reinforcing fibers, act as a glue to hold the fibers together, and protect the fibers from mechanical and environmental damages. Resins are divided into two major groups known as thermoset and thermoplastic. The matrix usually comprises 30%–40% of composite structure. The resin selection is crucial in the composite manufacturing process. The factors influencing the resin selection are

1. Processing temperature
2. Compatibility with fiber
3. Properties of resin
4. Cost
5. Viscosity
6. Process-related factors

Since the strengths of matrix materials are generally lower than fiber strengths by an order of magnitude or more, it is desirable to orient the fibers within a composite structure so that they will carry the major external loads. Matrix material properties can significantly affect how a composite will perform, particularly with respect to in-plane compression, in-plane shear, resistance to impact damage, and other interlaminar behavior, and especially when exposed to moisture and elevated temperatures.

13.5.1 Thermoplastic Matrices

Thermoplastic resins have gained importance in the recent years as the matrix material in composite manufacturing. The global thermoplastic composites market is burgeoning during last 5 years and is expected to reach $6.2 billion in 2014 with a global growth rate of 5.9% in the next 5 years. Linear molecules with no chemical linkages between the molecules are generally termed as "thermoplastic polymers". The molecules are held together by van der Waals or hydrogen bonding, that are readily deformed by the application of heat or pressure. Thermoplastic resins exhibit superior toughness (also called specific work of rupture) properties such as high impact strength and fracture resistance, but it is not linearly translated into properties of composites. The other features of this polymer are postformability, good shelf life at room temperature, ease of repair, and ease of handling (Table 13.3).

Among various thermoplastic matrices, many resins with excellent properties can be found, such as thermoplastic polyesters, polyamides, polysulfones, polyaryl ethers, thermoplastic polyimides, polyarylene sulfide,

TABLE 13.3

Properties of Resin

Resin Material	Density (g/cm³)	Tensile Modulus, GPa (10⁶ psi)	Tensile Strength, MPa (10³ psi)
Nylon	1.1	1.3–3.5 (0.2–0.5)	55–90 (8–13)
PEEK	1.3–1.35	3.5–4.4 (0.5–0.6)	100 (14.5)
PPS	1.3–1.4	3.4 (0.49)	80 (11.6)
Polyester	1.3–1.4	2.1–2.8 (0.3–0.4)	55–60 (8–8.7)
Polycarbonate	1.2	2.1–3.5 (0.3–0.5)	55–70 (8–10)
Acetal	1.4	3.5 (0.5)	70 (10)
Polyethylene	0.9–1.0	0.7–1.4 (0.1–0.2)	20–35 (2.9–5)
Teflon	2.1–2.3	—	10–35 (1.5–5.0)

and liquid crystalline polymers. Thermoplastic polymers are preferred in composites manufactured at high processing temperature (325°C–450°C) systems and higher temperature mold materials must be used.

13.5.2 Thermosetting Resins

Thermoset material is cured by heat or chemical reaction that is changed into an infusible and insoluble material. Thermosetting resins undergo irreversible chemical cross-linking reaction upon application of heat. These resins are recommended in the production of large and structural components while making use of lightweight–low cost tooling (low initial viscosity). Due to its brittle nature, it is normally used in conjunct with filler and reinforcement. The resin is processable at room temperature, which facilitates easy processability and better fiber impregnation in various processes such as filament winding, pultrusion, and RTM. The characteristic advantages of thermoset resins are greater thermal and dimensional stability, better rigidity, and higher electrical, chemical, and solvent resistance. The most widely used resins in thermoset composites are epoxy, polyester, vinylester, phenolics, cyanate esters, bismaleimides, and polyimides.

13.6 Prepregs and Preforms

13.6.1 Prepregs

Resin-impregnated fiber forms, commonly called prepregs, are manufactured by impregnating fibers with a controlled amount of resin (thermoset or thermoplastic), using solvent, hot-melt, or powder-impregnation technologies. Prepregs can be stored in "B-stage," or partially cured state, until they are needed for fabrication.

Prepregs provide consistent properties as well as consistent fiber/resin mix and complete wet-out. They eliminate the need for weighing and mixing resin and catalyst. Various types of drape and tack are provided with prepregs to meet various application needs. Drape is the ability of prepreg to take the shape of a contoured surface. For example, thermoplastic prepregs are not easy to drape, whereas thermoset prepregs are easy to drape. Thermoset prepregs have a limited shelf life and require refrigeration for storage. The advantages of prepreg materials over metals are their higher specific stiffness, specific strength, corrosion resistance, and faster manufacturing. The major disadvantage of prepreg materials is their higher cost.

13.6.2 Preforms

Textile preforming is a fiber placement method utilizing textile processes prior to the formation of composite structures. Textile preforms are the structural backbone of a composite analogous to the structural steel framework in a building. Preforms are near-net shape reinforcement forms designed for use in the manufacture of particular parts by stacking and shaping layers of chopped, unidirectional, woven, stitched, and/or braided fiber into a predetermined 3-D form. Complex part shapes can be approximated closely by careful selection and integration of any number of reinforcement layers in varying shapes and orientations. Preforms are feedstock for the RTM and SRIM processes, where a reinforcement in the form of a thick 2-D or 3-D fiber architecture is put in the mold cavity and then resin is injected into the cavity to obtain the composite part.

13.7 Composite Manufacturing Technologies

Fabricating a composite part is generally concerned with placing and retaining fibers in the direction and form that is required to provide specified characteristics while the part performs its design function. The fabrication of composites is a complex process, and it requires simultaneous consideration of various parameters such as component geometry, production volume, reinforcement and matrix types, tooling requirements, and process and market economics.

There are four basic steps involved in composite part fabrication: wetting/impregnation, layup, consolidation, and solidification. All composite manufacturing processes involve the same four steps, although they are accomplished in different ways. There are three general divisions of composite manufacturing processes: open molding (sometimes called contact molding), closed molding, and continuous mold process.

13.7.1 Open Mold Process

Open molding (contact molding) is the simplest method of fabrication of polymer matrix composites. Open molding is usually used for manufacturing large individual parts (swimming pools, boat bodies). Open molding method is mostly used for fabrication glass fiber-reinforcing polymers (fiberglasses) with polyester (sometimes epoxy or vinylester) matrix.

13.7.1.1 Hand Laminating or Hand Layup or Wet Layup

The most popular type of open molding is hand laminating process. Hand laminating is a primitive but effective method that is still widely used for prototyping and small batch production. Hand laminating using open molds has traditionally been used for making structures out of fiberglass and polyester. The hand laminating is used in the production of standard wind-turbine blades, production boats, and architectural moldings (Figure 13.15).

13.7.1.2 Spray Layup

In a spray layup method, the fiber is chopped in a handheld gun and fed into a spray of catalyzed liquid resin directed at the mold. The sprayed, catalyzed liquid resin will wet the reinforcement fibers, which are simultaneously chopped in the same spray gun. The deposited materials are left to cure under standard atmospheric conditions. Spray-up can be used to mold more complex shapes than hand layup (Figure 13.16).

13.7.1.3 Filament Winding

Filament winding is an automated process for creating parts of simple geometry wherein continuous resin-impregnated fibers are wound over a rotating male tool called mandrel. Successive layers of reinforcement are built up on the mandrel until the required thickness is achieved. The reinforcement can

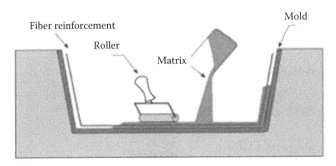

FIGURE 13.15
Hand lamination.

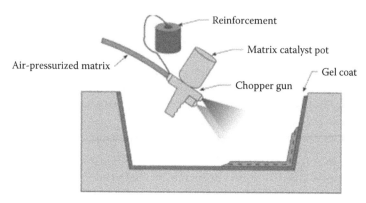

FIGURE 13.16
Spray layup process.

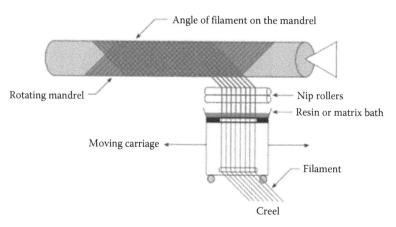

FIGURE 13.17
Filament winding.

be wound longitudinally, circumferentially, helically, or in a combination of two or more of these. The properties required from the finished article will often determine the angle of wind (Figure 13.17).

13.7.2 Closed Mold Process

Closed mold methods are used when mass production of identical parts with both smooth surfaces is required.

13.7.2.1 Compression Molding

Compression molding is one of the oldest manufacturing techniques in the composites industry. The recent development of high strength, fast cure,

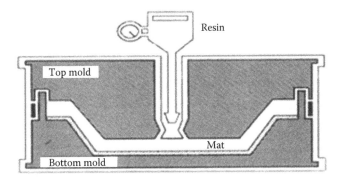

FIGURE 13.18
Resin transfer molding.

sheet molding compounds, bulk molding compounds, and advancement in press technology is making the compression molding process very popular for mass production of composite parts.

13.7.2.2 Resin Transfer Molding

RTM is a closed-mold, low-pressure process in which dry, preshaped reinforcement material is placed in a closed mold and a polymer solution or resin is injected at a low pressure, filling the mold and thoroughly impregnating the reinforcement to form a composite part. The RTM process is also known as a liquid transfer molding process. RTM offers the fabrication of near-net-shape complex parts with controlled fiber directions. Continuous fibers are usually used in the RTM process (Figure 13.18).

13.7.3 Continuous Mold Process

13.7.3.1 Pultrusion

Pultrusion is the only continuous, highly automated composite manufacturing process offering high-volume and good-quality production. Some of the advantages of using pultruded polymer-based composites are high specific strength, chemical and corrosion resistance, machinability, adhesive and mechanical joining, dimensional stability, nonmagnetic properties, and electromagnetic transparency (Figure 13.19).

13.8 Applications

Textile composites for high-performance applications are made from high-performance fibers such as carbon, glass, and aramid, which are bound together within a matrix of polymer material such as epoxy, polyester, etc.

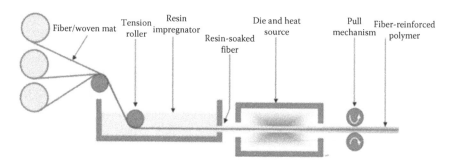

FIGURE 13.19
Pultrusion.

The inherent high strength–weight and stiffness–weight ratios and fatigue and corrosion resistance of composite materials give them the ability to produce components with significant weight and performance advantages over metal parts.

The superior impact and ballistic performance of 3-D woven composites has been utilized to produce innovative protective sports equipment such as shin guards and helmets and ballistic protection equipment for police personnel. Currently, the only reported application of 3-D woven composites in the building industry is for composite I-beams. The material consists of a 3-D woven carbon fiber/silicon carbide matrix sandwich, which will be used in the combustion chamber itself. The use of 3-D woven preforms by Delphi to produce automotive Class 8 composite bumpers results in improved resin infusion and better conformability to the complex shape.

Three-dimensional braided preforms have very high levels of conformability, drapability, torsional stability, and structural integrity, in general the highest of all the 3-D textile processes. The applications of 3-D braiding have primarily been in the area of composite components for rocket applications. Braiding processes are capable of forming quite intricately shaped preforms, and the process can be varied during operation to produce changes in the cross-sectional shape, as well as to produce tapers, bends, holes, and bifurcations in the final preform. It is used to produce composite truss section decking for use as a lightweight bridge deck, propeller blades for naval landing craft, propulsion shafts, and propellers on other vessels.

The knitting process can produce flat multilayer fabrics with highly curved fiber architectures. The more advanced industrial knitting machines are also capable of producing very complex, net-shaped structures at high production rates with little material wastage. Aircraft push rod fairing is manufactured with flat knitted glass fabric. The main disadvantage of knitting is that the fabric is not generally suited for structural composite applications owing to the highly curved nature of the yarn architecture. Noncrimp fabric is being used extensively for the manufacture of high-performance yachts and wind turbine blades. Its use is also increasing within the aerospace industry.

Bibliography

Akovali, G., *Handbook of Composite Fabrication*, Rapra Technology Ltd., Shrewsbury, U.K.

Campbell, F. C., *Manufacturing Technology for Aerospace Structural Materials,* Elsevier Ltd., the Netherlands, 2006.

Composite world, Fiber reinforcement forms, February 1, 2012, http://www.compositesworld.com/articles/fiber-reinforcement-forms (accessed May 29, 2013).

Daniel, I. M. and O. Ishai, *Engineering Mechanics of Composite Materials*, Oxford University Press, Oxford, U.K., 1994.

Hoa, S. V., *Principles of the Manufacturing of Composite Materials*, DEStech Publications, Inc., Lancaster, PA, 2009.

Jones, R. M., *Mechanics of Composite Materials*, Taylor & Francis, London, U.K., 1999.

Mazumdar, S. K., *Composites Manufacturing—Materials, Product and Process Engineering*, CRC Press, Boca Raton, FL, 2002.

Mutel, F., Executive Forum—Composites: A growing industry, JEC Group, *Textile World*, January 2013, http://www.textileworld.com/Articles/2013/January/January_February_issue/Executive_Forum.html (accessed May 29, 2013).

Netcomposites, http://www.netcomposites.com/guide (accessed May 29, 2013).

Pandey, P. C., Composite materials—Web based learning materials, NPTEL, http://ecourses.vtu.ac.in/nptel/courses/Webcourse-contents/IISc-BANG/Composite%20Materials/Learning%20material%20-%20composite%20material.pdf (accessed May 29, 2013).

Peters, S. T., *Handbook of Composites*, Chapman and Hall, London, U.K., 1998.

Rakutt, D., Fitzer, E., and Stenzenberger, H. D., The toughness and morphology spectrum of bismaleimide/polyetherimide carbon fabric laminates, *High Performance Polymers*, 3(2), 59 (1991).

Rapra, S., Tensile properties—In plane tension, www.rapra.net/composites/mechanical...testing/in-plane-tension.asp (accessed May 29, 2013).

Thermoplastic resins, in *Engineered Materials Handbook: Engineering Plastics*, Vol. 2, ASM International, Metals Park, OH, 1988, pp. 98–221.

Tuttle, M. E., *Structural Analysis of Polymeric Composite Materials*, Marcel Dekker, Inc., 2004.

U.S. Department of Defense, *Composite Materials Handbook, Department of Defense Handbook*, MIL-HDBK-17-1F, USA Vol. 1 of 5, 2002.

Index